CHEMISTRY AND BIOCHEMISTRY OF AMINO ACIDS, PEPTIDES, AND PROTEINS

A Survey of Recent Developments

◀ *Volume 2* ▶

Edited by

BORIS WEINSTEIN

DEPARTMENT OF CHEMISTRY
UNIVERSITY OF WASHINGTON
SEATTLE, WASHINGTON

MARCEL DEKKER, INC., New York

MARCEL DEKKER, INC.

270 Madison Avenue, New York, New York 10016

LIBRARY OF CONBRESS CATALOG CARD NUMBER: 75-142896

ISBN: 0-8247-1769-4

Current printing (last digit):
10 9 8 7 6 5 4 3 2 1

PRINTED IN THE UNITED STATES OF AMERICA

ABOUT THE SERIES

The amide bond is one of the less reactive organic functional groups, yet it serves as the cornerstone for the building of the many peptides and proteins found in living systems. The evolving science of molecular biology has served to stress again that the chemistry and biochemistry of amino acids, peptides, and proteins is interwoven into a complex pattern, which on closer examination is found to be dependent on a host of secondary factors associated with individual compounds. There has been a need for a new review series in this area, especially if the interrelationships between the various disciplines are to be discussed in a detailed fashion. In an ideal sense, each volume should contain some chapters on recent developments and applications of established techniques, whereas others might describe the background and problems for topics still under investigation. Too, the subjects encompassed here do permit a variety of treatments without undue duplication or specialization.

One need not remind the reader of the many life processes that are dependent upon specific amino acid, hormone, and enzyme systems. Each functions in a very unique fashion, yet, in the end, they must involve the reactions of fundamental organic chemistry. Sometimes this point is overlooked, and it will be restated in greater detail through the series. To balance the scale, the brief comment is made here that new protecting, labeling, and coupling agents are always desirable, but these must be put to the test by the synthesis or degradation of actual compounds, for which practical use exists in Nature.

It is anticipated that these volumes can be useful both to the specialist and nonspecialist, and may provide a reference point to those who may do research in a broad region, or to the active worker in a small field. Most importantly, these volumes can serve the general purpose of presenting various points of view on the amide bond to interested observers, who, at present, are unknown to one another.

BORIS WEINSTEIN

Seattle, Washington
December, 1970

iii

PREFACE

This volume, the second in a continuing series, covers a variety of topics of much interest to active investigators in the interrelated areas devoted to the biochemistry and chemistry of peptides.

The first chapter by Nobuo Izumiya and Tetsuo Kato describes the field of gramicidin antibiotics and the effect of structure on biological activity. The second by Haruaki Yajima and Hiroki Kawatani is a survey of the related effort given to the connection between structure and effect for adrenocorticotropic hormone. The next by K. Jankowski discusses the reactions between amino acid and small-ring organic compounds. A review by Darrell J. Woodman appraises the isoxazolium salt coupling procedure from the viewpoint of both mechanism and application. The last by Kaoru Harada summarizes the data now available to explain the prebiotic synthesis of amino acids and peptides.

The initial review was received in the late Fall of 1971, while the last was obtained in the Summer of 1972. As a result, the literature is covered through 1970, and, in several cases, selected citations are given for 1971. The authors are gratefully thanked for their contributions, and any errors, omissions, and delays are the responsibility of the editor.

It is hoped that the good reception given to the first volume in 1971 will continue through this and succeeding ones. Any suggestions as to format and content will be welcomed by the editor.

BORIS WEINSTEIN

Seattle, Washington
November, 1972

CONTRIBUTORS TO THIS VOLUME

Kaoru Harada, Institute for Molecular and Cellular Evolution, and Department of Chemistry, University of Miami, Coral Gables, Florida

Nobuo Izumiya, Laboratory of Biochemistry, Faculty of Science, Kyushu University, Fukuoka, Japan

K. Jankowski, Université de Moncton, Moncton, New Brunswick, Canada

Tetsuo Kato, Laboratory of Biochemistry, Faculty of Science, Kyushu University, Fukuoka, Japan

H. Kawatani, Faculty of Pharmaceutical Sciences, Kyoto University, Kyoto, Japan

Darrell J. Woodman, Department of Chemistry, University of Washington, Seattle, Washington

H. Yajima, Faculty of Pharmaceutical Sciences, Kyoto University, Kyoto, Japan

CONTENTS

About the Series   iii

Preface   v

Contributors to This Volume   vii

Contents of Volume 1   xi

1. CHEMISTRY AND BIOCHEMISTRY OF GRAMICIDIN S AND RELATED
   COMPOUNDS, *Tetsuo Kato* and *Nobuo Izumiya*   1

     I.   Introduction   2

     II.   Biosynthesis of Gramicidin S   4

     III.   Chemical Synthesis of Gramicidin S   6

     IV.   Chemical Synthesis and Biologic Activity
         of Peptides Related to Gramicidin S   11

     V.   Conformation of Gramicidin S   23

     VI.   Relationship between Structure and Biologic
         Activity   27

         References   31

2. SYNTHESIS OF ACTH-ACTIVE PEPTIDES AND ANALOGS,
   H. Yajima and *H. Kawatani*   39

     I.   Introduction   40

     II.   Effects of ACTH and Assay Methods   40

     III.   Isolation and Structure Determination   44

     IV.   Synthesis of ACTH Peptides   49

     V.   Structure-Function of ACTH Peptides   107

     VI.   ACTH Receptor   124

     VII.   ACTH Releasing Factor   127

         References   129

3. REACTION OF SMALL HETEROCYCLIC COMPOUNDS WITH AMINO
    ACIDS, *K. Jankowski*   145
        I.   Introduction   146
        II.   Reaction of Three-Membered Rings with
              Amino Acids   163
        III. Reaction of Four-Membered Rings with
              Amino Acids   181
        IV.   Illustrative Procedures   187
        V.   Tabular Survey of the Reactions   191
              References   200

4. THE ISOXAZOLIUM SALT METHOD OF PEPTIDE SYNTHESIS,
    Darrell J. Woodman   207
        I.   Introduction   208
        II.   Synthesis of 3-Unsubstituted Isoxazolium Salts   211
        III. Reaction with Carboxylic Acid Anions   217
        IV.   Enol Ester Stability and the Imide Rearrangement   233
        V.   Enol Ester Acylating Agents   238
        VI.   Acylation By-Products   240
        VII. Racemization   242
        VIII. Applications to Peptide Synthesis   250
        IX.   Special Applications   280
              References   287
5. SYNTHESES OF AMINO ACIDS AND PEPTIDES UNDER POSSIBLE
    PREBIOTIC CONDITIONS, *Kaoru Harada*   297
        I.   Formation of Amino Acids   298
        II.   Formation of Peptides and Polypeptides   316
              References   344

Author Index   353
Subject Index   373

CONTENTS OF VOLUME I

THE OPTICAL ANALYSIS OF AMINO ACID DERIVATIVES BY GAS CHROMATOG-
RAPHY, *John W. Westley*

THE CHEMISTRY OF CYCLOSERINE, *Charles H. Stammer*

THE USE OF HYDROGEN FLUORIDE IN PEPTIDE CHEMISTRY, *S. Sakakibara*

γ-GLUTAMYL PEPTIDES: CHEMISTRY AND IMMUNOCHEMISTRY, *D. E. Nitecki*
and *J. W. Goodman*

PEPTIDE ALKALOIDS, *Mary Païs* and *François-Xavier Jarreau*

CHAPTER 1

CHEMISTRY AND BIOCHEMISTRY OF

GRAMICIDIN S AND RELATED COMPOUNDS

Tetsuo Kato and Nobuo Izumiya

Laboratory of Biochemistry
Faculty of Science
Kyushu University
Fukuoka, Japan

I.   INTRODUCTION.   2
II.   BIOSYNTHESIS OF GRAMICIDIN S.   4
III. CHEMICAL SYNTHESIS OF GRAMICIDIN S.   6
IV.   CHEMICAL SYNTHESIS AND BIOLOGIC ACTIVITY OF
      PEPTIDES RELATED TO GRAMICIDIN S. 11
      A. Linear Peptides 11
      B. Analogs of Gramicidin S 15
      C. Peptides with Smaller and Larger
          Ring Structures 21
      D. Tyrocidines 22
V.   CONFORMATION OF GRAMICIDIN S. 23
VI.   RELATIONSHIP BETWEEN STRUCTURE AND BIOLOGIC ACTIVITY. . . 27
      REFERENCES. 31

1

I. INTRODUCTION

In 1939, Dubos isolated a crude antibiotic preparation, named tyrothricin, from cultures of <u>Bacillus</u> <u>brevis</u> [1]. It was recognized afterward that tyrothricin was composed of two crystalline components [2]. The first, a neutral fraction, was called gramicidin, and contained several linear polypeptide derivatives, while the second material consisted of a mixture of basic peptides. The latter, the tyrocidines, were separated by Battersby and Craig with the aid of countercurrent distribution techniques [3].

In search of a similar antibiotic from Russian soil, Gause and Brazhnikova carried out extensive microbiologic screening tests and discovered another peptide antibiotic from a strain of <u>B. brevis</u> [4]. This compound was named gramicidin S. After purification by crystallization, its homogeneity was established by classic analytical [5], diffusion [6], and countercurrent distribution methods [7]. Amino acid [5], sequence [8, 9] and 2,4-dinitrophenylamino acid determinations [10] indicated gramicidin S to be a cyclodecapeptide containing the pentapeptide sequence -Val-Orn-Leu-D-Phe-Pro-, which is repeated twice to form a 30-membered macrocyclic structure (<u>1</u>). The results of x-ray crystallographic studies [11, 12], diffusion measurements [6], and a molecular weight determination [13] provided further evidence for this structure.

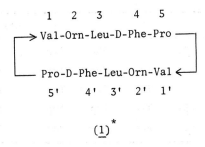

$$\underline{(1)}^{*}$$

Pure gramicidin S dihydrochloride, m.p. 277-278°C, $[\alpha]_D$ = -289° ± 10° (c 0.43, 70% ethanol) [14], kills Staphylococci at a concentration of 3 μg/ml, and E. coli at 50 μg/ml in a nutritive medium [4].

Both the Craig and the Kurahashi groups [15-19] have determined the structures 2 to 6 of a number of tyrocidines. The results indicate that tyrocidines are also cyclic decapeptides and contain the same pentapeptide sequence as gramicidin S, but the other half of the molecule is built from different sequences. As a result, it may not be inappropriate to conclude that gramicidin S is just another particular variant of the tyrocidine family [20]. A typical example of the neutral gramicidins of Dubos, valine-gramicidin A, has the structure N-formylpentadecapeptide ethanolamide (7) [21].

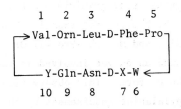

$$\underline{(2)}-\underline{(6)}$$

---

*Abbreviations used are: Z, benzyloxycarbonyl; Trt, triphenylmethyl; Tos, p-toluenesulfonyl; Z(OCH₃), p-methoxy-benzyloxycarbonyl; Boc, tert-butyloxycarbonyl; OMe, methyl ester; ONp, p-nitrophenyl ester; TFA, trifluoroacetic acid; DCC, N,N'-dicyclohexylcarbodiimide. Amino acid symbols denote the L configuration unless otherwise indicated by D or DL.

		W	X	Y	References
Tyrocidine A	(2)	Phe	Phe	Tyr	15
B	(3)	Trp	Phe	Tyr	16
C	(4)	Trp	Trp	Tyr	17
D	(5)	Trp	Trp	Trp	18
E	(6)	Phe	Phe	Phe	19

HCO-Val-Gly-Ala-D-Leu-Ala-D-Val-Val-D-Val-Trp-D-Leu-

Trp-D-Leu-Trp-D-Leu-Trp-NHCH$_2$CH$_2$OH

(7)

II. BIOSYNTHESIS OF GRAMICIDIN S

Ample evidence indicates that the biosynthesis of peptide
antibiotics is independent of the ribosomal RNA-dependent process,
such as protein synthesis; rather, it is dependent on a purely
enzymatic process. Several oligopeptides that are possible intermediates in the biosynthesis of gramicidin S have been isolated in
cell-free extracts of B. brevis; for example, H-D-Phe-Pro-OH [22],
H-D-Phe-Pro-Val-OH [23], and H-D-Phe-Pro-Val-Orn-OH [24]. Yamada
and Kurahashi [25] found an enzyme involved in the racemization of
L-phenylalanine to the D- form. This observation explains the
reason for the L-phenylalanine requirement by the bacteria.

Recently, Lipmann [26] discovered a multiple enzyme system
composed of two active fractions, I and II, involved in the
biosynthesis of gramicidin S. Fraction I is a complex capable of
activating the four constituent L-amino acids of gramicidin S.
Fraction II activates, racemizes, and incorporates phenylalanine
into the product whenever this particular amino acid residue has
the D- configuration. Further, II catalyzes a condensation

between the carboxyl group of D-phenylalanine, bound to an enzyme
sulfhydryl group, and the free imino group of L-proline, which is
one of the four L-amino acids all linked by their carboxyl functions
to separate sulfhydryl groups within fraction I. Successive reactions within the enzyme system result in the generation of peptide-
active thioester chains and ultimately in the formation of gramicidin
S. However, there is no evidence for the existence of chains between
six and nine residues long. A doubling cyclization reaction between
two antiparallel pentapeptide units might then form gramicidin S in
this biosynthetic procedure [26].

Within the past year, Laland [27] showed that the cyclization
of two pentapeptides takes place on the same enzyme molecule in the
form of a head-to-tail condensation and suggested a scheme for the
cyclization reaction involving a 4'-phosphopantetheine arm. The
phosphopantetheine appears to transfer each intermediate peptide to
the corresponding enzyme in order to make a thioester linkage, as in
the case of fatty acid biosynthesis [28].

On the other hand, a three-component enzyme system active in
tyrocidine biosynthesis was prepared from B. brevis ATCC 8185. Each
fraction activates phenylalanine, proline, and the remaining tyrocidine constituent amino acids [29]. Nine enzyme-bound intermediate
peptides were isolated from cell-free systems [30]. All the peptides,
including H-Phe-Pro-OH,H-Phe-Pro-Phe-OH, and others, up to the linear
decapeptide H-Phe-Pro-Phe-Phe-Asn-Gln-Phe-Val-Orn-Leu-OH, were bound
to the enzyme proteins by thioester linkages, as in the case of
gramicidin S. An interesting difference between the gramicidin S
and the tyrocidine [29] biosynthetic processes is that cyclization
in the former seems to be a rather fast process, but in the latter
case, cyclization probably is a rate-limiting process.

Pollard et al. [31] found another possible precursor of
gramicidin S in cell-free extracts of B. brevis ATCC 9999. A linear
decapeptide formula has been assigned to this product (8).

HCO-D-Phe-Pro-Val-Orn-Leu-D-Phe-Pro-Val-Orn-Leu-NHCH$_2$CH$_2$OH

(8)

Since the N-formylpentadecapeptide ethanolamide structure (7)
of the linear gramicidins of Dubos resembles this possible intermediate (8), the two families of antibiotics found in Bacillus
species, the linear gramicidins and the cyclic tyrocidines, appear
to be more closely related to each other than their apparent
structures would suggest. However, there is no direct evidence
for compound (8) being a precursor of gramicidin S. This polyamide
might be a mere artifact of the cell-free system in which the
peptide is found free [31].

III. CHEMICAL SYNTHESIS OF GRAMICIDIN S

Schwyzer and Sieber achieved the synthesis of gramicidin S
in 1957 [14]. This was the first chemical preparation of a cyclopeptide antibiotic. The reactions are described in Fig. 1.
The protected pentapeptide methyl ester (9) derived from
Z-Val-Orn(Tos)-N$_3$ and H-Leu-D-Phe-Pro-OMe was converted to the
corresponding trityl derivative (11). After saponification, the
protected pentapeptide acid (12) was coupled with the pentapeptide
ester (10) through the use of dicyclohexylcarbodiimide to yield the
decapeptide derivative (13). Saponification formed the protected
decapeptide acid (14), which was converted to the corresponding
p-nitrophenyl ester (15) by the action of di-p-nitrophenyl sulfite
[32]. After cleavage of the trityl group, the decapeptide ester
trifluoroacetate (16) was subjected to cyclization in a large
volume of pyridine. Purification on the ion-exchange resins and
crystallization gave ditosylcyclodecapeptide (17) in 28% yield.
Removal of the tosyl groups by treatment with sodium in liquid
ammonia afforded the desired cyclodecapeptide as a dihydrochloride
salt (I·2HCl).

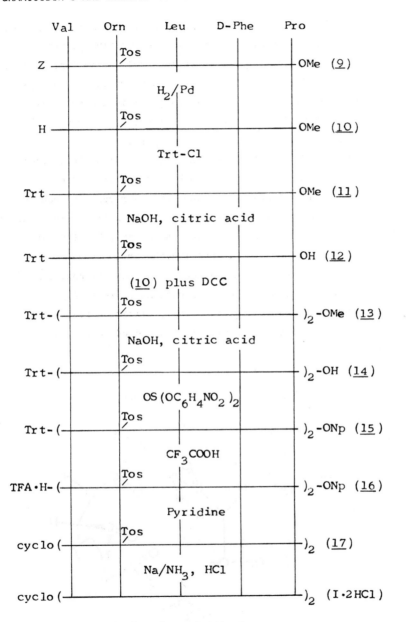

FIG. 1.   Synthesis of gramicidin S.

The final cyclopeptide was found to be identical with natural gramicidin S on the basis of comparison of melting points, R_f values on paper chromatograms in various solvent systems, infrared spectra, x-ray diffraction patterns, and microbiologic assays.

Schwyzer and Sieber synthesized gramicidin S not only from the corresponding decapeptide active ester, but also with the pentapeptide active ester H-Val-Orn(Tos)-Leu-D-Phe-Pro-ONp [33]. Two molecules of the latter combined to give ditosylgramicidin S through a doubling cyclization. Schwyzer explained the mode of this reaction as follows [34]: When the L-tripeptide sequence -Val-Orn-Leu- of the two molecules takes the antiparallel β-pleated sheet structure proposed by Pauling and Corey [35], the C-terminal active esters will be placed opposite the amino group of another molecule because the peptide backbone kinks at the position of D-phenylalanine and L-proline residues, as shown in Fig. 2. This orientation will lead to the doubling cyclization reaction and will afford the structure for gramicidin S depicted in Fig. 3.

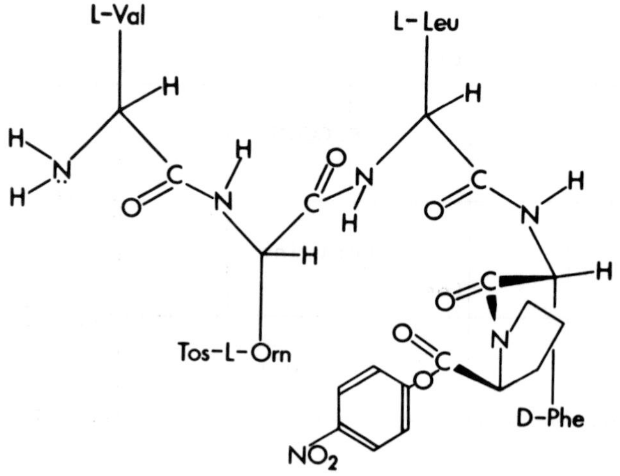

FIG. 2. A possible conformation of the pentapeptide active ester in a doubling cyclization reaction.

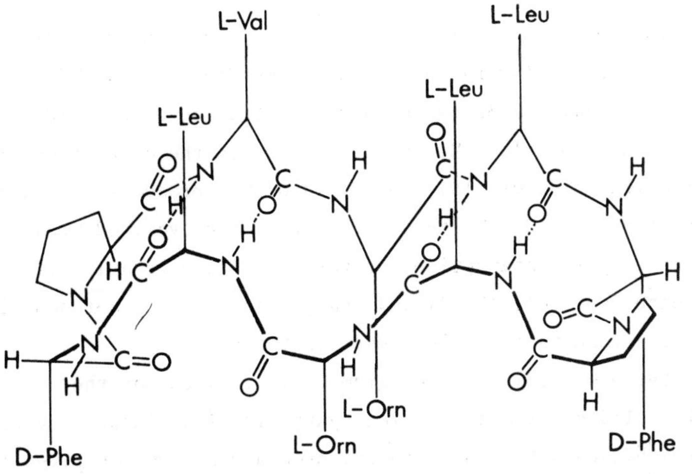

FIG. 3. A possible β-pleated sheet structure of gramicidin S.

Waki and Izumiya [36] subjected the p-nitrophenyl ester of

H-Val-Orn(Z)-Leu-D-Phe-Pro-OH

(18)

to a cyclization reaction and observed the formation of a mixture of
two products, protected cyclic penta- and decapeptides (see Fig. 4).
On Sephadex LH-20 column chromatography, the dimerized cyclodecapeptide derivative was eluted quickly, while the monomeric
cyclopentapeptide was obtained from the more slowly moving fractions.
They indicated the molar ratio of the monomer to the dimer was
29:71, 32:68, and 45:55 when the concentration of the active ester
derived from (18) was 30×10^{-3} M, 3×10^{-3} M, and 0.3×10^{-3} M,
respectively.

Matsuura and Izumiya [37] examined the influence of various
coupling methods on the cyclization of the same pentapeptide (18).
When the N-hydroxysuccinimide ester of (18) was used, the yield of

the protected monomer and dimer was 14 and 38% at a concentration
of 3×10^{-3} M. The molar ratio of the monomer to dimer was found
to be 16:84, 26:74, and 31:69 in the concentration of 30×10^{-3} M,
3×10^{-3} M, and 0.3×10^{-3} M, respectively. In the case of cyclization from the azide of (18), the yield of the products decreased
to 18 and 4% at 3×10^{-3} M; that is, most of the cyclized product
was monomeric cyclopentapeptide derivative. The alteration of amino
acid sequence (see Sec. IV, B) and changes in coupling methods also
influence the results found in these reactions.

    Klostermeyer [38] prepared gramicidin S by use of the
Merrifield solid-phase method. The synthesis of a linear decapeptide-
resin was achieved with two different procedures. One involved the
use of dicyclohexylcarbodiimide as the coupling reagent with Boc-
amino acids, and the other employed activated Boc-amino acid N-
hydroxysuccinimide esters. After being cleaved from the supporting
polymer resin, pure

$$H-(Val-Orn(Tos)-Leu-D-Phe-Pro)_2-OH$$

(19)

could be obtained only by the former procedure. Cyclization of
structure 19 with a large amount of dicyclohexylcarbodiimide
produced the desired ditosylcyclodecapeptide. After deprotection,
the synthetic product was identical with an authentic sample made
by Schwyzer and Sieber [14].

    A combination of the furfuryloxycarbonyl group for
α-aminoprotection and a resinol-type resin was applied to another
solid-phase synthesis of gramicidin S. The yield and biologic
activity of the product were similar to those obtained by the usual
Merrifield method [39]. Ohno et al. [40] has reported an improved
synthesis of gramicidin S. α-p-Methoxybenzyloxycarbonyl-δ-p-
nitrobenzyloxycarbonyldecapeptide hydrazide was obtained almost
quantitatively from the resin by hydrazinolysis [41]. The purity of

the intermediate compounds was determined spectrophotometrically by
measuring the absorption of the p-nitrobenzyloxycarbonyl group.
Cyclization via the azide method in pyridine furnished a protected
cyclodecapeptide in a good yield. The hydrogenolyzed product
exhibited the same activity as natural gramicidin S.

IV.   CHEMICAL SYNTHESIS AND BIOLOGIC ACTIVITY
OF PEPTIDES RELATED TO GRAMICIDIN S

A. Linear Peptides

The relation between the chemical structure and the biological
activity of gramicidin S has attracted the prolonged attention of
several investigators. The cyclic structure of the molecule and the
presence of an amino acid of D- configuration, D-phenylalanine, as
well as another amino acid rarely encountered in a natural protein,
L-ornithine, are notable and significant features of this peptide.
Furthermore, the specific pentapeptide sequence found in gramicidin
S and the tyrocidines seems to play a role in producing antibacterial
activity. In order to study the specific structural features that
might affect the activity of the molecule, several linear di-, tri-
[42], tetra-, and pentapeptides [43], e.g., H-Val-DL-Orn-Leu-D-Phe-
Pro-OH, were synthesized and tested. However, all of them proved
to be only negligibly active. These findings indicate that growth
inhibitory activities cannot be correlated with either the mere
presence of D-amino acids in the peptide or the specific sequence
of the peptide. Harris and Work [42] suggested peptide antibiotics
containing D-amino acids are active not because they have this
character in common, but rather by virtue of the individual structure of the peptide.

In contrast to the negligible activity of these smaller peptides, the linear decapeptide analogs (20-22) of gramicidin S exhibit considerable activity against E. coli, as shown in Table 1 [44, 45]. This result implies that the cyclic structure is not indispensable for the appearance of antibacterial activity. By contrast, the chain length of the linear peptides seems to be important in exhibiting biologic activity. Since some linear decapeptides reveal synergistic activity with gramicidin S against E. coli, the mode of antibacterial action for such linear peptides may differ from gramicidin S [45].

Makisumi et al. [46] synthesized several linear penta- and decapeptide analogs, including the supposed precursor (8) suggested by Pollard [31]. Some of the decapeptide analogs, such as the formyldecapeptide ethanolamide, were fairly active against B. subtilis; however, the activity of the hypothetical precursor was extremely low (see Table 2).

The preparation of various amino acid polymers and copolymers containing appropriate amino acid residues was carried out by Katchalski et al. [47, 48]. The polymers were synthesized from the corresponding amino acid N-carboxyanhydride in anhydrous dioxane. The δ-amino group of ornithine was protected with the

TABLE 1

Minimum Inhibitory Concentration for

E. coli by Linear Decapaptide Analogs of Gramicidin S

Compound	Concentration (μg/ml)
H-(Val-Orn-Leu-D-Phe-Pro)$_2$-OH (20)	15
H-(Val-Orn-Leu-D-Tyr-Pro)$_2$-OH (21)	107
H-(Val-Lys-Leu-D-Phe-Pro)$_2$-OH (22)	66.7
Gramicidin S (GS)	1.85
GS + (20)	0.5 + 3.8
GS + (22)	0.5 + 12.8

TABLE 2

Minimum Inhibitory Concentration for
Microorganisms by Linear Decapeptide Derivatives

Compound[a]	Concentration (µg/ml)		
	S. aureus (FDA 209P)	B. subtilis (ATCC 6633)	E. coli (IAM 1253)
Gramaicidin S	1.6	0.8	25
For-I-Ea (7)	100	50	100
For-II-Am	25	3.1	100
Ac-II-Am	50	6.3	100
For-II-EA	25	6.3	100
Ac-II-EA	50	12.5	100
H-II-OH	25	3.1	25

[a] I, (D-Phe-Pro-Val-Orn-Leu)$_2$; II, (Val-Orn-Leu-D-Phe-Pro)$_2$; For, HCO; EA, NHCH$_2$CH$_2$OH; Ac, CH$_3$CO; Am, NH$_2$.

benzyloxycarbonyl group, and after polymerization, the protecting group was cleaved by the action of hydrogen bromide in acetic acid. These copolymers, usually containing the same amino acids as those present in gramicidin S or alanine and sarcosine residues, possessed a striking resemblance to the activity of gramicidin S.

The minimum concentrations of gramicidin S and some of the copolymers that inhibit the growth of E. coli and Micrococcus pyogenes in synthetic media are summarized in Table 3. A random copolymer composed of the amino acids of gramicidin S in the proper ratio inhibits the growth of the organism tested at nearly the same concentration as the natural one. An alteration in the configuration of phenylalanine residue from D to L causes little change in antibacterial activity. The substitution of lysine for ornithine yields an equally active compound. The elimination of proline, or its substitution by sarcosine, results in no loss of activity. Copolymers

TABLE 3

Minimum Inhibitory Concentration for Microorganisms by
Amino Acid Polymers and Copolymers Related to Gramicidin S

Amino acid composition and optical configuration					DP^a	Concentration (µg/ml)	
Val	Orn	Leu	Phe	Pro		E. coli	M. pyogenes
L	L	L	D	L	30	2.5 - 5	2.5 - 5
L	L	L	L	L	40	10 - 15	10 - 15
L	Lys L	L	D	L	40	2.5 - 5	2.5 - 5
L	L	L	D	Sar	25	5 - 10	5 - 10
L	L	L	D		20	2.5 - 5	2.5 - 5
	L	L	Ala_3 DL		25	2.5 - 5	5 - 10
	L	L	Ala L		45	2.5 - 5	2.5 - 5
	L	L			23	0.3 - 0.5	2.5 - 5
		L	Ala_3 L		26	>250	>250
	L				73	1.5 - 2.5	1.5 - 2.5
			Ala DL		35	>150	>150
Gramicidin S						2.5 - 5	0.5 - 1

^{a}Average number of amino acid residues for a molecule.

of ornithine with L-leucine or DL-alanine are all pronouncedly
active, and one of them, the (Orn, Leu) copolymer, even surpassed
the natural antibiotic in activity against E. coli. Marked antibacterial inhibition is found in all the copolymers containing
ornithine, as well as in polyornithine. By contrast, the polymers
and copolymers without basic residues are inactive. These results
reveal the importance of basic amino acids and of amino acids with
a hydrophobic side chain.

Linear analogs of the tyrocidines have never been made, although such analogs may show antimicrobial action. Gramicidin A (7) was synthesized by Sarges and Witkop by the usual solution method [49], and the same peptide is under preparation by Fontana and Gross [50] with the solid-phase method.

B. Analogs of Gramicidin S

Schwyzer and Sieber prepared bis-homogramicidin S, which contains lysine in place of ornithine, by a cyclization reaction from the corresponding decapeptide active ester [51]. The cyclization reaction with the pentapeptide active ester was also performed in order to obtain the dimerized cyclodecapeptide [33]. The replacement of ornithine by lysine did not produce a significant loss in antibacterial activity.

Izumiya's group has done extensive synthetic work [36, 52-63] on the substituted analogs of gramicidin S. In all cases, the cyclization reaction was done with the corresponding deca- or pentapeptide intermediates. The use of decapeptide active esters yielded cyclic decapeptide exclusively. By contrast, cyclization with pentapeptide active esters produced three types of products: dimerized cyclodecapeptide, monomeric cyclopentapeptide, and a mixture of cyclic deca- and pentapeptides. In the last cases, the ratio of the cyclic dimer to the monomer varies with the sequence of each pentapeptide. In other words, the primary structure of the reacting pentapeptide active esters mediates the mode of cyclization and probably the conformation of the resultant cyclic peptides.

The steps necessary to separate a monomeric cyclopeptide and a dimerized cyclopeptide from a reaction mixture is diagrammed in Fig. 4 as an example of these experiments [59]. The p-methoxy-benzyloxycarbonyl group was used as the α-N-protecting group in the initial peptide p-nitrophenyl ester, while the benzyloxycarbonyl group was used for δ-N-protection. The former was cleaved

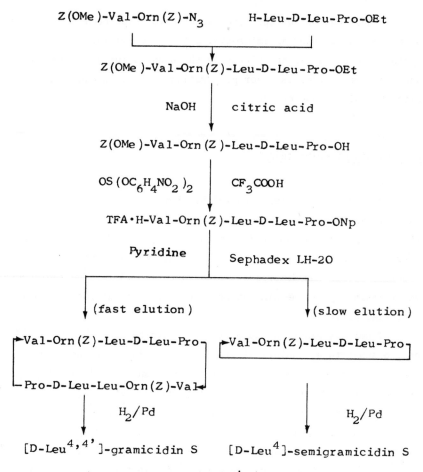

FIG. 4. Synthesis of [D-Leu$^{4,4'}$]-gramicidin S and its semiform.

selectively prior to cyclization by the action of trifluoroacetic acid. Splitting of the latter protecting group was effected by catalytic hydrogenation. The separation of cyclic penta- and decapeptides was achieved by column chromatography on Sephadex LH-20 for protected cyclic peptides and on carboxymethylcellulose for naked cyclic peptides. The molar ratios of the cyclic monomer to the dimer derived from several pentapeptide esters are summarized in Table 4.

TABLE 4

Ratio of Protected Monomer and Dimer
after Cyclization of Linear Pentapeptide Active Esters

p-Nitrophenyl ester of[a]					Ratio of compounds in product[b]		References
					Z-cyclic monomer	Z-cyclic dimer	
1	2	3	4	5			
Val	Orn(Z)	Leu	D-Phe	Pro	32	68	36
Gly					100	0	52
Gly				Gly	100	0	53
Ala					91	9	52
Leu					78	22	54
	Dab(Z)[c]				30	70	55
	Lyz(Z)				29	71	55
		Gly			59	41	56
		Ala			43	57	56
			Gly		0	100	57
			D-Ala		25	75	58
			D-Val		17	83	59
			D-Leu		15	85	59
				Gly	79	21	60, 61
				Sar	85	15	62
				β-Ala	89	11	63
Orn(Z)	Leu	D-Phe	Gly	Gly	100	0	64
D-Phe	Leu	Orn(Z)	Val	Pro	53	57	64
(Retrogramicidin S pentapeptide)							

[a]After the first compound listed, only variations of residue are shown.

[b]The concentration of linear pentapeptide p-nitrophenyl esters in pyridine was approximately 3×10^{-3} M. The ratios shown are weight/weight.

[c]α,γ-Diaminobutyric acid.

Schwyzer proposed that the act of acquiring the pleated sheet conformation in the transition state or in the product might force the reaction in such a desired direction. If so, then the peptides with odd numbers of amino acid residues would be expected to be especially prone to doubling cyclization [34]. However, Table 4 indicates that the presence of L-valine residue in the amino terminus is the decisive factor in dimerization and that the proline residue in the carboxyl end probably plays an important role in the doubling reaction [52, 53]. When glycine occupied the amino terminal of the peptide, no doubling reaction occurred; by contrast, with L-alanine and its hydrophobic side chain, the ratio of dimerized product decreased to below 10% [52]. Further, the cyclized dimer had the same activity as gramicidin S (see Table 5). Inexplicably, the substitution of glycine for D-phenylalanine caused exclusive dimerization. There is still no reasonable explanation for this finding.

All of the above cyclic peptides were subjected to biologic assay. None of the cyclic pentapeptides showed antibacterial activity. The specific activity of the cyclic decapeptides are given in Table 5.

The two aliphatic side chains of Val[1] and Leu[3] can be replaced with other aliphatic amino acids--for example, L-alanine--without intensive decrease of the activity. However, when the substituent amino acid is glycine, this conversion results in the complete loss of activity. The chain length of the basic amino acid can be varied without significant effect on the activity [55]. The substitution of aliphatic D-amino acids for D-phenylalanine did not produce a bad effect [58, 59], yet glycine again caused substantial loss of activity [57]. Interestingly, proline could be exchanged for glycine without loss of activity [60]. The activity of [Sar[5,5']]-gramicidin S was identical to the natural one [62], but the substitution of β-alanine for proline afforded an inactive product [63].

TABLE 5

Minimum Inhibitory Concentration for

Microorganisms by Analogs of Gramicidin S[a]

Amino acid sequence of cyclodecapeptide[b]					Concentration (µg/ml)			References
					B. subtilis	S. aureus	E. coli	
1	2	3	4	5				
Val	Orn	Leu	D-Phe	Pro	5	5	>100	36
Gly					100	100	>100	52
Gly				Gly	>100	>100	>100	53
Ala					5	5	>100	52
Leu					5	10	>100	54
	Dab				5	10	>100	55
	Lys				10	10	>100	55
		Gly			>100	>100	>100	56
		Ala			20	50-100	>100	56
			Gly		50	>100	>100	57
			D-Ala		25	50		58
			D-Val		10	10	>100	59
			D-Leu		5	10	>100	59
				Gly	20		>100	60, 61
				Sar	5	10	>100	62
				-Ala	1C0	100	>100	63
Retrogramicidin S					50	50	>100	64
Natural gramicidin S					5	5	>100	36

[a]Modified Stephenson-Whetham's medium was used.

[b]Only the varied residues are shown other than the first compound

An all-L-gramicidin S was prepared by Rothe and Eisenbeiss
[65]. They made a linear pentapeptide, H-Val-Orn(Boc)-Leu-Phe-Pro-OH,
by stepwise elongation with benzyloxycarbonylamino acid N-hydroxysuccinimide esters. Cyclization of the linear pentapeptide by

o-phenylene chlorophosphite as a coupling reagent gave a mixture of
cyclic monomer and dimer in 24% and 36% yield, respectively. The
all-L-gramicidin S was inactive against gram negative bacteria.
The synthesis of cyclo-(Val-Lys-Leu-Gly-Pro)$_2$ was done in a similar
manner and the compound had weak activity (40 µg/ml) against
S. aureus [66]. In a patent [67], the preparation and some antibacterial characteristics were described for three analogs: cyclo-
(Leu-Leu-Leu-Phe-Pro)$_2$, cyclo-(Val-Tyr-Leu-Phe-Pro)$_2$, and cyclo-
(Val-Lys-Leu-Phe-Pro)$_2$. Okawa et al. [68] mentioned the synthesis
of cyclo-(Ser(Bzl)-Orn-Ser(Bzl)-D-Phe-Pro)$_2$ and reported a positive
antibacterial activity for the compound.

    Retrogramicidin S, a cyclodiastereomer having the amino acid
sequence of gramicidin S in the reverse direction, was made by
Waki and Izumiya [64] in order to investigate the contribution of
the peptide bonds in gramicidin S to antibacterial activity.
Cyclization through the corresponding pentapeptide active ester
furnished a mixture of cyclic monomer and dimer in a weight ratio
of 53:47. Retrogramicidin S exhibits weak activity (50 µg/ml)
against B. subtilis. Shemyakin et al. [69] synthesized retro-
[Gly$^{5,5'}$]-gramicidin S and retroenatio-[Gly$^{5,5'}$]-gramicidin S.
These peptides were obtained by a doubling reaction involving the
glycine C-terminated linear pentapeptide p-nitrophenyl esters. In
turn, these units were built stepwise from the C-terminus in which
the δ-amino group of ornithine was protected by a phthaloyl
residue. The two peptides had high antibacterial activity. This
result suggests that the topochemical approach has considerable
potential in structure-activity studies.

    [Gly$^{5,5'}$]-Gramicidin S was constructed with the solid-phase
method by Halstrøm and Klostermeyer [70]. The stepwise elongation
of the peptide chain to furnish a decapeptide-resin was achieved
by the use of dicyclohexylcarbodiimide. Release of the decapeptide

H-(Val-Orn(Tos)-Leu-D-Phe-Gly)$_2$-OH

(23)

and cyclization by dicyclohexylcarbodiimide and N-hydroxysuccinimide, followed by deprotection, furnished the desired peptide, which exhibited the same activity as that reported by Aoyagi et al. [60]. Shemyakin's group [71] prepared [Gly$^{5,5'}$]-gramicidin S by the application of a soluble polymeric carrier. After the same decapeptide (23) was isolated from the soluble decapeptide-resin, it was cyclized in pyridine through the p-nitrophenyl ester of (23). After deprotection, the resulting cyclodecapeptide posessed strong antibacterial activity.

C. Peptides with Smaller and Larger Ring Structures

Izumiya's group has made a series of peptides with rings smaller than the 30-membered one found in gramicidin S. In Sec. III, B, a number of cyclic pentapeptides were described, but these showed no antibacterial activity. Furthermore, several cyclic di- [72], hexa- [73], and heptapeptides [74] having appropriate amino acid sequences were prepared, yet none had any activity. Some of these structures are shown in (24).

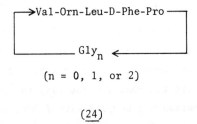

$$\begin{array}{l} \rightarrow \text{Val-Orn-Leu-D-Phe-Pro} \\ \quad\quad\quad\quad \text{Gly}_n \leftarrow \end{array}$$

(n = 0, 1, or 2)

(24)

[β-Ala$^{5,5'}$]-Gramicidin S, by virtue of the two additional methylene groups, has an internal 32-membered ring, but exhibited no activity (see Table 5). Sesqui- (the 45-membered ring) and digramicidin S (the 60-membered ring) were synthesized in the same laboratory, and these compounds revealed appreciable activities.

Here, the minimum inhibitory concentrations of sesqui-, digramicidin S, and natural gramicidin S for B. subtilis were 50, 10, and 5 µg/ml, respectively [75]. A penetrating insight into the exact relationship between molecular conformation and biologic activity would be required to secure further progress in this area.

D. Tyrocidines

The Craig and Kurahashi groups have elucidated structures (2)-(6) for the various components of the tyrocidine complex (see Sec. I). Recently, Izumiya's group has published details of the syntheses for each component with the exception of tyrocidine D [76-79].

The procedure for the synthesis of tyrocidine A (2) followed earlier schemes (76). Thus, the azide coupling of Z(OMe)-Phe-D-Phe-NHNH$_2$ and H-Asn-Gln-Tyr-OEt gave the pentapeptide ethyl ester. The corresponding hydrazide was joined to another pentapeptide fragment H-Val-Orn(Z)-Leu-D-Phe-Pro-OH (18), and the resultant decapeptide was cyclized by the same procedure as described for gramicidin S. Both the synthetic cyclodecapeptide and the natural tyrocidine A were identical to each other in terms of all physical and spectral characteristics.

In the case of tyrocidine B (3) [77] and C (4) [78], the usual azide procedure was avoided because it may have caused oxidation of the tryptophan residue. As a result, Z(OMe)-Asn-Gln-Tyr-NHNH$_2$ was coupled with the pentapeptide (18) to form an octapeptide, which upon hydrogenation was coupled with Z(OMe)-Trp-D-Phe(or D-Trp)-OH by the action of dicyclohexylcarbodiimide and N-hydroxysuccinimide. Cyclization and hydrogenation gave the desired cyclic decapeptides, which showed the expected molecular weights, as well as the expected elemental and amino acid analyses and reasonable antibacterial activities.

Kurahashi et al. [19] isolated a new substance, tyrocidine E, in a medium containing phenylalanine, but devoid of tyrosine and

tryptophan. The peptide was assigned the primary structure (6) on
the basis of an amino acid analysis. The synthesis of this peptide
was achieved by a procedure similar to that described for tyrocidine A [79].

The preparation of some amino acid substituted analogs of the
tyrocidines has been tried in the authors' laboratory. Four analogs,
[Gly^6]-, [Gly^7]-, [Gly^{10}]-, and [Gly^{6,7,10}]-tyrocidine A, where
glycine has replaced the aromatic residue, were prepared with the
solid-phase method [80]. The analogs showed reasonable antibacterial
activity. For example, the minimum inhibitory concentration of
[Gly^7]-tyrocidine A for S. aureus was 0.63 µg/ml and of gramicidin
S, 0.39 µg/ml.

V. CONFORMATION OF GRAMICIDIN S

Although studies on the conformational analyses of polypeptides
have made much progress, similar investigations of the gramicidin S
molecule have not yet led to unambiguous conclusions. In fact, there
are several incompatible views on the precise conformation of this peptide. Various molecular models for gramicidin S have been proposed,
with or without experimental evidence [12, 81-86] and some of these
studies have been reviewed briefly [86, 87].

In 1953, Abbott and Ambrose [81] proposed an initial model for
gramicidin S on the basis of infrared dichroism work. The peptide
backbone forms an α_{11} ribbon structure with one hydrogen bond in a
seven-membered ring. It is important to note that this conformation
does not have a genuine twofold axis of symmetry.

In 1957 Hodgkin [12] reported preliminary results from the
x-ray analysis of gramicidin S derivatives and mentioned that this
peptide should have a twofold axis of symmetry. From this data,
several possible models for the molecule were proposed. One of them

is based on the β-pleated sheet structure (see Fig. 3), which seems
to be by far the most likely. Indeed, this specific pattern was
employed by Schwyzer [34] to explain the apparent doubling cyclization
reaction as mentioned in Sec. III, although this hypothesis now seems
to be in doubt [36]. The β-sheet model has been supported by the
results of microbial assays of amino acid substituted analogs, as is
shown in Sec. VI. Another structure suggested is an α,β mixed type
structure and is compatible with the optical data [87].

Warner's model [82] is simply two fused hexagons, similar to
naphthalene, and is strain-free. It employs the usual bond-angles
and fulfills Kauzmann's suggestion [88] that close proximity of the
two aromatic rings would yield the most desirable hydrophobic arrangement. However, this particular conformation requires all peptide
bonds to be in the cis conformation. By contrast, Balasubramanian
[87] proposed, on the basis of infrared experiments, that at least
six peptide bonds of gramicidin S should be in the trans conformation.

Liquori et al. [83] calculated the potential energy of the
peptide units found in gramicidin S and proposed a conformation
consisting of two antiparallel single turns of a right-handed
α-helix bridged by proline and phenylalanine residues. Each
helical section contains one hydrogen bond between the carboxyl
oxygen of valine and the amino hydrogen of phenylalanine. This
model appeared to be compatible with the results of nuclear
magnetic resonance (NMR) studies [89], optical rotatory dispersion
(ORD), and circular dichroism (CD) measurement [90]. Yet, Bradbury
et al. [91] claimed considerable difficulties were involved if the
NMR chemical shift of the α-CH peak, as customarily used for conformation analysis of polypeptides, is compared with the peak
position of homopolypeptides in various solvent systems. Furthermore,
the apparent similarity of the ORD and the CD curves for gramicidin
S to those of various α-helical polypeptides must be interpreted
carefully because of possible confusion with other conformations
[87, 92, 93]. Lately, Conti has revised his earlier attempt to
correlate the NMR spectra and the conformation of gramicidin S, so
this approach remains an open problem [94].

Scheraga's group [84] has suggested several structures on the basis of energy minimization calculation. Their model was revised twice [95] and the last one has no distinct hydrogen bond, but contains three conformationally different types of exchangeable hydrogen atoms. The potential energy of this conformation is found to be the lowest among those known for gramicidin S [96]. It must be noted that their suggestions are deduced from purely theoretical work and have, as yet, received no experimental support.

Schwyzer and Ludescher [97] prepared N,N'-diphthaloylgramicidin S to investigate the NMR spectra of the molecule and demonstrated that the phthaloyl groups on the ornithine side chains are in close contact with the phenyl groups of the phenylalanine residues. Only one model satisfies this condition; it consists of six L-amino acid residues in the antiparallel β-pleated sheet structure.

Another proposal comes from Craig's group [85]. It is based on differential dialysis [20, 98], ORD [98], NMR [85], ^{13}C-NMR [99], and deuterium or tritium exchange experiments [100]. They conclude that above all, the NMR investigation requires that gramicidin S must have: (a) C_2 axis of symmetry; (b) a small (trans) dihedral angle (ϕ) for ornithine, leucine, and valine; and (c) a large (cis) dihedral angle for phenylalanine. They propose a model having two antiparallel L-tripeptide sequences with four intramolecular hydrogen bonds. The high-field chemical shift of the valine amide protons could be explained well by this model as well as by a shielding effect from the carbonyl groups of phenylalanine and/or valine. At least six of the peptide carbonyl groups and the corresponding amide hydrogen bonds are in the plane perpendicular to the C_2 axis, while two phenylalanine peptide carbonyl groups and amide hydrogen bonds are almost parallel to this axis. Moreover, this molecule may exhibit infrared dichroism in the carbonyl and amide regions, as shown by Abbott and Ambrose [81].

Ohnishi and Urry [101] measured the temperature-dependent chemical shifts for the amide protons in gramicidin S. The hydrogen-bonded amide protons of valines and leucines exhibit a lower slope with temperature than the non-hydrogen-bonded amide protons of

ornithines and phenylalanines. This result can be taken as another
confirmation of the β-sheet structure with four intramolecular
hydrogen bondings, including the amide protons of valines and
leucines.

Pysh [102] introduced a new method, optical calculation, to
screen a large number of hypothetical structures obtained from
energy minimization considerations. By the use of this procedure,
conformations that yield a spectrum different from the observed
spectrum could be excluded. Three of the structures found as local
minimums in the energy surface were used to define the final set of
conformations to be screened. Only one of the three geometries, the
structure proposed by Craig [85], leads to a significantly large
negative b_0 value [103]. This result is additional evidence of a
β-sheet-type structure for gramicidin S.

On the basis of conformational calculations made by Camilletti
et al. [104] the presence of a dyad axis, four hydrogen bonds, and
the proximity of the δ-amino groups of the two ornithine residues
were suggested in the cyclic peptide. To confirm this point, they
prepared the copper complex of bis-salicyl-aldiminato-gramicidin S.
A comparison of the ultraviolet spectra and the results of x-ray
analyses for this and related compounds confirmed both the rigidity
of the molecule and the over-all similarity between the shapes of
the derivatives.

Thus, the most plausible geometry for the gramicidin S molecule
seems to be the structure having an antiparallel L-tripeptide sequence
with four hydrogen bondings. With respect to the Phe-Pro segment,
however, the dihedral angles assigned by Craig [85], for example
ϕ = 150°, and ψ = 130° for phenylalanine, were criticized by the
Russian workers [86].

Ovchinnikov et al. [86] prepared N,N'-diacetylgramicidin S and
proposed that this derivative should contain six hydrogen-bonded
amide groups. Two of these are formed by the α-amino groups of
ornithines, while the $^3J_{NH-CH}$ constant of 4.1 Hz for the phenylalanine residues corresponds to the gauche conformation. The
structure derived from these findings has ϕ and ψ values, as shown

in Table 6 [86]. They made N,N'-carbonylgramicidin S and showed
that the spectral characteristics of the compound was similar to
those of gramicidin S and diacetylgramicidin S.

Since gramicidin S, the tyrocidines, and the amino acid-
substituted analogs of gramicidin S posessing antibacterial
activities show related ORD [87, 105] and CD [92] curves, this
suggests that the loci of similar conformations may be correlated
to the biological activity of these molecules.

VI. RELATIONSHIP BETWEEN STRUCTURE AND BIOLOGIC ACTIVITY

Ample knowledge has been accumulated on the structure and
biologic activity of natural antibiotic polypeptides. Yet the
precise mechanism of action of these compounds still remains un-
clear. This specific situation may be even worse than the one
just discussed for conformational analysis.

As shown by Harris and Work [42, 43], the presence of
several unusual amino acids in an appropriate peptide sequence is
insufficient to generate satisfactory antibacterial activity.
Katchalski revealed the remarkable activities of amino acid copolymers, especially those containing ornithine residues. The
occurrence of ornithine in gramicidin S and the tyrocidines gives
the basic character to these molecules and causes an increase in

TABLE 6

Values of ϕ and ψ in Gramicidin S

	Val	Orn	Leu	D-Phe	Pro
ϕ	60	70	60	235	120
ψ	300	290	290	70	140

the permeability of the cell membrane, thereby leading to leakage
of the cell constituents. The mode of action of the synthetic amino
acid copolymer is similar to that of the natural antibiotic, because
a clear cross-resistance relationship was found between an (Orn, Leu,
Ala) copolymer and gramicidin S [47]. The (Orn, Leu) or (Orn, Leu,
Ala) copolymers and the natural antibiotic possess common structural
features characteristic of cationic detergents. Both contain hydrophilic cationic groups and lipophilic aliphatic side chains, the
former promoting electrostatic attraction to the negatively charged
bacteria, and the latter facilitating the anchoring of the polypeptide
in the lipid layer of the cell membrane [106].

Znamenskaya and Belozerskii [107] tested the antibacterial
activity of gramicidin S derivatives on Micrococcus pyogenes var.
aureus. The substitution of guanidines for the δ-amino groups, and
the alkylation and nitration of gramicidin S, did not produce any
noticeable changes in activity. However, blocking of the δ-amino
groups with acyl residues formed inactive compounds. The presence
of a free δ-amino group in gramicidin S seems to be essential for
biologic activity as in the case of the amino acid copolymer.

Stepanov et al. [108] prepared several δ-aminoacyl and
δ-peptidyl derivatives of gramicidin S. The antibacterial activity
of these peptides slowly decreased as the chains were lengthened in
the peptidyl portions. On the other hand, when the δ-amino groups
of ornithines were changed to carboxylates, the activity of these
derivatives fell to almost negligible values [109].

The antibacterial activity of the amino acid substituted
analogs of gramicidin S can be explained nicely if we assume that
the antibiotic molecule is in the β-pleated sheet structure. Then
the peptide backbone lies on a planer surface, one side being
hydrophilic and another hydrophobic. Substitution of alanine or
leucine for valine will not cause a substantial change in the
nature of the hydrophobic surface of the molecule. Furtheimore,
the replacement of D-phenylalanine by another aliphatic amino acid
of D- configuration will not disturb the conformation of the mole-

cule as a whole. Also, variation in the chain length of the basic
amino acids will not cause any change in the shape of the molecule.
Thus it is quite acceptable that these substitutions do not result
in a decrease of activity. The weak activity of retrogramicidin S
[64] can be explained similarly, because the hydrophobic pyrrole
rings of the proline residues will face the hydrophilic ornithine
side, and this may hinder a possible electrostatic interaction
between the ornithine side chain and the cell surface.

On the other hand, the ORD experiments have shown that there
is a remarkable difference between the ORD curve of the biologically
active cyclic peptides and the inactive analogs [105], as shown in
Fig. 5. This finding suggests not only that a considerable change
of conformation takes place on such a substitution, but also that
the resulting inactivation must be dependent on the tertiary shape
of the molecule.

There are several interesting reports concerned with the
interaction of antibiotics and cell surfaces. Uemura et al. [110]
incubated ^{14}C-labeled gramicidin S with sensitive strains of
B. subtilis or S. aureus and found that over 80% of the radioactive
gramicidin S was absorbed by the cells. Later, when the cells had
been disrupted and centrifuged, 70% of the activity was found in
the ribonucleoprotein fraction. If gramicidin S-resistant strains
of these bacteria were used, then only 25 to 30% of the antibiotic
was absorbed by the cell, and the radioactivity was found exclusively
in the bacterial cell wall. Miki [111] observed that gramicidin
S-sensitive strains of Micrococci release nucleotides instantly at
0°C, even with the antibiotic in the low concentration of 1 μg/ml.
This result has also been confirmed by Bulgakova et al. [112].

Hiramatsu [113] reported the protoplasts of Micrococci
lysodeiktici were lysed by the action of gramicidin S in a concentration of 10 μg/ml. By contrast, the protoplasts of non-
sensitive E. coli were lysed only slightly, and the protoplasts of
gramicidin S-producing strains of B. brevis were not affected by
the antibiotic at the high concentration of 100 μg/ml.

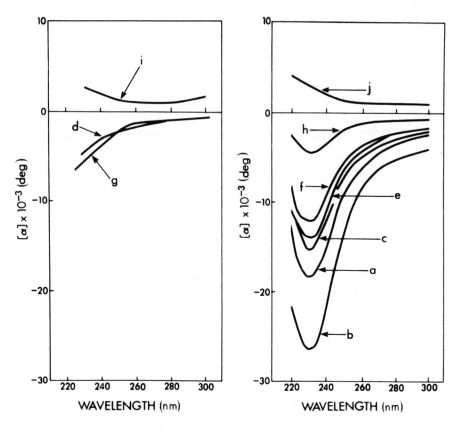

FIG. 5. Optical rotatory dispersions of cyclic penta- and decapeptides related to gramicidin S (GS) in ethanol. a, GS; b, [Dab$^{2,2'}$]-GS; c, [Lys$^{2,2'}$]-GS; d, semiGS; e, [D-Ala$^{4,4'}$]-GS; f, [Gly$^{4,4'}$]-GS; g, retrosemiGS; h, retroGS; i, [β-Ala^{5}]-semiGS; j, [β-Ala$^{5,5'}$]-GS.

These observations support the speculation that gramicidin S penetrates the cell wall of the sensitive microorganism and interacts with the cell membrane, probably changing the permeability of the membrane and releasing nucleotides and other intracellular components [111, 112]. Presumably, the gramicidin S-resistant strains may be able to repulse the penetration of the antibiotic through the cell wall [109]. As discussed above, the cationic detergent

theory has been a useful working hypothesis and did prompt
several noteworthy investigations. Unfortunately, we have not yet
been able to explain decisively the mode of interaction between
the gramicidin S molecule and the cell membrane.

Some new approaches to the structure-activity relationship
were published recently. The topochemical investigations begun
by Shemyakin have given fruitful results, especially in the field
of cyclic depsipeptides. The remarkable antibacterial activity
of a retroenatiomer of $[Gly^{5,5'}]$-gramicidin S could also be
rationalized in a topochemical fashion [114].

It has been reported that valinomycin and enniatin react
as selective cation carriers [115]. A similar ion-carrying
activity of tyrocidine B was recently demonstrated [116]. Unlike
valinomycin and enniatin, gramicidin S does not have an internal
cavity capable of accommodating metal cations, as shown in Sec. V.
This may account for its inability to induce cation transport
effects [117]. Thus, the mode of interaction between gramicidin S
and a biologic membrane seems to be different from that of the
tyrocidines, in spite of the similar amino acid sequence and
biological origin of both antibiotics. Precise knowledge on the
tertiary structure of these peptides, especially on the perturba-
tion of conformations in the solution state, is absolutely
necessary for a full understanding of the structure-activity
relationships in gramicidin S and related compounds.

REFERENCES

1. R. J. Dubos, J. Exptl. Med., 70, 1, 249 (1939); through CA,
   33, 6902, 8678 (1939).
2. R. D. Hotchkiss and R. Dubos, J. Biol. Chem., 132, 791, 793
   (1940); 136, 803 (1940).

3. A. R. Battersby and L. C. Craig, J. Am. Chem. Soc., 74, 4019, 4023 (1952).

4. G. F. Gause and M. G. Brazhnikova, Am. Rev. Soviet Med., 2, 134 (1944); through CA, 39, 1893[3] (1945); G. F. Gause and M. G. Brazhnikova, Nature, 154, 703 (1944).

5. R. L. M. Synge, Biochem. J., 39, 363 (1945).

6. K. O. Pederson and R. L. M. Synge, Acta Chem. Scand., 2, 408 (1948).

7. A. R. Battersby and L. C. Craig, J. Am. Chem. Soc., 73, 1887 (1951).

8. R. Consden, A. H. Gordon, A. J. P. Martin, and R. L. M. Synge, Biochem. J., 41, 596 (1947).

9. A. J. P. Martin and R. Mittelman, Biochem. J., 43, 353 (1948).

10. F. Sanger, Biochem. J., 40, 261 (1946).

11. G. M. J. Schmidt, D. C. Hodgkin, and B. M. Oughton, Biochem. J., 65, 744 (1957).

12. D. C. Hodgkin and B. M. Oughton, Biochem. J., 65, 752 (1957).

13. A. N. Belozerskii and T. S. Paskhina, Biokhimiya, 10, 344 (1945); through CA, 40, 1557[8] (1946).

14. R. Schwyzer and P. Sieber, Helv. Chim. Acta, 40, 624 (1957).

15. A. Paladini and L. C. Craig, J. Am. Chem. Soc., 76, 688 (1954).

16. T. P. King and L. C. Craig, J. Am. Chem. Soc., 77, 6627 (1955).

17. M. A. Ruttenberg, T. P. King, and L. C. Craig, Biochemistry, 4, 11 (1965).

18. M. A. Ruttenberg and B. Mach, Biochemistry, 5, 2864 (1966).

19. K. Fujikawa, Y. Sakamoto, T. Suzuki, and K. Kurahashi, Biochim. Biophys. Acta, 169, 520 (1968).

20. L. C. Craig, Science, 144, 1093 (1964).

21. R. Sarges and B. Witkop, J. Am. Chem. Soc., 86, 1861, 1862 (1964).

22. S. Otani, T. Yamanoi, Y. Saito, and S. Otani, Biochem. Biophys. Res. Commun., 25, 590 (1966).

23. S. Tomino and K. Kurahashi, Biochem. Biophys. Res. Commun., 17, 288 (1964).

24. H. Holm, L. D. Frhølm, and S. Laland, Biochim. Biophys. Acta, 115, 361 (1966).

25. M. Yamada and K. Kurahashi, J. Biochem. (Tokyo), 63, 59 (1968).

26. W. Gevers, H. Kleinkauf, and F. Lipmann, Proc. Natl. Acad. Sci. U.S., 63, 1335 (1969).

27. E. Stoll, Ø. Frøyshov, H. Holm, T. L. Zimmer, and S. G. Laland, FEBS Letters, 11, 348 (1970).

28. C. C. Gilhuus-Moe, T. Kristensen, J. E. Bredesen, T.-L. Zimmer, and S. G. Laland, FEBS Letters, 7, 287 (1970).

29. R. Roskoski, Jr., W. Gevers, H. Kleinkauf, and F. Lipmann, Biochemistry, 9, 4839 (1970).

30. R. Roskoski, Jr., H. Kleinkauf, W. Gevers, and F. Lipmann, Biochemistry, 9, 4846 (1970).

31. L. W. Pollard, N. V. Bhagavan, and J. B. Hall, Biochemistry, 7, 1153 (1968).

32. B. Iselin and R. Schwyzer, Helv. Chim. Acta, 43, 1760 (1960).

33. R. Schwyzer and P. Sieber, Helv. Chim. Acta, 41, 2186 (1958).

34. R. Schwyzer, Amino Acids and Peptides with Antimetabolic Activity (CIBA Foundation Symposium), Churchill, London, 1958, p. 171.

35. L. Pauling and R. B. Corey, Proc. Natl. Acad. Sci. U.S., 39, 247 (1953).

36. M. Waki and N. Izumiya, J. Am. Chem. Soc., 89, 1278 (1967); M. Waki and N. Izumiya, Bull. Chem. Soc. Japan, 40, 1687 (1967).

37. S. Matsuura and N. Izumiya, 24th Annual Meeting of Chemical Society of Japan, Osaka, April 1971.

38. H. Klostermeyer, Chem. Ber., 101, 2823 (1968).

39. G. Losse and K. Neubert, Tetrahedron Letters, 1970, 1267.

40. M. Ohno, K. Kuromizu, H. Ogawa, and N. Izumiya, J. Am. Chem. Soc., 93, 5251 (1971).

41. M. Ohno and C. B. Anfinsen, J. Am. Chem. Soc., 89, 5994 (1967).

42. J. I. Harris and T. S. Work, Biochem. J., 46, 196 (1950).

43. J. I. Harris and T. S. Work, Biochem. J., 46, 582 (1950).

44. B. F. Erlanger and L. Goode, Nature, 174, 840 (1954).

45. B. F. Erlanger and L. Goode, Science, 131, 669 (1960).

46. S. Makisumi and N. Izumiya, Tetrahedron Letters, 1970, 227;
    S. Makisumi, M. Waki, and N. Izumiya, Bull. Chem. Soc. Japan,
    44, 143 (1971); S. Makisumi, S. Matsuura, M. Waki, and
    N. Izumiya, Bull. Chem. Soc. Japan, 44, 210 (1971).

47. E. Katchalski, L. Bichowski-Slomnitzki, and E. Valcani,
    Biochem. J., 55, 671 (1953).

48. L. Bichowsky-Slomnicki, A. Berger, J. Kurtz, and E. Katchalski,
    Arch. Biochem. Biophys., 65, 400 (1956).

49. R. Sarges and B. Witkop, J. Am. Chem. Soc., 87, 2020 (1965).

50. A. Fontana and E. Gross, personal communication, 1970.

51. R. Schwyzer and P. Sieber, Helv. Chim. Acta, 41 1582 (1958).

52. M. Kondo and N. Izumiya, Bull. Chem. Soc. Japan, 40, 1975 (1967).

53. M. Kondo, H. Aoyagi, T. Kato, and N. Izumiya, Bull Chem. Soc.
    Japan, 39, 2234 (1966).

54. M. Kondo and N. Izumiya, Bull. Chem. Soc. Japan, 43, 1850 (1970).

55. M. Waki, O. Abe, R. Okawa, T. Kato, S. Makisumi, and N. Izumiya,
    Bull. Chem. Soc. Japan, 40, 2904 (1967).

56. O. Abe and N. Izumiya, Bull. Chem. Soc. Japan, 43, 1202 (1970).

57. R. Nagata, M. Waki, M. Kondo, H. Aoyagi, T. Kato, S. Makisumi,
    and N. Izumiya, Bull. Chem. Soc. Japan, 40, 963 (1967).

58. S. Lee, R. Ohkawa, and N. Izumiya, Bull. Chem. Soc. Japan, 44,
    158 (1971).

59. H. Aoyagi, T. Kato, M. Waki, O. Abe, R. Okawa, S. Makisumi,
    and. N. Izumiya, Bull. Chem. Soc. Japan, 43, 782 (1969).

60. H. Aoyagi, T. Kato, M. Ohno, and N. Izumiya, J. Am. Chem.
    Soc., 86, 5700 (1964); H. Aoyagi, T. Kato, M. Ohno, M. Kondo,
    M. Waki, S. Makisumi, and N. Izumiya, Bull. Chem. Soc. Japan,
    38, 2139 (1965).

61. H. Aoyagi, M. Kondo, T. Kato, S. Makisumi, and N. Izumiya,
    Bull. Chem. Soc. Japan, 40, 1685 (1967).

62 H. Aoyagi and N. Izumiya, Bull. Chem. Soc. Japan, 39, 1747
    (1966).

63. S. Matsuura, M. Waki, S. Makisumi, and N. Izumiya, Bull. Chem.
    Soc. Japan, 43, 1197 (1970).

64. M. Waki and N. Izumiya, Tetrahedron Letters, 1968, 3083;
    M. Waki and N. Izumiya, Bull. Chem. Soc. Japan, 41, 1909 (1968).
65. M. Rothe and F. Eisenbeiss, Angew. Chem., 80, 907 (1968).
66. M. Rothe and F. Eisenbeiss, Z. Naturforsch., 21b, 814 (1966).
67. R. Schwyzer (to CIBA Ltd.), Ger. Pat. 1, 112, 525 (1962);
    through CA, 75, 949f (1962).
68. K. Okawa, S. Hase, and A. Uno, 18th Annual Meeting of Chemical
    Society of Japan, Tokyo, April 1965.
69. M. M. Shemyakin, Yu. A. Ovchinnikov, V. T. Ivanov, and I. D.
    Ryabova, Experientia, 23, 326 (1967); M. M. Shemyakin, Yu. A.
    Ovchinnikov, and V. T. Ivanov, Agnew. Chem. Intern. Ed. Engl.,
    8, 492 (1969).
70. J. Halstrøm and H. Klostermeyer, Liebigs Ann. Chem., 715, 208
    (1968).
71. A. A. Kiryushkin, Yu. A. Ovchinnikov, I. V. Kozhevnikova, and
    M. M. Shemyakin, in Peptides (H. C. Beyerman, A. van de Linde,
    and W. Maassen van den Brink, eds.), North-Holland, Amsterdam,
    1967, p. 100.
72. N. Izumiya, T. Kato, Y. Fujita, M. Ohno, and M. Kondo, Bull.
    Chem. Soc. Japan, 37, 1809 (1964).
73. T. Kato, M. Kondo, M. Ohno, and N. Izumiya, Bull. Chem. Soc.
    Japan, 38, 1202 (1965).
74. O. Abe, K. Kuromizu, M. Kondo, and N. Izumiya, Bull. Chem. Soc.
    Japan, 43, 914 (1970).
75. S. Matsuura and N. Izumiya, Experientia, 28, 1402 (1972).
76. M. Ohno and N. Izumiya, J. Am. Chem. Soc., 88, 376 (1966);
    M. Ohno, T. Kato, S. Makisumi, and N. Izumiya, Bull. Chem.
    Soc. Japan, 39, 1738 (1966).
77. K. Kuromizu and N. Izumiya, Experientia, 26, 587 (1970); Bull.
    Chem. Soc. Japan, 43, 2199 (1970).
78. K. Kuromizu and N. Izumiya, Tetrahedron Letters, 1970, 1471;
    Bull. Chem. Soc. Japan, 43, 2944 (1970).
79. N. Mitsuyasu and N. Izumiya, Experientia, 26, 476 (1970);
    N. Mitsuyasu, S. Matsuura, M. Waki, M. Ohno, S. Makisumi,
    and N. Izumiya, Bull. Chem. Soc. Japan, 43, 1829 (1970).

80. K. Okamoto, K. Kuromizu, M. Waki, T. Kato, and N. Izumiya, 24th Annual Meeting of Chemical Society of Japan, Osaka, April 1971.

81. N. B. Abbott and E. J. Ambrose, Proc. Roy. Soc. (London), A219, 17 (1953).

82. D. T. Warner, Nature, 190, 120 (1961).

83. A. M. Liquori, P. De Santis, A. L. Kovacs, and L. Mazzarella, Nature, 211, 1039 (1966).

84. G. Vanderkooi, S. J. Leach, G. Némethy, R. A. Scott, and H. A. Scheraga, Biochemistry, 5, 2991 (1966).

85. A. Stern, W. A. Gibbons, and L. C. Craig, Proc. Natl. Acad. Sci. U.S., 61, 734 (1968).

86. Yu. A. Ovchinnikov, V. T. Ivanov, V. F. Bystrov, A. I. Miroshnikov, E. N. Shepel, N. D. Abdullaev, E. S. Efremov, and L. B. Senyavina, Biochem. Biophys. Res. Commun., 39, 217 (1970).

87. D. Balasubramanian, J. Am. Chem. Soc., 89, 5445 (1967).

88. W. Kauzmann, Advan. Protein Chem., 14, Academic, New York, 1959, p. 33.

89. A. M. Liquori and F. Conti, Nature, 217, 635 (1968).

90. F. Quadrifoglio and D. W. Urry, Biochem. Biophys. Res. Commun., 29, 785 (1967).

91. E. M. Bradbury, B. G. Carpenter, C. Crane-Robinson, and H. W. E. Rattle, Nature, 220, 69 (1968).

92. S. Laiken, M. Printz, and L. C. Craig, J. Biol. Chem., 244, 4454 (1969).

93. G. D. Fasman, H. Hoving, and S. N. Timasheff, Biochemistry, 9, 3316 (1970).

94. F. Conti, Nature, 221, 777 (1969).

95. R. A. Scott, G. Vanderkooi, R. W. Tuttle, P. M. Shames, and H. A. Scheraga, Proc. Natl. Acad. Sci. U.S., 58, 2204 (1967).

96. F. A. Momany, G. Vanderkooi, R. W. Tuttle, and H. A. Scheraga, Biochemistry, 8, 744 (1969).

97. R. Schwyzer and U. Ludescher, Biochemistry, 7, 2514, 2519 (1968).

98. L. C. Craig, Proc. Natl. Acad. Sci. U.S., 61, 152 (1968).

99. W. A. Gibbons, J. A. Sogn, A. Stern, L. C. Craig, and L. F. Johnson, Nature, 227, 840 (1970).

100. S. L. Laiken, M. P. Printz, and L. C. Craig, Biochemistry, 8, 519 (1969).

101. M. Ohnishi and D. W. Urry, Biochem. Biophys. Res. Commun., 36, 194 (1969).

102. E. S. Pysh, Science, 167, 290 (1970).

103. P. J. Urnes and P. Doty, Advan. Protein Chem., 16, Academic, New York, 1961, p. 401.

104. C. Camiletti, P. De Santis, and R. Rizzo, Chem. Commun., 1970, 1073.

105. T. Kato, M. Waki, S. Matsurra, and N. Izumiya, J. Biochem. (Tokyo), 68, 751 (1970).

106. M. Sela and E. Katchalski, Advan. Protein Chem., 14, Academic, New York, 1959, p. 439.

107. M. P. Znamenskaya and A. N. Belozerskii, Antibiotiki, 2, 36 (1957); through CA, 51, 16888f (1957).

108. V. M. Stepanov, A. B. Silaev, and A. N. Polin, Antibiotiki, 3, 49 (1958); through CA, 53, 2476h (1959).

109. V. M. Stepanov and A. B. Silaev, Zh. Obshch. Khim., 31, 3811 (1961); through CA, 57, 16738f (1962).

110. I. Uemura, C. J. Cherng, T. Saito, and Y. Hashimoto, Osaka Shiritsu Daigaku Igaku Zasshi (in Japanese) 9, 3625 (1960); through CA, 55, 10587C (1961).

111. Y. Miki, Osaka Shiritsu Daigaku Igaku Zasshi (in Japanese), 9, 4005 (1960); through CA, 55, 10587g (1961).

112. V. G. Bulgakova and A. N. Polin, Antibiotiki, 11, 714 (1966); through CA, 65, 15823e (1966).

113. T. Hiramatsu, Osaka Shiritsu Daigaku Igaku Zasshi (in Japanese), 10, 267 (1961); through CA, 56, 14720h (1962).

114. Yu. A. Ovchinnikov, Vestn. Akad. Nauk SSSR, 38, 43 (1968); through CA, 69, 83431q (1968).

115. Yu. A. Ovchinnikov, V. T. Ivanov, A. V. Evstratov, V. F.
     Bystrov, N. D. Abdullaev, E. M. Popov, G. M. Lipkind,
     S. F. Arkhipova, E. S. Efremov, and M. M. Shemyakin, Biochem.
     Biophys. Res. Commun., 37, 668 (1969).
116. M. C. Goodall, Biochim. Biophys. Acta, 219, 28 (1970).
117. D. C. Tosteson, T. E. Andreoli, M. Tieffenberg, and P. Cook,
     J. Gen. Physiol., 51, 373S (1968).

CHAPTER 2
SYNTHESIS OF ACTH-
ACTIVE PEPTIDES AND ANALOGS

H. Yajima and H. Kawatani

Faculty of Pharmaceutical Sciences
Kyoto University
Kyoto, Japan

I.   INTRODUCTION.   40
II.   EFFECTS OF ACTH AND ASSAY METHODS   40
III.   ISOLATION AND STRUCTURE DETERMINATION   44
IV.   SYNTHESIS OF ACTH PEPTIDES.   49
        A. Hydrogen Bromide Procedure.   50
        B. Sodium in Liquid Ammonia Procedure.   55
        C. Dilute Hydrochloric Acid Procedure.   62
        D. Trifluoroacetic Acid Procedure.   69
        E. Hydrogen Fluoride Procedure   91
        F. Solid-Phase Procedure   100
V.   STRUCTURE-FUNCTION OF ACTH PEPTIDES   107
        A. Chain Length and Activity   108
        B. Synthetic Analogs and Activity.   113
        C. Stereoisomers and Activity.   119
        D. Des-Chain Peptides and Activity   122
VI.   ACTH RECEPTOR   124
VII.   ACTH RELEASING FACTOR   127
        REFERENCES.   129

I. INTRODUCTION

ACTH (Adrenocorticotropic hormone or corticotropin) is one of
the peptide hormones of anterior pituitary origin. Regardless of
species variation, the mammalian ACTHs known today are straight-
chain polypeptides; each of these is composed of 39 amino acid
residues. The secretion of ACTH is presumably under the hypo-
thalamic control of CRF (corticotropin releasing factor), though
its chemical nature is not yet clearly characterized. The target
organ of ACTH is the adrenal cortex, where the production of corticoid
hormone is stimulated. Through this adrenal cortex action, the hormone
seems to participate in the defense mechanism of mammals. Since glu-
cocorticosteroids are known to exert an anti-inflammatory action,
expecially against rheumatism, this antiphlogistic effect in terms
of production is therapeutically important. The chemical synthesis
of ACTH-active peptides has continued in various laboratories
throughout the world and the synthetic hormone is now available
for clinical use.

Several review articles are available on the biologic aspects
of ACTH: Astwood et al. in 1952 [1]; Hays and White in 1954
[2]; Engel in 1961 [3]; Liddle et al. in 1962 [4]; Phillips and
Bellamy in 1963 [5]; and Lebovitz and Engel in 1964 [6]. With
respect to synthetic aspects, see: Hofmann and Yajima in 1962
[7]; Hofmann in 1962 [8], 1963 [9, 10]; Li in 1956 [11], 1959
[12], 1962 [13, 14]; Schwyzer in 1963 [15], 1964 [16], 1966 [17];
and Schröder and Lübke in 1965 [18].

II. EFFECTS OF ACTH AND ASSAY METHODS

ACTH is produced by the basophilic cells of the anterior lobe
of the pituitary. When administered to hypophysectomized animals,
ACTH exerts a number of characteristic physiological responses:

maintenance of the normal histology and weight of the adrenal
cortex, depletion of adrenal cortical lipid, cholesterol, and
ascorbic acid, and an increase in the level of circulating plasma
glucocorticoids. In connection with steroidogenesis in the
adrenal cortex, ACTH participates in protein metabolism and
gluconeogenesis, as well as fat metabolism. Water and salt
retention is seen, too; this causes Cushing's disease after prolonged administration of ACTH. As for extra-ACTH activity, several
comprehensive review articles are available [3, 14], and some
references are available concerning the adipokinetic activity
of ACTH [19-23].

In addition to the general effects on steroidogenesis and
general metabolism, ACTH has the ability to disperse melanin
granules of amphibia, reptiles, and humans. The MSH activity of
ACTH is about 1/100 of α-melanocyte-stimulating hormone (MSH), the
most potent melanotropic agent in the pituitary. The structure of
this latter compound is the same as the first 13 amino acid
sequence of ACTH, although the N-terminus is acetylated and the
C-terminus is protected in the form of an amide [24].

Among the many physiologic effects produced by ACTH, two
specific responses furnish the basis for standard quantitative
assays, namely, the depletion of ascorbic acid in the adrenal
cortex and steroidogenesis as measured in vivo and in vitro
systems [25].

Ascorbic Acid Depletion Test. This method was developed by
Sayers et al. [26] in 1949 and adopted by the Pharmacopeia of the
United States [27]. Hypophysectomized male rats are anesthetized
for 21-24 hr and the left adrenal is removed. The test solution
is given intravenously, and after 1 hr, the right adrenal is
removed. The ascorbic acid depletion occurring between the two
adrenals is a function of the administered ACTH dose.

In Vivo Steroidogenetic Assay. This method was devised by
Guillemin et al. [28] in 1959. The ACTH activity can be determined
by measuring the rise in plasma hydrocortisone level in 24-hr hypo-

physectomized rats after 15 min of administration of a compound to
be tested. Alternatively, a total of 17-hydroxysteroids can be
measured by the rise of Porter-Silber chromogens.

In Vitro Steroidogenetic Assay. This determination was introduced by Saffran and Schally [29] in 1955. A rat adrenal cortex is
incubated with a solution of ACTH to be tested and the corticoids
produced, after extraction with dichloromethane, are measured
spectrometrically. It will be noted, however, that the potency
of ACTH determined by the in vitro method does not parallel the
results obtained by the ascorbic acid depletion test. One should
keep in mind that the exact role of ascorbic acid in terms of the
production of corticoids is not known. After hypophysectomy, the
ability of the rat adrenal to secrete corticosteroids in response
to ACTH administration is lost within a few days. Yet the effect
on ascorbic acid content can be observed for a much longer period
[30].

Improved in Vitro Steroidogenetic Assay. The international
standard of ACTH was first determined in 1950 using the ACTH preparation according to Munson [27]. Initially, the activity
produced by 1 mg of this product (La-1-A) was defined as 1 IU.
Later, in 1955, the effect corresponding to 0.88 mg was defined
as 1 IU. With this international standard, the activity of highly
purified ACTH preparations is around 80-150 IU/mg. Synthetic
peptides have been assayed by one or more of the three methods
mentioned above. However, they were often found to have varying
potencies, depending upon the assay method employed, and none of
these techniques are accurate enough to measure the very low
potency of synthetic peptides. Therefore a much more sensitive
and specific method, preferably an in vitro method, is required
for precise work in this area.

Along these lines, attempts were made to disperse the adrenal
tissue by various proteolytic enzymes [31-35]. In 1971, Sayers
et al. [36] developed a more sensitive in vitro assay system. Here,
rat adrenal tissue was dispersed by tryptic digestion. Bovine serum
albumin was added to prevent excessive cell damage during this treat-

ment and the dispersed cells were collected by centrifugation. The
use of soybean trypsin inhibitor was recommended to terminate the
undesired proteolytic partial digestion [36, 37]. The cell suspension
was incubated with a given quantity of ACTH for 2 hr and the reaction
was stopped by the addition of dichloromethane. Corticosterone was
then determined by the fluorescent procedure, in which the minimum
effective dose is equivalent to a 0.1 microunit (1.0 pg) of ACTH.
In addition, the cells are highly selective and do not respond to
relatively large quantities (10^7 pg) of insulin, glucagon, oxytocin,
vasopressin, or angiotensin II. Using this method, Sayers et al.
[36] demonstrated that 10^7 pg of α-MSH exhibited a weak action
corresponding to that found for 10 pg of ACTH. Nakamura and Tanaka
[37] evaluated the ACTH activity of α-MSH as 0.0043 IU/mg, though it
had been estimated as about 1/400 of ACTH-[1-24]-OH by the earlier
assay [30]. As a result of this new technique, many synthetic peptides with low activity should be reexamined for a better understanding
of the relationship between structure and function in ACTH.

Radioimmunoassay. As stated above, the potency of ACTH was
defined as a function of effects occurring in the adrenal cortex.
However, a newly developed radioimmunossay now makes it possible to
measure a minute amount of ACTH directly in plasma.

The concentration of plasma ACTH has been determined by the
method of Lipscomb and Nelson [38], but relatively large amounts
of blood and skillful techniques are required to obtain accurate
results. An advancement in the field of immunoassay was promised
when sufficiently pure ACTH preparations became available for
antigen formation. In 1963, Felber et al. [39] introduced the
radioimmunoassay of ACTH by applying the antigen principle devised
originally by Berson and Yalow [40]. Because of the low molecular
weight of ACTH, some difficulty was encountered in preparing an
antigen with a high specific activity. Consequently the value
(7000-25,000 pg/ml) obtained initially by these authors must be
judged as too high. Later, Yalow and Berson [41] were able to
prepare ^{131}I-ACTH with a high specific activity (1000 mCi/mg) and

reported that the concentration of ACTH in plasma (0.2 ml) could
be determined with confidence. The normal value of circulating
ACTH in plasma is estimated as around 20-80 pg/ml. By contrast,
the value obtained earlier by a hemagglutination inhibition
immunoassay [42] seems in error (120,000-260,000 pg/ml). A further
improvement in the radioimmunoassay of ACTH has yielded even
greater accuracy and sensitivity [43]. It seems highly probable
that the C-terminal portion of ACTH represents the immunological
activity [44], since a steroidogenetically inert fraction corresponding to the 22 to 39-amino-acid sequence of sheep ACTH is
immunologically more active on a molar basis than the entire
molecule [45].

III. ISOLATION AND STRUCTURE DETERMINATION

Starting from the ACTH concentrate prepared by the hydrochloric
acid-acetone procedure of Lyons [46], two groups of investigators,
Li et al. [47] with sheep, and Sayers et al. [48] with hog, isolated,
in apparently homogeneous form, a protein of molecular weight about
20,000, isoelectric point 4.7, respectively. The peptide nature of
ACTH was first disclosed in 1951, when Astwood et al. [49, 50]
separated a low-molecular-weight substance from hog pituitary
extract by adsorption on oxycellulose. By this single oxycel
treatment, the activity of 1-2 U.S.P. U/mg was increased to approximately 80 U.S.P. U/mg.

Based on this observation, Bell et al. [51, 52] submitted the
substance extracted from hog pituitary to further purification by
countercurrent distribution in the solvent system of n-butanol-6%
acetic acid containing 3.5% sodium chloride. They separated, among
other active fractions, the most abundant and homogeneous material,
a so-called β-corticotropin with an approximate molecular weight
of 4500. This compound exhibited an adrenal ascorbic acid-depleting

activity of 80-100 U.S.P. U/mg and a steroidogenetic potency of
94.5 + 10.6 U/mg [53]. The product was then submitted to standard
peptide analysis.

In 1954, Bell et al. announced the structure of β-corticotropin
[54, 55], which is composed of 39 amino acid residues, as shown in
Fig. 1. In their work, fragments derived from proteolysis by either
pepsin, trypsin, or chymotrypsin were separated by a micro countercurrent distribution machine, or by paper chromatography, when
necessary. The peptide sequences of these fragments were performed
by end-group analysis and stepwise Edman degradation.

Independently, White et al. [56, 57] purified the oxycel
concentrates derived from pig pituitary by column chromatography
on Amberlite XE-97, followed by countercurrent distribution in
the solvent system of sec-butanol-0.2% trichloroacetic acid. A
homogeneous sample designated as corticotropin A was isolated and
the structure was published in 1955 [58-63]. A minor discrepancy
was noted between the arrangement of the amino acid residues at
positions 25 to 29 in their structure and β-corticotropin. Since
these two materials are from the same source, Harris [64] repeated
the sequence analysis, and the result was in favor of the formula
of Bell et al. [55].

Li and his associates [65-67] isolated a homogeneous sample
designated as α-corticotropin from sheep pituitary after submitting
oxycel concentrates to zone electrophoresis on starch, IRC-50
column chromatography, and subsequently to countercurrent distribution in the solvent system of 2-butanol and 0.2% trichloroacetic acid.
Its structural elucidation was published in 1955 [68-70].

Subsequently, Li et al. [71, 72] were able to determine the
total structure of ACTH from bovine origin. In 1959, the amino
acid sequence of human ACTH was proposed by Lee et al. [73, 74].
The low activity of human ACTH may be due to inactivation occurring
during the isolation procedure.

The amino acid compositions and sequences of pig, sheep, beef,
and human ACTH are shown in Table 1 and Fig. 1, respectively. As
far as amino acid compositions are concerned, sheep, beef, and

TABLE 1

Amino Acid Composition of Pig, Sheep, Beef, and Human ACTH

Amino acid	Pig	Sheep	Beef	Human	Amino acid	Pig	Sheep	Beef	Human
Ser	2	3	3	3	Trp	1	1	1	1
Tyr	2	2	2	2	Gly	3	3	3	3
Met	1	1	1	1	Lys	4	4	4	4
Glu	5	4	4	5	Pro	4	4	4	4
Gln	0	1	1	0	Val	3	3	3	3
His	1	1	1	1	Asp	1	2	1	1
Phe	3	3	3	3	Ala	3	3	3	3
Arg	3	3	3	3	Leu	2	1	1	1

human ACTH are identical. They differ from pig ACTH by one more serine and one less leucine. Isoleucine is absent and one mole of methionine represents the only sulfur-containing amino acid in ACTH. It is noteworthy that all of these hormones possess the same amino acid sequence in positions 1 to 24 and species variation occurs only between residues 25 to 33. The C-terminal portions are again identical.

In addition to the ACTHs listed in Fig. 1, one must keep in mind the possibility that more than one active component can be separated from a species. Dixon and Stack-Dunne [75] reported the presence of several active components in oxycel concentrates of pig pituitary. At least two components, corticotropin A_1 [Gln^{30}-ACTH] (later revised as Asn^{25}-ACTH by Gráf et al. See footnote of Fig. 1.), were separable by ion-exchange chromatography on IRC-50 (XE-64). Corticotropin A_1 can be converted to A_2 (a deaminated compound, later assigned as Asp^{25}-ACTH) on standing for several hours at pH 11.3. Even with these minor differences in their structures, they are distinguishable on starch gel electrophoresis because of the high anodic migration of A_2. When corticotropin A and

FIG. 1. Amino Acid Sequences of Pig, Sheep, Beef, and Human ACTH.*

```
H-Ser-Tyr-Ser-Met-Glu-His-Phe-Arg-Trp-Gly-Lys-Pro-Val-Gly-Lys-Lys-Arg-Arg-Pro-Val-
   1           5                  10                15                 20

Lys-Val-Tyr-Pro ──── species variation ──── Ala-Phe-Pro-Leu-Glu-Phe-OH
 21         25                      33    35                     39
```

Species	Amino acid sequences between positions 25 and 33	ACTH activity (U/mg)		MSH activity (U/g)	Reference
		Ascorbic acid depletion	in vitro steroidogenesis		
Pig	1---Asp-Gly-Ala-Glu-Asp-Gln-Leu-Ala-Glu---39 25 30 33	100-150 80-100	94.5 ± 10.6	1.7×10^8	56 52,53
Sheep	1---Ala-Gly-Glu-Asp-Asp-Glu-Ala-Ser-Gln---39	150	177	0.6×10^8	57
Beef	1---Asp-Gly-Glu-Ala-Glu-Asp-Ser-Ala-Gln---39	---	140	0.5×10^8	62
Human	1---Asp-Ala-Gly-Glu-Asp-Gln-Ser-Ala-Glu---39	26	52	0.4×10^8	64

*Since this chapter was written several revised amino acid sequences for ACTHs were reported[a,b,c,d] and human ACTH was synthesized[d]:

Species	Revised amino acid sequences between positions 25 to 33
Pig[a,b]	1---Asn-Gly-Ala-Glu-Asp-Glu-Leu-Ala-Glu---39
Sheep[c]	1---Asp-Gly-Ala-Glu-Asp-Glu-Ser-Ala-Gln---39
Beef[c]	1---Asn-Gly-Ala-Glu-Asp-Glu-Ser-Ala-Gln---39
Human[d]	1---Asn-Gly-Ala-Glu-Asp-Glu-Ser-Ala-Glu---39

a. L. Gráf, S. Bajusz, A. Patthy, E. Baráti, G. Cseh, Acta Biochim. Biophys. Acad. Sci. Hung., 6, 415 (1971).
b. B. Riniker, P. Sieber, W. Rittel, H. Zuber, Nature New Biol., 235, 114 (1972).
c. H. Li, Biochem. Biophys. Res. Commun., 49, 835 (1972).
d. P. J. Bennett, P. J. Lowry, and C. McMartin, Biochem. J., 133, 11 (1973); P. Sieber, W. Rittel, B. Riniker, Helv. Chim. Acta, 55, 1234 (1972).

β-corticotropin were examined by this sensitive technique [76], it
was found that the former contains mainly A_1 and a small amount of
A_2, whereas the latter contains mainly A_2 and little A_1. These
minor differences must take place during the isolation procedure.

In addition to β-corticotropin, Shepherd et al. [52] obtained
other minor components with some ACTH activity from pig pituitary:
α_1 (11.2 ± 2.1 U/mg), α_3 (34.0 ± 2.8 U/mg), α_4 (25.8 ± 5.1 U/mg),
δ_1 (11.6 ± 2.0 U/mg), δ_2 (21.7 ± 4.1 U/mg), and δ_3 (27.6 ± 6.9 U/mg).
The chemical nature of these active components remained to be
elucidated at a future date.

Later, Lande et al. [77] isolated from pig pituitary extracts
three peptides with high steroidogenic potency in vivo. They
reported that these components are similar in molecular weight
to pig ACTH as known today, but differ from the usual amino acid
composition in molar ratios of aspartic acid, threonine, serine,
glycine, valine, methionine, and phenylalanine. A sequence analysis
of these fractions and a comparison with the minor components of
Shepherd et al. [52] are not available at the present time.

A heterogeneity in ACTH from sheep pituitary was reported by
Pickering et al. [78]. Pig-type ACTH was isolated from sheep
pituitary and it appears to represent about 10% of the ACTH
present. In addition, they offered the evidence, though the
results are inconclusive, that sheep pituitary extracts contain a
peptide having the sequence of ACTH, but lacking the N-terminal
hexapeptide, Ser-Tyr-Ser-Met-Glu-His, and the C-terminal amino acid.

The presence of two active components in bovine pituitary was
also recorded by White and Peters [79].

Considering the heterogeneity of adrenocorticotropically active
components in pituitary glands, Li [80] proposed the following
terminology for polypeptides with ACTH activity as isolated from
sheep, porcine, and bovine: α_{sheep}-, α_{pig}-, α_{beef}- (or briefly,
α_s-, α_p-, α_b-)-ACTH. Whenever more than one active component is
isolated, as in the case for sheep or pig, the terms β_s-, γ_s- or
β_p-, γ_p-ACTH were to be employed.

IV. SYNTHESIS OF ACTH PEPTIDES

The remarkable achievements in the field of ACTH synthesis
were provided initially by observations that came from modification
attempts, especially some studies on partial enzymatic digestion.
Although the peptide nature of adrenocorticotropic hormone was
proven by inactivation on exposure to trypsin and pepsin [81, 82],
it was found later that the biological activity was not completely
destroyed when an ACTH concentrate was treated with pepsin under
controlled conditions [83]. From such a peptic hydrolysate, a
partially purified material, designated as corticotropin B, with
a reported adrenal ascorbic acid-depleting activity of 300 U.S.P.,
was isolated [57], but this was not a pure substance.

When pure β-corticotropin was submitted to partial digestion
with pepsin, Bell et al. [54] were able to isolate several active
components by countercurrent distribution. These compounds,
designated as peptides P_2 (27.5 ± 7.9 U/mg), P_3 (31.4 ± 4.7 U/mg),
and P_4 (21.5 ± 10.0 U/mg), correspond to the N-terminal 31, 30,
and 28 amino acid residues of the β-corticotropin molecule,
respectively. Moreover, they noted that on acid hydrolysis, four
C-terminal amino acid residues may be removed from the shortest
peptide fragment, P_4, without affecting biologic activity.
However, the expected tetracosapeptide was neither isolated nor
characterized. It may be recalled that this tetracosapeptide
portion is the structural element that is identical in all of the
ACTHs known today.

It should be emphasized here that the above observation made
by Bell et al. [54] offered the first evidence, although it was
indirect, for an important concept in peptide chemistry — namely
hormones that the entire molecule of certain peptide hormones may
not be required for biologic activity. These tests on the peptic
degradation products of β-corticotropin implied that the N-terminal
24 amino acid sequence is the physiologically important section.

As a result, the initial work on the synthesis of ACTH was conducted in order to elucidate the smallest portion of the molecule,
possibly smaller than the tetracosapeptide, which is fully active.

As is mentioned later, the first synthesis of a peptide with
some biologic activity was announced by Boissonnas and Guttmann in
1958. However, their synthetic eicosapeptide, ACTH-[1-20]-OMe,
was reported to be only 2-3% as active as the natural product.
Li et al. in November, 1960, reported simultaneously the synthesis
of ACTH-[1-19]-OH and Gln^5- ACTH-[1-19]-OH, respectively. Both
peptides exhibited an activity of approximately 30-35%. The first
synthesis of a peptide, ACTH-[1-23]-OH, possessing essentially full
biologic activity was achieved by Hofmann and one of the present
authors (H.Y.) in January, 1961. The total synthesis of porcine
ACTH was accomplished by Schwyzer et al. in 1963, and that of
human ACTH by Bajusz et al. in 1968.

Numerous efforts still continue to identify the essential
structural elements in ACTH on the one hand, and to improve the
synthetic methodology of this clinically important pituitary
hormone on the other. The majority of the synthetic activities
in this field have been done by the research groups of Hofmann
(University of Pittsburgh), Li (University of California),
Schwyzer (Ciba, Switzerland), Boissonnas (Sandoz, Switzerland),
Otsuka (Shionogi, Japan), Geiger (Hoechst, Germany), Fujino
(Takeda, Japan), Bajusz and Medzihradszky (Eötvös University,
Hungary), and Kisfaludy (Gedeon Richter, Hungary).

The authors summarize the synthetic methodologies discussed
here by subdividing the deblocking procedures used on the final
synthetic peptides, since this step determines the main strategy
in peptide synthesis.

A. Hydrogen Bromide Procedure

In 1956, Boissonnas et al. [84] published the synthesis of the
eicosapeptide methyl ester, ACTH-[1-20]-OMe, according to the route

illustrated in Scheme 1-a. They employed Lys(Z) at positions 11, 15, and 16, and Glu(OBzl) at position 5. The guanidino function of Arg was not protected. Coupling reactions between synthetic sub-units, as well as the synthesis of these subunits, were done with the dicyclohexylcarbodiimide (DCC) procedure, except for the synthesis of the N-terminal pentapeptide, Z-Ser-Tyr-Ser-Met-Glu(OBzl)-OH, which was prepared by the reaction of Z-Ser-Tyr-Ser-azide with H-Met-Glu(OBzl)-OH. In order to prepare the latter dipeptide, Pht-Met-Glu(OBzl)-OH was treated with one equivalent of hydrazine, as illustrated in Scheme 1-b.

For the preparation of the middle sequence, Tri-His(Tri)-Phe-Arg-Trp-OH (positions 6-9), these authors united Tri-His(Tri)-Phe-OH and H-Arg-Trp-OMe by DCC. Because of the steric hindrance of the bulky trityl group, the former compound had to be prepared from Z-His-Phe-OMe, as illustrated in Scheme 1-c. The tetrapeptide, Tri-Gly-Lys(Z)-Pro-Val-OH (positions 10-13), was prepared by a 2 + 2 coupling with DCC, as shown in Scheme 1-d [85].

The C-terminal heptapeptide, Tri-Gly-Lys(Z)-Lys(Z)-Arg-Arg-Pro-Val-OMe (positions 14-20), was obtained by condensation of Tri-Gly-Lys(Z)-Lys(Z)-OH with H-Arg-Arg-Pro-Val-OMe, as illustrated in Scheme 1-e and 1-f [85]. This is a rare example of the preparation of an Arg-peptide using N^G-unprotected Arg as a carboxyl component.

The trityl group was removed selectively from the protected heptapeptide methyl ester (positions 14-20) by hot acetic acid and the product was converted to the corresponding hydrochloride, which, after neutralization with tributylamine, was submitted to the next coupling reaction with Tri-Gly-Lys(Z)-Pro-Val-OH (positions 10-13). Similar reactions were repeated in order to construct the entire amino acid sequence of the eicosapeptide, as shown in Scheme 1-a. At the final stage, the protecting groups, Z and Bzl ester, were removed by HBr in a mixture of acetic acid and methyl ethyl sulfide in order to avoid the possible alkylation of the Met residue. The final product was isolated by precipitation from its aqueous solution at pH above 6.

During the course of this synthesis, in order to cleave the
α-amino protecting group without affecting the Z group attached as
the ε-amino function of the Lys residue, the investigators used the
trityl group, which is removable by acetic acid. However, because
of the steric hindrance associated with this protecting group,
Tri-Gly-OH is the only amino acid that can be introduced to the
peptide bond by either mixed anhydride or DCC procedure. Consequently they had to sacrifice the racemization-free peptide
condensations available at the Gly residues that lie at positions
10 and 14 in the ACTH molecule.

The full details of this synthesis and the purification of
the final product are unavailable, but it is not surprising that
the first synthetic, but racemic, ACTH peptide had activity of
only 2-3 U/mg (by Saffran and Schally in vitro test). It should
be noted that even in the synthesis of small peptides, racemization
will take place when the carboxyl group of an acyldipeptide is
activated by DCC.

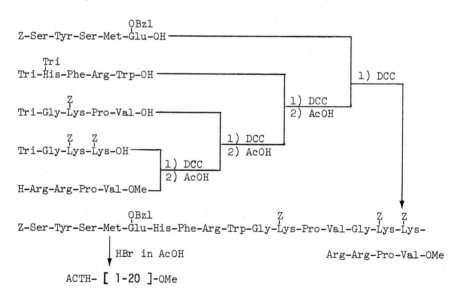

Scheme 1-a.   Synthetic outline of ACTH-[1-20]-OMe,
              Boissonnas et al. [84].

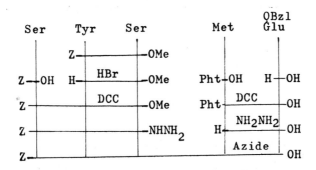

Scheme 1-b.   Positions 1-5.

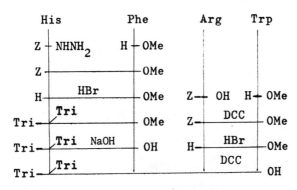

Scheme 1-c.   Positions 6-9.

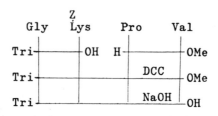

Scheme 1-d. Positions 10-13.

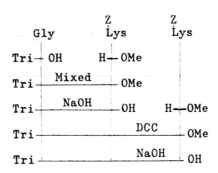

Scheme 1-e. Positions 14-16.

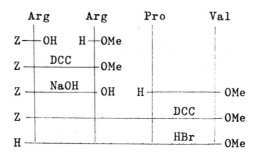

Scheme 1-f. Positions 17-20.

The eicosapeptide amide, ACTH-[1-20]-NH$_2$, was later synthesized by Hofmann and one of the present authors (H.Y.) in 1962 by a method different from that used by Boissonnas et al. This synthetic peptide amide is nearly as active as the natural ACTH, as is mentioned in another section.

Later, Boissonnas et al. adopted Lys(Boc) instead of Lys(Z) and prepared a number of ACTH peptides: D-Ser^1-Nle^4- Val^{25}-ACTH-[1-25]-NH$_2$ (ascorbic acid depleting test, 625 + 130 IU/mg) [86]; NVal^4-Val^{25}-ACTH-[1-25]-NH$_2$ (active as natural ACTH) [87]; NLe^4-Lys^{17}-Lys^{18}-Val^{25}-ACTH-[1-25]-NH$_2$ (active as natural ACTH) [87]; NLe^4-Lys^{17}-Lys^{18}-D-Val^{25}-ACTH-[1-25]-NH$_2$ (highly active) [87]; and NLe^4- Val^{25}-ACTH-[1-25]-NH$_2$ (highly active) [87].

B. Sodium in Liquid Ammonia Procedure

Li and his associates are the only group that routinely removes all protecting groups at the final stage under basic conditions. They adopted the sodium in liquid ammonia procedure, as described by du Vigneaud et al., in the synthesis of oxytocin [88] and vaso-pressin [89]. Thus His(Bzl), Lys(Tos), and Arg(Tos) are required for starting materials. Among those, Arg(Tos) was introduced into peptide synthesis by Schwyzer and Li [90] and its preparation was reported in detail by Schnabel and Li [91], and Ramachandran and Li [92]. The tosylation of Z-Arg-OH was achieved in acetone-water solution, while the pH was maintained at 11-11.5 with sodium hydroxide during the reaction. Z-Arg(Tos)-OH could not be obtained in crystalline form; however, the cyclohexylamine salt is a crys-talline compound.

Reaction 1. Preparation of Z-Arg(Tos)-OH,
Schnabel and Li [91] and Ramachandran and Li [92].

As an example, the synthesis of the nonadecapeptide [93, 94],
ACTH-[1-19]-OH, accomplished in 1960, is given in Scheme 2-a.

As shown in Scheme 2-e, the C-terminal pentapeptide (positions
15-19) was prepared in a stepwise manner starting with H-Pro-OMe.
Condensation of appropriate Z-amino acids by the DCC or p-nitrophenyl
ester procedures and subsequent removal of the Z group by catalytic
hydrogenation were employed in the elongation of the peptide chain.
After saponification of the protected pentapeptide methyl ester by
alkali, the desired pentapeptide, Z-Lys(Tos)-Lys(Tos)-Arg(Tos)-Arg(Tos)-
Pro-OH, was purified by countercurrent distribution in the solvent
system chloroform-toluene-methanol-water (5:5:8:2). In order to
establish the optical purity, all protecting groups were removed
by sodium in liquid ammonia and the resulting product, H-Lys-Lys-
Arg-Arg-Pro-OH, was characterized as a helianthate in a crystalline
form [95].

The tetrapeptide, Z-Lys(Tos)-Pro-Val-Gly-OH (positions 11-14),
was prepared according the Scheme 2-d. For some reason, the dipeptide, Z-Pro-Val-OH, was condensed by DCC with H-Gly-OMe. The
N^{α}-protecting group was removed by hydrogenation and the resulting
tripeptide amine was combined with Z-Lys(Tos)-OH by the p-nitrophenyl ester procedure to give Z-Lys(Tos)-Pro-Val-Gly-OMe, which
was then saponified.

As shown in Scheme 2-c, in order to construct the hexapeptide,
Z-Glu(OBzl)-His(Bzl)-Phe-Arg(Tos)-Trp-Gly-OH (positions 4-10), the
dipeptide, Z-His(Bzl)-Phe-OH, was converted to the corresponding
p-nitrophenyl ester, which was allowed to react with H-Arg(Tos)-

Trp-Gly-OMe. The product, after saponification, was hydrogenated
to give H-His(Bzl)-Phe-Arg(Tos)-Trp-Gly-OH. During hydrogenation,
the N^{im}-benzyl groups suffered cleavage to the extent of 30%. However, subsequent coupling with Z-Glu(OBzl)-ONP was performed and
the product was extensively purified by countercurrent distribution
in the toluene system mentioned above.

The N-terminal tetrapeptide hydrazide, Z-Ser-Tyr-Ser-Met-NHNH$_2$
(positions 1-4), was prepared by the azide coupling of 2 + 2 subunits,
as shown in Scheme 2-b.

Assembling these four units was done, as shown in Scheme 2-a.
The C-terminal pentapeptide obtained above, Z-[15-19]-OH, was
hydrogenated and the resulting pentapeptide was condensed with the
p-nitrophenyl ester of Z-[11-14]-OH to yield the protected nonapeptide, Z-[11-19]-OH (yield 91%). This nonapeptide was prepared
alternatively by saponification of Z-[11-19]-OMe, which was obtained by coupling of Z-[11-14]-OH with H-Lys(Tos)-Lys(Tos)-Arg(Tos)-
Arg(Tos)-Pro-OMe by either the p-nitrophenyl ester method (yield 61%)
or the mixed anhydride method (yield 88%). The saponification of
Z-[11-19]-OMe was not complete within a short time period, while
prolonged alkali treatment caused considerable damage to the
material. The impure nonapeptide, Z-[11-19]-OH, was then hydrogenated to give the partially protected nonapeptide, H-[11-19]-OH.

Next, a coupling reaction between the two subunits, Z-[5-10]-OH
and H-[11-19]-OH, was achieved by a mixed anhydride procedure using
isobutyl chloroformate. The protected pentadecapeptide, Z-[5-19]-OH,
was purified by countercurrent distribution in the toluene system
(yield 35%). Hydrogenolysis of this product cleaved the Z group,
as well as the benzyl groups attached at Glu and His, to give the
partially protected pentadecapeptide, H-[5-19]-OH, which was then
submitted to the final coupling reaction with the N-terminal
tetrapeptide, Z-[1-4]-azide. The resulting fully protected nonadecapeptide was purified by precipitation from methanol with
ethyl acetate (yield 56%).

The tosyl groups were then removed from Z-[1-19]-OH by
sodium in liquid ammonia. Sodium was added in small pieces over

an interval of 30 min until a blue color persisted. The crude
product was desalted by IRC-50 and then purified by countercurrent
distribution in the solvent systems 0.1% acetic acid-n-butanol-
pyridine (11:5:3) and 2-butanol-0.5% aqueous trichloroacetic acid.
From 200 mg of the protected nonadecapeptide, a homogeneous sample
of 26 mg (in vitro steroidogenetic ACTH potency 39.8 U/mg) was
obtained.

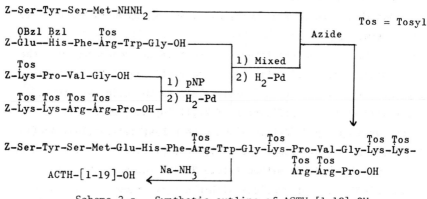

Scheme 2-a.   Synthetic outline of ACTH-[1-19]-OH
              (Li et al. [93, 94]).

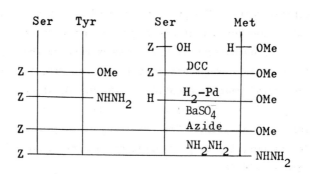

Scheme 2-b.   Positions 1-4.

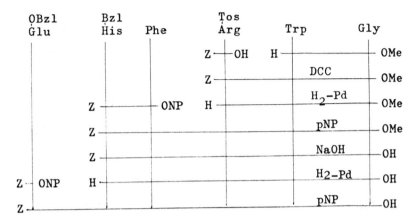

Scheme 2-c. Positions 5-10.

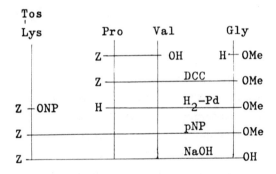

Scheme 2-d. Positions 11-14.

Throughout this synthesis, saponification of the methyl ester of a relatively large peptide containing the Arg derivative, Z-[11-19]-OMe, proved disadvantageous. Therefore, in 1964, Li et al. [96] improved the synthesis of this nonadecapeptide by taking H-Pro-OBu^t as the C-terminus, in which the protecting group is removable by trifluoroacetic acid. Subsequently, Lys(Tos) was replaced by Lys(Boc). However, Arg(Tos) was still kept in this synthesis, as shown in Scheme 2-f.

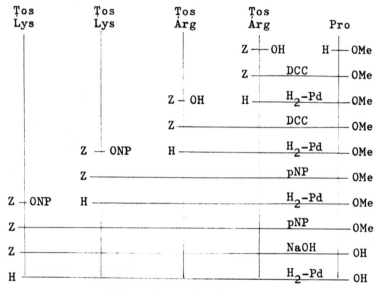

Scheme 2-e. Positions 15-19.

The final coupling reaction was performed between Z-[1-10]-OH and H-[11-19]-OBu^t by DCC and the fully protected nonadecapeptide was purified by countercurrent distribution in the solvent system of carbon tetrachloride-chloroform-methanol-0.01 M ammonium acetate (1:3:3:1). The pure, blocked derivative was first treated with trifluoroacetic acid to remove both the Boc and Bu^t ester groups, plus the benzyl group at Glu^5. The Arg(Tos) derivative of the nonadecapeptide was purified by countercurrent distribution in the solvent system of n-butanol-acetic acid-water (4:1:5), yield 38%.

Finally, the remaining protecting moieties, the Z and Tos groups, were removed by reduction with sodium in liquid ammonia. The desired peptide was isolated by desalting on IRC-50 resin followed by chromatography on CM-cellulose using continuous gradient elution with ammonium acetate (final concentration 0.2 M, pH 6.7). After a second chromatography on CM-cellulose, the nonadecapeptide, H-[1-19]-OH (yield 43%), was shown to be homogeneous by electrophoresis on paper and on polyacrylamide gel (in vitro steroidogenetic potency 33 U/mg).

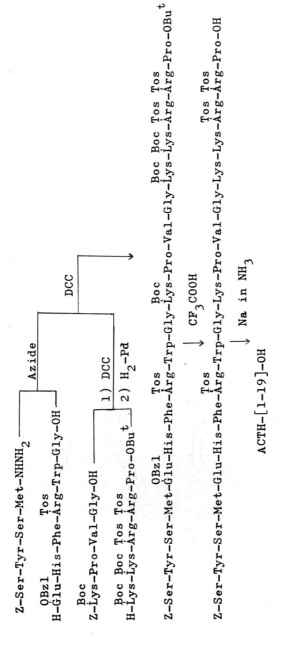

Scheme 2-f. Improved synthesis of ACTH-[1-19]-OH
(Li et al. [96]).

The latter synthetic procedure was then utilized to synthesize
a number of ACTH-active peptides:

Peptide	Activity	Reference
	(Steroidogenesis in vitro)	
ACTH-[1-17]-OH	5.2 U/mg	97, 98
ACTH-[1-17]-NH2	42 U/mg	99
ACTH-[1-18]-NH2	33 U/mg	99
ACTH-[1-19]-NH$_2$	35 U/mg	99
ACTH-[1-26]-OH	96 U/mg	101
Prolinol^{19}-ACTH-[1-19]	338 U/μM	100
Lys^8-ACTH-[1-17]-NH	1.0 U/mg	102
Gly^{12}-Gly^{13}-ACTH-[1-17]-NH$_2$	1.0 U/μM	103
Gly^{11}-Gly^{13}-ACTH-[1-17]-NH$_2$	0.5 U/μM	103
Gly^{11}-Gly^{12}-Gly^{13}-ACTH-[1-19]-OH	0.3 U/μM	103
ACTH-[1-10]-[15-19]-OH	0.4 U/mg	104, 105

As seen in the two synthetic outlines discussed above, and as
is noted in later examples, an improvement in yield at the final
deblocking procedure is usually an extremely difficult task in
the synthesis of complex peptides. Incomplete deblocking of one
or another protecting groups may result in the formation of a
mixture of peptides, which makes purification very difficult.
In addition, unexpected side reactions may, of course, decrease the
yield of the desired compound. Finally, in the synthesis under
discussion, cleavage of the Pro peptide bond by sodium in liquid
ammonia [106-109] may be responsible for lowering the yield to
some extent.

C. Dilute Hydrochloric Acid Procedure

ACTH activity as measured both in crude extracts and in pure
corticotropin is remarkably stable to acid treatment. It is known

that prolonged heating at 100°C in 0.1 N hydrochloric acid exerts
little effect on the biologic activity of these preparations [54].
Based on these observations, in 1961, Hofmann and one of the present
authors (H.Y.) [110, 111] assembled the 1-23 amino acid sequence in
a protected form using protecting groups, which are removable by
exposure to dilute hydrochloric acid. The formyl group was selected
as a protecting group for the ε-amino function of Lys in order to
fulfill this demand. A successful synthesis of a fully active tricosapeptide, ACTH-[1-23]-OH, was achieved, as shown in Scheme 3-a.

For the preparation of N^{ε}-formyllysine[Lys(For)-OH], the copper
chelate of lysine was exposed to ethyl formate at pH of 8-9 to
afford the less soluble chelate of H-Lys(For)-OH, which, on
decomposition with hydrogen sulfide, gave the formyl compound:

$$
\begin{array}{ccccc}
\text{NH}_2 & & \text{NH-CHO} & & \text{NH-CHO} \\
| & & | & & | \\
(\text{CH}_2)_4 & \xrightarrow[\text{NaOH}]{\text{HCOOC}_2\text{H}_5} & (\text{CH}_2)_4 & \xrightarrow{\text{H}_2\text{S}} & (\text{CH}_2)_4 \\
| & & | & & | \\
\text{NH}_2-\text{CH}-\text{COO}^{\ominus} & & \text{NH}_2-\text{CH-COO}^- & & \text{NH}_2-\text{CH-COOH} \\
1/2\ \text{Cu}^{2+} & & 1/2\ \text{Cu}^{2+} & &
\end{array}
$$

(2)

Reaction 2. Preparation of H-Lys(For)-OH,
Hofmann et al. [112].

Prior to the synthesis of ACTH-[1-23]-OH, some model experiments
were performed. First, H-His-Phe-Arg-Trp-Gly-Lys(For)-Pro-Val-NH$_2$
on treatment with 0.1 N HCl gave the corresponding Lys peptide without noticeable destruction of Trp [112]. It was found further that
the same treatment converted Ac-Ser-Tyr-Ser-Met-Gln-OH, used previously for the synthesis of α-MSH derivatives [113-115], into
H-Ser-Tyr-Ser-Met-Glu-OH with no major cleavage of the peptide
bond [109]. Thus it became feasible to employ available partial
MSH sequences in the synthesis of ACTH peptides.

The synthetic route to the C-terminal heptapeptide amide,
H-Arg-Arg-Pro-Val-Lys(For)-Val-Tyr-NH$_2$ (positions 17-23), is illus-

trated in Scheme 3-e. The peptide chain was elongated in a stepwise manner starting from H-Tyr-OMe by the mixed anhydride procedure. The Z group was used extensively for the protection of the α-amino function and, after addition of one amino acid residue, this unit was removed by hydrogenation without affecting the formyl group. The mixed anhydride procedure is particularly preferable in condensing Z-Arg(NO$_2$)-OH to H-Pro-Val-Lys(For)-Val-Tyr-NH$_2$, while the resulting protected tetrapeptide amide in acetic acid was hydrogenated over a Pd catalyst for 12 hr. The guanidino function of the Arg residue in the resulting tetrapeptide amide was protonated in the form of an acetate salt. Incorporation of Arg at position 17 in the above tetrapeptide was performed in essentially the same manner, and the product was again hydrogenated. The resulting partially protected heptapeptide amide, H-[17-23]-NH$_2$, was purified by column chromatography on CM-cellulose using ammonium acetate as eluent.

The middle sequence (positions 11-16) was prepared by coupling Z-Lys(For)-Pro-Val-Gly-OH and H-Lys(For)-Lys(For)-OMe with carbonyl-diimidazole. An alternative attempt to unite these two fragments with DCC failed, probably because of predominant acylurea formation. The resulting protected hexapeptide methyl ester was converted to the corresponding hydrazide Z-Lys(For)-Pro-Val-Gly-Lys(For)-Lys(For)-NHNH$_2$, as shown in Scheme 3-d.

Two additional pentapeptides, Ac-Ser-Tyr-Ser-Met-Gln-NHNH$_2$ (positions 1-5) [113] and H-His-Phe-Arg-Trp-Gly-OH (positions 6-10) [116], were intermediates used in the α-MSH synthesis, as seen in Scheme 3-b and 3-c, respectively. For the synthesis of·the latter pentapeptide, a coupling reaction in the style of 2 + 2 was performed. When Z-Phe-Arg(NO$_2$)-OH and H-Trp-Gly-OMe were condensed by the mixed anhydride procedure, fortunately, an optically pure tetrapeptide, Z-Phe-Arg(NO$_2$)-Trp-Gly-OH, could be obtained after a single recrystallization.

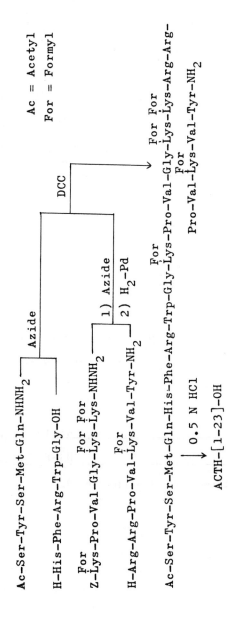

Scheme 3-a. Synthetic outline of ACTH-[1-23]-OH
(Hofmann and Yajima et al. [110, 111]).

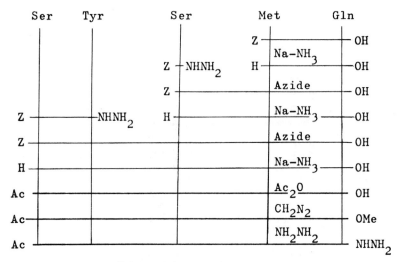

Scheme 3-b. Positions 1-5.

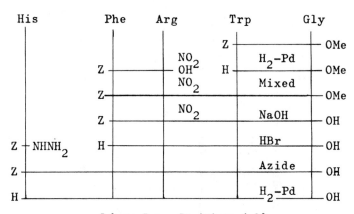

Scheme 3-c. Positions 6-10.

The N-terminal decapeptide, Ac-[1-10]-OH, was obtained after condensation of the above two subunits by the azide procedure. The product was purified by column chromatography on CM-cellulose. Thus from 111 mg of H-His-Phe-Arg-Trp-Gly-OH, 118 mg of Ac-[1-10]-OH was obtained, which was then converted into the corresponding hydrochloride.

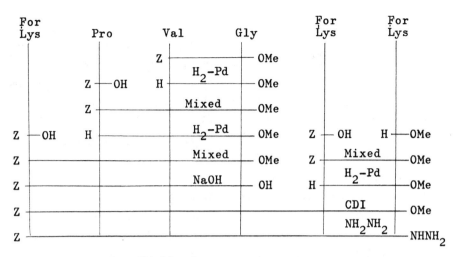

CDI = Carbonyldiimidazole

Scheme 3-d.   Positions 11-16.

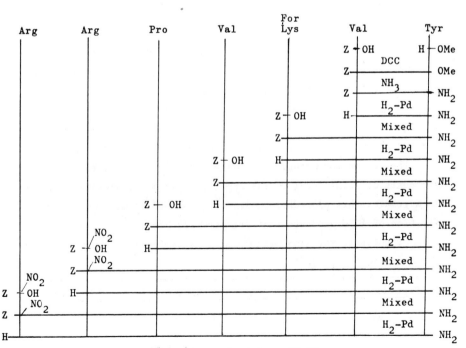

Scheme 3-e.   Positions 17-23.

The construction of the entire sequence of the tricosapeptide is illustrated in Scheme 3-a. The heptapeptide amide, H-[17-23]-NH$_2$, and the hexapeptide hydrazide, Z-[11-16]-NHNH$_2$, were combined via the azide procedure. The product, after hydrogenolysis, was purified by column chromatography on CM-cellulose to give the partially protected tridecapeptide amide, H-[11-23]-NH$_2$ (yield 74%) [117]. This compound, after conversion to the corresponding hydrochloride, was condensed with Ac-[1-10]-OH by DCC and the resulting fully protected tricosapeptide amide, Ac-[1-23]-NH$_2$, was obtained in an analytically pure form by purification on CM-cellulose (yield 46%).

The purified peptide was then treated with 0.5 N HCl containing 0.2% thioglycolic acid in a boiling water bath for 80 min to remove the formyl groups from the various Lys residues, the acetyl group from the N-terminus, and the amide from Gln. The material was submitted to purification by column chromatography on CM-cellulose. The chromatographic pattern indicated that the peptide bonds following Gly^{10} and Gly^{14} were partially cleaved by the acid treatment. It was possible, however, to separate the tricosapeptide (20-30% yield) that possessed the essential full biological activity of ACTH (by ascorbic acid-depleting test, 103 ± 10.4 U/mg) from the lower-molecular-weight peptides. Examination of the product by leucine aminopeptidase and carboxypeptidase digestion revealed that the Gln^5 was converted to Glu, as expected, and the amide of the C-terminal Tyr was also cleaved to the free carboxylic acid.

The column chromatographic system used in 1961 for the synthesis of ACTH-[1-23]-OH was applied later by other investigators. As a result, this procedure became the standard method for the purification of ACTH peptides that are strongly basic.

Essentially the same methodology was employed to synthesize the following ACTH peptides:

Peptide	Activity (U/mg)[a]	Reference
ACTH-[1-13]-NH$_2$	< 0.1	109
ACTH-[1-16]-OH	< 0.1	118
ACTH-[1-20]-NH$_2$	111	119,120

[a]Ascorbic acid-depletion test

Several other analogs were synthesized by the trifluoroacetic acid procedure, which are mentioned later:

Peptide	Activity (U/mg)[a]	Reference
α-Aminobutyric acid[4]-Gln[5]-ACTH-[1-20]-NH$_2$	31	121, 122
Gln[5]-pyrazolylalanine[6]-ACTH-[1-20]-NH$_2$	50	123, 124
Gln[5]-Phe[9]-ACTH-[1-20]-NH$_2$	2	124
Gln[5]-[^{14}C]-Phe[7]-[1-20]-NH$_2$ 150% of corticotropin A$_1$		125

[a]Ascorbic acid-depleting test

The synthetic strategy discussed here is the method of choice, in which the acid-basic properties of the natural product are purposely utilized. Moreover, the usefulness of H-Lys(For)-OH in the synthesis of complex peptides was demonstrated for a complex example. Since many Lys(For)-containing peptides are water soluble, this property permits one to purify every synthetic intermediate by ion-exchange chromatography. However, the partial cleavage of the Gly bond under acidic conditions seems to be a limitation of this approach. Recently, the authors have reanalyzed the conditions for the removal of the formyl group and have found that this group can be cleaved under mild conditions by treatment with hydrazine acetate or hydroxylamine hydrochloride in pyridine at pH 6.0. The improved procedure was applied to the synthesis of α-MSH [126], monkey [127, 128], and human β-MSH [129, 130]. Geiger et al. [131, 132] have also discussed the aminolysis of the formyl group in detail.

D. Trifluoroacetic Acid Procedure

1. Procedure Employed by Schwyzer et al.

Schwyzer et al. introduced Lys(Boc) and similar derivatives for the synthesis of ACTH peptides. This protecting group can be removed quantitatively by treatment with trifluoroacetic acid (TFA),

but is stable to catalytic hydrogenation. The Lys(Boc) compound
was first prepared by Schwyzer and Rittel [133], who allowed tert-
butyl azidoformate to react with Lys copper chelate to form H-Lys
(Boc)-OH copper complex. This latter product was then decomposed
with hydrogen sulfide to generate the desired intermediate.

$$
\begin{array}{ccccc}
NH_2 & & CO\text{-}O\text{-}C(CH_3)_3 & & Boc \\
| & & | & & | \\
(CH_2)_4 & \xrightarrow{(CH_3)_3C\text{-}OCO\text{-}N_3} & NH & \xrightarrow{H_2S} & NH \\
| & & (CH_2)_4 & & (CH_2)_4 \\
NH_2\text{—}CH\text{—}COO^{\ominus} & & NH_2\text{—}CH\text{—}COO^- & & NH_2\text{—}CH\text{-}COOH \\
& & & & \\
1/2\ Cu^{2+} & & 1/2\ Cu^{2+} & &
\end{array}
$$

$$(\underline{3})$$

In 1960, Schwyzer et al. announced the synthesis of Gln^5- ACTH-
[1-19]-OH [134], but later Gln^5 was replaced with $Glu(OBu^t)$. Since
the tert-butyl ester is known to be cleaved by TFA, it thus became
possible to prepare ACTH peptides in a protected form using both
Boc and tert-butyl ester protecting groups.

· a. Synthesis of ACTH-[1-24]-OH. Among many ACTH peptides prepared by Schwyzer et al., a synthetic outline to ACTH-[1-24]-OH
[135, 136], as done in 1961, is now discussed here in detail. In
terms of a general strategy, synthetic subunits were selected in
such a way that racemization-free condensation at a peptide bond
can take place at Gly and Pro, as seen in Scheme 4-a.

The synthetic route to the C-terminal pentapeptide, H-Val-Lys
(Boc)-Val-Tyr-Pro-OBu^t (positions 20-24), is illustrated in Scheme
4-f [137]. During the synthesis of this pentapeptide, it was found
that the DCC condensation involving Z-Val-Lys(Boc)-OH with
H-Val-Tyr-Pro-OBu^t produced about 25% racemization of the Lys
residue. The all-L-protected pentapeptide ester, Z-Val-Lys(Boc)-
Val-Tyr-Pro-OBu^t, was isolated by column chromatography on silica
using chloroform-ethyl acetate (2:1) as the eluent. Alternatively,
H-Val-Tyr-Pro-OBu^t was elongated in the stepwise manner by first
adding Z-Lys(Boc)-OH with DCC, and then Z-Val-OH via the mixed
anhydride procedure. The protected pentapeptide ester, prepared
by two alternate routes, was hydrogenated to give the partially

protected pentapeptide ester, H-[20-24]-OBu^t, in an analytically
pure form.

As shown in Scheme 4-e, for the synthesis of the partially
protected pentapeptide ester, H-Lys(Boc)-Lys(Boc)-Arg(NO$_2$)-Arg(NO$_2$)-
Pro-OMe (positions 15-19) [138], the coupling of two peptide units,
Tri-Lys(Boc)-Lys(Boc)-OH and H-Arg(NO$_2$)-Arg(NO$_2$)-Pro-OMe, was performed with either DCC (yield 64%) or N-ethyl-5-phenylisoxazolium-
3'-sulfonate (yield 81%). No analytical data were presented for the
product; however, the optical purity of the isolated compound was
confirmed by trypsin digestion of the TFA-treated sample. The
trityl moiety was removed from the product by 75% acetic acid to
give the partially protected pentapeptide ester, H-[15-19]-OMe.

Because of the known acid sensitivity of the trityl unit, one
presumably could not apply a racemization-free azide coupling
technique to obtain the pentapeptide ester.

The synthesis of the tetrapeptide hydrazide, PZ-Lys(Boc)-Pro-
Val-Gly-NHNH$_2$ (positions 11-14) [133], is illustrated in Scheme 4-d.
In this scheme and in the above preparation of H-[20-24]-OBu^t, the
useful phenylazo-benzyloxycarbonyl (PZ) group was demonstrated for
the first time. This colored protecting group, like the older Z
group, can be removed by catalytic hydrogenation. PZ-Lys(Boc)-Pro-
Val-Gly-OEt was prepared in a stepwise manner starting with
H-Gly-OEt, and then converted to the corresponding hydrazide in
the usual manner.

The N-terminal decapeptide (positions 1-10) [139] was obtained
by condensation of two subunits, a tetrapeptide unit (positions 1-4)
and a hexapeptide unit (positions 6-10). To prepare the former
tetrapeptide subunit, Boc-Ser-Tyr-Ser-azide was allowed to react
with H-Met-OMe, and the resulting protected tetrapeptide ester was
converted to the corresponding hydrazide, Boc-Ser-Tyr-Ser-Met-NHNH$_2$
[140]. The latter hexapeptide subunit, H-Glu(OBu^t)-His-Phe-Arg-Trp-
Gly-OH [141], was formed by the reaction of Z-Glu(OBu^t)-His-azide
with H-Phe-Arg(NO$_2$)-Trp-Gly-OMe, followed by saponification and
subsequent hydrogenation. During the latter treatment, the nitro
group of the Arg residue was reductively removed. Since the tert-

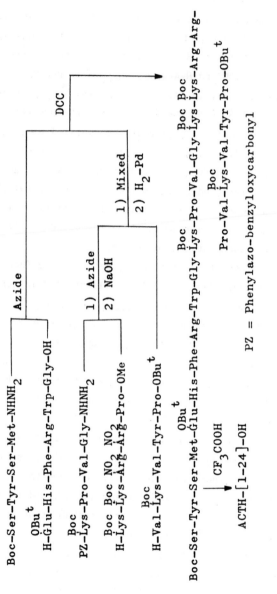

Scheme 4-a. Synthetic outline of ACTH-[1-24]-OH
(Schwyzer et al. [135, 136]).

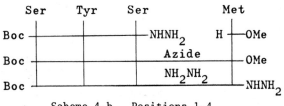

Scheme 4-b. Positions 1-4.

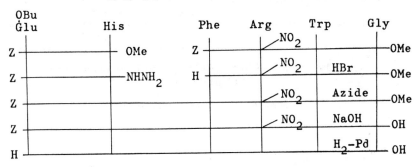

Scheme 4-c. Positions 5-10.

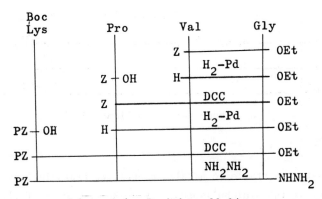

Scheme 4-d. Positions 11-14.

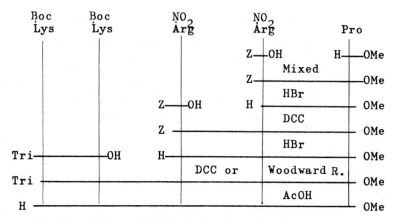

Woodward R. = N-Ethyl-5-phenylisoxazolium-
3'-sulfonate

Scheme 4-e. Positions 15-19.

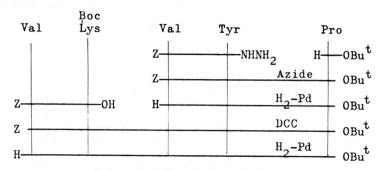

Scheme 4-f. Positions 20-24.

butyl ester is stable to alkali, as well as hydrazine, the azide
coupling of the Glu(OBu^t)-peptide was thereby successfully achieved,
as shown in Scheme 4-c.

As seen in Scheme 4-a, Boc-[1-4]-NHNH$_2$ was converted to the
corresponding azide (IR 4.75 μ), which was then allowed to react

with H-[5-10]-OH to give the N-terminal decapaptide, Boc-Ser-Tyr-Ser-Met-Glu(OBu^t)-His-Phe-Arg-Trp-Gly-OH (yield 70-85%). The product was recrystallized from 90% methanol [139].

Next, condensation of subunits, PZ-[11-14]-NHNH$_2$ and H-[15-19]-OMe, was performed. The product, PZ-[11-19]-OMe, was then purified by precipitation from chloroform with ether and recrystallized from hot acetonitrile (yield 75%). This material was then saponified in 75% dioxane to give PZ-Lys(Boc)-Pro-Val-Gly-Lys(Boc)-Lys(Boc)-Arg(NO$_2$)-Arg(NO$_2$)-Pro-OH. The free acid was obtained by precipitation from alkali solution with HCl and acetic acid. From 1.34 g of the methyl ester, 1.26 g of PZ-[11-19]-OH was obtained. This result seems almost too good in comparison with the similar saponification reactions of base sensitive Arg-peptides made by Li et al. [94].

PZ-[11-19]-OH thus isolated was used for the next coupling reaction with H-[20-24]-OBu^t via the mixed anhydride procedure with isobutyl chloroformate. The product, PZ-[11-24]-OBu^t, was purified by column chromatography on alumina employing chloroform-methanol (95:5) as the eluent. It is recorded that from 990 mg of PZ-[11-19]-OH, 1.14 g of the product was obtained after crystallization from acetonitrile. This compound was then hydrogenated over a Pd catalyst at a pressure of 5 atm in 90% acetic acid for more than 20 hr to remove the PZ and the nitro groups from Arg. The compound, H-[11-24]-OBu^t, H-Lys(Boc)-Pro-Val-Gly-Lys(Boc)-Lys(Boc)-Arg-Arg-Pro-Val-Lys(Boc)-Val-Tyr-Pro-OBu^t, was purified by column chromatography on CM-cellulose with ammonium acetate as the eluent (yield 70%).

The final coupling reaction between Boc-[1-10]-OH and H-[11-24]-OBu^t·tritosylate was performed in pyridine by means of DCC, and the resulting fully protected tetracosapeptide, Boc-[1-24]-OBu^t, was purified by countercurrent distribution in the solvent system of methanol-buffer-chloroform-carbon tetrachloride (8:4:5:2) (buffer: 285 ml of acetic acid and 192.5 g of ammonium acetate in 10 liters of water, yield 55%).

The final protecting groups were removed from tetracosapeptide by treatment with 90% TFA at 20° for 1 hr. The deblocked peptide, after conversion to the corresponding acetate (hexaacetate), was obtained directly in an analytically pure form (activity 106 ± 14 U.S.P. U/mg) in 90% yield.

Thus the structural element of ACTH that is common to all ACTHs known today was synthesized for the first time. The peptide prepared in this fashion is now available for clinical use as "Synacthen." Most important, the allergic phenomena that are often found with the use of natural ACTH has not been observed with the laboratory material.

b.   Total Synthesis of Porcine ACTH (1956 formula). As seen in the synthesis of ACTH-[1-24]-OH, the condensation of peptide subunits, as well as the synthesis of various subunits, was done by either the azide, the mixed anhydride, or the DCC procedure. In addition to these three tools, the newly developed active ester method aided Schwyzer and Sieber [142, 143] in accomplishing by 1963 the total synthesis of porcine ACTH, whose structure was first presented by Bell et al. [54, 55].

The synthetic route to porcine ACTH is illustrated in Scheme 4-g. The C-terminal, partially protected pentadecapeptide ester, H-[25-39]-OBu^t, was prepared in a stepwise manner by the p-nitrophenyl ester method starting from H-Phe-OBu^t (over-all yield 6.9%). A combination of the Z group (for α-amino protection) and the Bu^t ester (for side chain protection of Asp and Glu) allowed the elongation of the peptide chain in a stepwise fashion.

Next, some available partial sequences of ACTH-[1-24]-OH were slightly modified for a new approach in this area. In order to pre-pare the protected octapeptide, Z-Arg-Arg-Pro-Val-Lys(Boc)-Val-Tyr-Pro-OH (positions 17-24), the corresponding Bu^t ester was treated with TFA and the exposed free ε-amino group of the Lys residue was again masked with tert-butyl azidoformate. The protected octapep-tide, after protonation of the Arg residue by hydrochloric acid, was coupled with the partially protected pentadecapeptide,

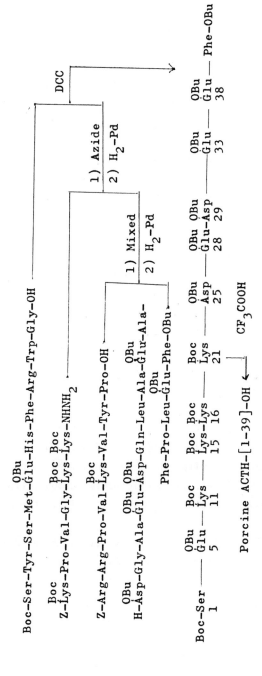

Scheme 4-g. Synthetic outline of porcine ACTH (1956 formula) (Schwyzer and Sieber [142, 143]).

H-[25-39]-OBu^t, by the mixed anhydride procedure using pivaloyl
chloride. The resulting protected tricosapeptide ester, Z-[17-39]-
OBu^t, was purified by countercurrent distribution in the solvent
system of chloroform-carbon tetrachloride-methanol-0.5 M ammonium
acetate (8:8:17:7) (yield 76%). Hydrogenation then gave H-[17-39]-
OBu^t, since by using the free Arg-peptide instead of the blocked
Arg(NO$_2$), the hydrogenation conditions could be reduced immensely
in terms of time and effort.

The protected hexapeptide hydrazide, Z-Lys(Boc)-Pro-Val-Gly-Lys
(Boc)-Lys(Boc)-NHNH$_2$ (positions 11-16), was prepared by using earlier
techniques. As seen in the synthesis of ACTH-[1-24]-OH, the azide
derived from the Gly-terminal tetrapeptide, Z-Lys(Boc)-Pro-Val-Gly-
NHNH$_2$, was employed again. The formyllysine derivative of this
hexapeptide unit had been used previously by Hofmann et al. [110,
111]. Condensation of the hexapeptide unit, Z-[11-16]-NHNH$_2$, with
H-[17-39]-OBu^t was performed by the standard azide procedure.
Again, countercurrent distribution in a similar ammonium acetate-
chloroform system was necessary in order to purify the product,
Z-[11-39]-OBu^t (yield 35%). Hydrogenation in 80% acetic acid gave
the partially protected nonacosapeptide ester, H-[11-39]-OBu^t,
isolated as the sulfate.

A DCC coupling united the hydrochloride form of the N-terminal
decapeptide, Boc-Ser-Tyr-Ser-Met-Glu(OBu^t)-His-Phe-Arg-Trp-Gly-OH,
and H-[11-39]-OBu^t in pyridine at 50° for 93 hr. The crude material,
isolated by precipitation with ether, was subjected to countercurrent
distribution in the solvent system chloroform-carbon tetrachloride-
methanol-0.1 M ammonium acetate (8:8:17:7) (yield in the purification
step 46%).

The protected nonatriacontapeptide ester was treated with
concentrated TFA at room temperature for 15 min, and the product,
after conversion to the corresponding acetate by Amberlite IRA-400,
was incubated with thioglycolic acid at 50° for 15 hr. Treatment
of the final product with a thiol compound, as demonstrated pre-

viously by Hofmann et al. [110, 111], is very effective in preventing
possible oxidation of the Met residue. The incubated sample was then
purified by CM-Sephadex using a gradient of pH 6.5 ammonium acetate
buffers of 0.05 and 0.8 M (the yield was not given). Thin-layer
chromatographically pure porcine ACTH was characterized as both an
acetate and a hydrate. This sample exhibited ACTH activity of
110 U/mg.

By essentially the same procedure, Schwyzer, Sieber, Kappeler,
Rittel, Riniker, and their associates at Ciba synthesized a number
of ACTH peptides:

Peptide	Activity, U/mg	Reference
ACTH-[1-16]-OMe	1	144
Orn^{17}-Orn^{18}-ACTH-[1-24]-OH	equiv to 1-24	145
D-Ser^1-ACTH-[1-24]-OH	750.	146
D-Ser^1-ACTH-[1-13]-NH$_2$	0.1	147
D-Ser^1-ACTH-[1-16]-NH$_2$	1.4	147
D-Ser^1-ACTH-[1-19]-OH	72.4	147
D-Ser^1-ACTH-[1-39]-OH	140	147
D-Ser^1-D-Tyr^2-ACTH-[1-24]-OH	equiv to 1-24	146
D-Ser^1-D-Tyr^2-D-Ser^3-D-Met^4-ACTH-[1-24]-OH	equiv to 1-24	146
D-Glu^5-D-His^6-D-Phe^7-D-Arg^8-ACTH-[1-24]-OH	inactive	146
D-Ala^1-ACTH-[1-24]-OH	{3 times more active than 1-24}	146
D-Ser^1-NVa^{17}-NVa^{18}-ACTH-[1-19]-NH$_2$	about ½ of 1-19-NH$_2$	148
D-Ser^1-NLe^{17}-NLe^{18}-ACTH-[1-19]-NH$_2$	about ½ of 1-19-NH$_2$	148
D-Ser^1-Lys^{17}-Lys^{18}-ACTH-[1-18]-NH$_2$	high & prolonged activ.	149

(They also synthesized some lipophilic-substituted [1-19] analogs
[148].)

Through these preparations, the useful nature of Lys(Boc),
Glu(OBu^t), and Asp(OBu^t) was demonstrated for the synthesis of
highly complex peptides. The most advantageous feature of the TFA
procedure is that no peptide bond cleavage takes place during the

acid treatment. However, the tert-butyl cation [150] that forms
during the reaction with TFA may cause undesired side reactions.
This can be reduced by the use of 90% TFA, as demonstrated by
Kappeler and Schwyzer [135, 136] in the synthesis of ACTH-[1-24]-OH,
or it may be minimized by addition of anisole.

The total synthesis of porcine ACTH was accomplished by a combination of two amino protecting groups, namely, the Z group for
α-amino protection and the Boc group for side-chain protection.
Moreover, the location of the Met residue near the N-terminus in
the ACTH molecule made this situation possible. If this amino
acid is found near the C-terminal portion in the molecule, then
elongation of the peptide chain is impossible, because the Z group
can no longer be removed under catalytic hydrogenation, due to the
poisonous nature of the sulfur atom in Met. If this is true, and
the TFA procedure is still pursued, then other amino protecting
groups, such as o-nitrophenylsulfenyl [151, 152] or 2-(p-biphenyl)-
isopropoxycarbonyl [153-155], which possess different properties
from the above two groups, will be needed, as demonstrated in the
synthesis of glucagon [156, 157] and calcitonin [158, 159]. In
this respect, the synthesis of ACTH peptides is a lucky target.

2. *Procedure Employed by Geiger et al.*

An improved synthesis of ACTH-[1-23]-NH$_2$ was reported by
Geiger et al. [160] in 1964. Regarding the deblocking procedure
that is employed at the final step, their synthetic method can be
classified as a TFA procedure.

As shown in Scheme 5, the tricosapeptide amide was prepared
by uniting three main subunits: H-Arg-Arg-Pro-Val-Lys(Boc)-Val-
Tyr-NH$_2$ (positions 17-23), Z-Lys(Boc)-Pro-Val-Gly-Lys(Boc)-Lys(Boc)-
NHNH$_2$ (positions 11-16), and Boc-Ser-Tyr-Ser-Met-Glu(OBu^t)-His-Phe-
Arg-Trp-Gly-OH (positions 1-10). These subunits are essentially
the same as those adopted by Hofmann et al. [110, 111] in the

synthesis of ACTH-[1-23]-OH, although the protecting groups are
different from each other, and they were prepared in a somewhat different fashion.

The C-terminal heptapeptide was obtained by a racemization-
free condensation of Z-Arg(NO_2)-Arg(NO_2)-Pro-OH and H-Val-Lys(Boc)-
Val-Tyr-NH_2 via the mixed anhydride method (yield 63%) and the product was catalytically hydrogenated to give the partially protected
heptapeptide amide, H-[17-23]-NH_2 [161].

The middle sequence was prepared by condensation of Z-Lys(Boc)-
Pro-Val-Gly-OH with H-Lys(Boc)-Lys(Boc)-OMe by DCC, and the product
was converted into the corresponding hydrazide, Z-[11-16]-$NHNH_2$,
which is identical with the subunit used by Schwyzer and Sieber
[142, 143] in the synthesis of porcine ACTH. The results of the
DCC coupling may cast some doubt, since Hofmann et al. [117] observed predominant acyl urea formation in a similar coupling, while
Rittel [138] applied the azide procedure to the Gly-terminal tetrapeptide.

The azide coupling of Z-[11-16]-$NHNH_2$ with H-[17-23]-NH_2 gave
a crude Z-[11-23]-NH_2, which, without purification, was subjected
to catalytic hydrogenation. The product was then purified by
column chromatography on CM-cellulose using gradient of ammonium
acetate buffers (0.02 to 0.22 M). The partially protected tridecapeptide amide, H-[11-23]-NH_2, was obtained as a triacetate (yield
36%) [161].

In order to prepare the N-terminal decapeptide, Z-Glu(OBu^t)-
$NHNH_2$ was utilized in the synthesis of Z-Glu(OBu^t)-His-Phe-Arg-Trp-
Gly-OH, which, after hydrogenation, was condensed with Boc-Ser-Tyr-
Ser-Met-$NHNH_2$ [162]. It is known that the DCC activation of a
large carboxyl component produces a considerable amount of acylurea
as a side-reaction product and this makes it difficult to isolate
the desired compound. Therefore these authors converted the N-terminal decapeptide to the corresponding p-nitrophenyl ester; this latter
product was then coupled with the hydrochloride form of H-[11-23]-NH_2
in pyridine. Although an exact comparison between the various pro-

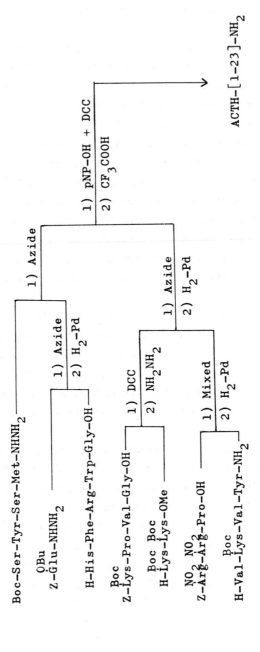

Scheme 5. Synthetic route to ACTH-[1-23]-NH$_2$ (Geiger et al. [160]).

cedures was not made, the crude product was purified by Sephadex
G-25. From 2.9 g of H-[11-23]-NH$_2$, 4.4 g of the crude peptide was
obtained, while column chromatography separated 930 mg of pure
Boc-[1-23]-NH$_2$ from 2.0 g of the crude material.

The protected tricosapeptide amide was treated with TFA at room
temperature for 45 min. One drop of thioglycolic acid was added to
the reaction mixture to prevent possible oxidation of the Met residue.
The deblocked peptide was purified by column chromatography on CM-
cellulose using ammonium acetate as eluent, and then by continuous
high-voltage electrophoresis. From 1 g of the protected peptide,
there was obtained 315 mg of pure H-[1-23]-NH$_2$ with an activity of
100 U/mg.

Later, a large-scale coupling reaction between Boc-[1-10]-OH
(16.5 g) and H-[11-23]-NH$_2$ (25 g) was performed to give Boc-[1-23]-
NH$_2$ (29 g) [163]. In 1970, König and Geiger recommended the use of
1-hydroxy-benzotriazole and DCC as a coupling procedure in peptide
synthesis [164, 165]. This could bring a more satisfactory result
to this specific reaction.

By using various synthetic subunits that were well characterized,
Geiger et al. synthesized a number of ACTH peptides modified around
the N-terminus:

Peptides	Activity, U/mg	Reference
Ala^3-ACTH-[1-23]-NH$_2$	50	166
Des-Ser^1-ACTH-[2-23]-NH$_2$	51	166
Gly^1-ACTH-[1-23]-NH$_2$	104	166
Phe^2-ACTH-[1-23]-NH$_2$	66	166
Pro^1-ACTH-[1-23]-NH$_2$	48	163
β-Ala^1-ACTH-[1-23]-NH$_2$	478	163
Leu^4-ACTH-[1-23]-NH$_2$	66	163
β-Ala^1-Leu^4-ACTH-[1-23]-NH$_2$	73	163

Peptides	Activity, U/mg	Reference
β-Ala^1-Lys^{17}-ACTH-[1-17]-NH-(CH$_2$)$_4$-NH$_2$	807	167
β-Ala^1-ACTH-[1-16]-NH-(CH$_2$)$_4$-NH$_2$	136	167
p-HO-C$_6$H$_4$-CH$_2$-CH$_2$-CO-Ser^3- Lys^{17}-ACTH-[3-17]-NH-(CH$_2$)$_4$-NH$_2$	149	167
CH$_3$-(CH$_2$)$_3$-O-CO-Glu^5-Lys^{17}-ACTH-[5-17]- NH-(CH$_2$)-NH$_2$	92	167

3. *Procedure Employed by Otsuka et al.*

Otsuka and Inouye [168] synthesized ACTH-[1-18]-OH and the corresponding amide in 1965 by making the protecting groups removable by TFA. They prepared a number of ACTH peptides using the N-hydroxysuccinimide ester method for the condensation steps.

For the synthesis of ACTH-[1-18]-NH$_2$, the construction of the C-terminal octapeptide amide (positions 11-18) was achieved by condensation of the pentapeptide unit (positions 11-15) and the tripeptide unit (positions 16-18). Thus the condensation was made between Lys^{15} and Lys^{16}, as shown in Scheme 6-a. The Z-Lys(Boc)-Pro-Val-Gly-Lys(Boc)-azide derived from the corresponding hydrazide [169] was allowed to react with H-Lys(Boc)-Arg(NO$_2$)-Arg(NO$_2$)-NH$_2$, and the resulting peptide (yield 84%) was subsequently hydrogenated to give H-Lys(Boc)-Pro-Val-Gly-Lys(Boc)-Lys(Boc)-Arg-Arg-NH$_2$ [170].

The N-terminal decapeptide, Boc-Ser-Tyr-Ser-Met-Glu(OBu^t)-His-Phe-Arg-Trp-Gly-OH (yield 80%) [171], was prepared by uniting the two subunits, Boc-Ser-Tyr-Ser-Met-NHNH$_2$ [171, 172] and H-Glu(OBu^t)-His-Phe-Arg-Trp-Gly-OH [173]. This type of coupling reaction was also done by Schwyzer et al. [139] and Geiger et al. [162]. During the course of the hexapeptide synthesis, Z-Glu(OBu^t)-His(Z)-OMe was exposed to hydrazine. The Z group attached at N^{im} of the His residue [174] was removed by this reagent to give Z-Glu(OBu^t)-His-NHNH$_2$.

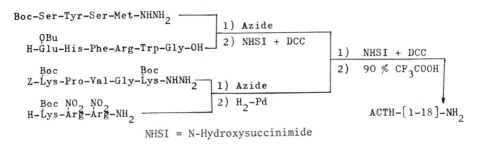

Scheme 6-a.   Synthetic route to ACTH-[1-18]-NH$_2$

(Otsuka and Inouye [168]).

The decapeptide, Boc-[1-10]-OH, was converted to the corresponding N-hydroxysuccinimide ester, instead of the p-nitrophenyl ester,
as used by Geiger et al. [160], and this active ester was allowed to
react with H-[11-18]-NH$_2$.   The resulting protected octadecapeptide
amide was purified by column chromatography on CM-cellulose using a
mixture of tert-butanol and 2 M ammonium acetate.   The purified
sample was then treated with 90% TFA and the desired octadecapeptide
amide (in vivo steroidogenetic activity 39.6-58 U/mg) was obtained
in an analytically pure form after purification by column chromatography on CM-cellulose.

Following the synthesis of ACTH-[1-18]-NH$_2$, Otsuka et al. [175]
reported a preliminary synthesis of human-type ACTH-[1-27]-OH (1959
formula).   The C-terminal octapeptide, H-Val-Lys(Boc)-Val-Tyr-Pro-Asp-
(OBu^t)-Ala-Gly-OBu^t (positions 20-27), was prepared by either the DCC
or the p-nitrophenyl ester method in a stepwise manner starting from
H-Gly-OMe.   Z-Lys(Boc)-Arg(NO$_2$)-Arg(NO$_2$)-Pro-OH was condensed with this
partially protected octapeptide via the N-hydroxysuccinimide ester
method, and the product, after hydrogenation, was joined with the
available pentapeptide hydrazide, Z-Lys(Boc)-Pro-Val-Gly-Lys(Boc)-
NHNH$_2$.   After hydrogenation of the product, the partially protected
heptadecapeptide, H-[11-27]-OBu^t, was obtained.   The N-terminal
decapeptide carrying the Z and benzyl ester groups, Z-Ser-Tyr-Ser-

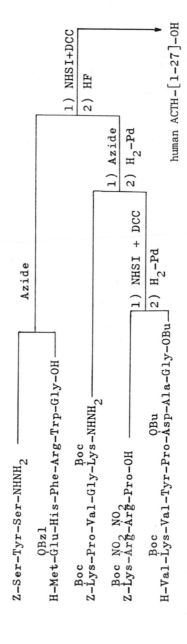

Scheme 6-b. Synthetic route to human ACTH-[1-27]-OH
(Otsuka et al. [175]).

Met-Glu(OBzl)-His-Phe-Arg-Trp-Gly-Oh, was coupled to this unit
by N-hydroxysuccinimide ester method, as shown in Scheme 6-b.

In the synthesis of this Met-containing peptide, the Z and
benzyl ester groups were used in combination with the Boc and tert-
butyl ester groups. In order to remove all protecting groups in one
step, the protected heptacosapeptide was treated with hydrogen
fluoride, according to Sakakibara et al. The heptacosapeptide is
reported to exhibit in vivo steroidogenetic activity of 267 U/mg.
This procedure can be classified as a combination of both the TFA
and the hydrogen fluoride deblocking techniques.

Many ACTH peptides were prepared by this group using the
standard TFA procedure:

Peptides	Activity, U/mg	Reference
ACTH-[1-18]-OH	9.4	168
Gly^1-ACTH-[1-14]-OH	active	176
Gly^1-[1-18]-NH_2	27	177
Gly^1-Gly^{11}-[1-18]-NH_2	1.1	178
Gly^1-Gly^{11}-Gly^{13}-[1-18]-NH_2	0.003	178
β-Ala^1-[1-18]-NH_2	237	179
β-Ala^1-Orn^{15}-[1-18]-NH_2	50	178
β-Ala^1-D-Phe^7-Orn^{15}-[1-18]-NH_2	9.6	178
Ibu^1-[1-18]-NH_2	83	180
Ibu^1-Orn^{15}-[1-18]-NH_2	137	178
Ibu^1-[1-24]-OMe	236	176
Ibu^1-[1-27]-OH	325	176
[Des-Lys^{15}-Lys^{16}]-Gly^1-[1-14]- [17-18]-OH	0.13	181
[Des-Lys^{11}-Gly^{14}]-Gly^1-[1-10]- [12-13]-[15-18]-NH_2	0.05	178
[Des-Lys^{15}]-Gly^1-[1-14]- [16-18]-NH_2	2.1	182

4. *Procedure Employed by the Hungarian Groups*

In 1967, Bajusz and Medzihradszky et al. [183, 184] reported
the synthesis of the nonatriacontapeptide corresponding to the
entire amino acid sequence of human ACTH, the structure of which
was proposed by Lee and Lerner [73, 74]. Since protecting groups
derived from tert-butanol were applied here, their procedure
is similar to that of the preparation of porcine ACTH performed by
Schwyzer and Sieber [142, 143]. In addition to Lys(Boc), Asp(OBu^t),
and Glu(OBu^t), Tyr(OBu^t) was also applied to this synthesis.

Peptides and peptide derivatives were routinely purified by
column chromatography on silica. Suitable solvents for elution
were selected by thin-layer chromatography, since separation on
thin-layer silica gel was quite similar to that done on column
chromatography. Usually, ethyl acetate-pyridine-water-methanol-
acetic acid or formic acid mixtures were recommended; in addition,
chloroform-water-methanol or acetone mixtures were found to be
effective for the purification work.

Peptide fragment condensations were performed by the pentachlorophenyl active ester method. Acylurea formation involving
DCC could be suppressed by using pentachlorophenol and DCC in the
proportion of 3:1. The molecular ratio of the acid and the amino
components in a reaction was 1:1. Thus an average yield of 50-60%
was obtained from each fragment condensation reaction. N-Hydroxysuccinimide esters were also applied for similar purposes.

The synthetic route to human ACTH (1959 formula) is illustrated
in Scheme 7.

The C-terminal dodecapeptide (positions 28-39) and the tripeptide, Z-Asp(OBu^t)-Ala-Gly-OH (positions 25-27), were coupled by
either the mixed anhydride or pentachlorophenyl ester methods. The
resulting pentadecapeptide (positions 25-39), after hydrogenolysis,
was allowed to react with the pentachlorophenyl ester of Z-[15-24]-OH,
which was prepared by condensation of Z-[15-19]-OH with H-[20-24]-OH
via the same active ester procedure. All pentachlorophenyl esters

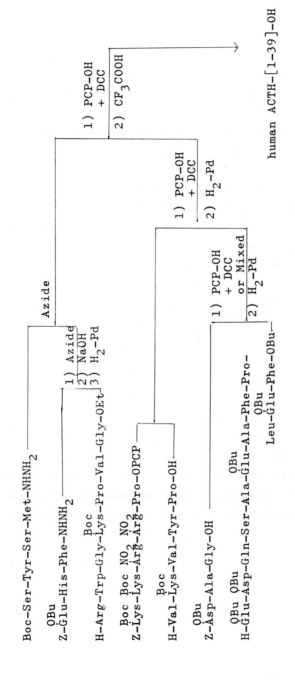

Scheme 7.   Synthetic route to human ACTH (1959 formula)
(Bajusz et al. [183, 184]).

used in these syntheses were characterized by complete elemental
analysis.

The pentacosapeptide ester, Z-[15-39]-OBu^t, was made alternatively
by the N-hydroxysuccinimide ester procedure. The partially protected
dodecapeptide ester, H-[28-39]-OBu^t, was condensed with the N-hydroxysuccinimide ester of Z-[20-27]-OH, which was prepared by uniting the
two fragments, Z-[20-24]-OH and H-[25-27]-OH, via the same active
ester. The resulting Z-[20-39]-OBu^t, after hydrogenation, was allowed
to react with the pentachlorophenyl ester of Z-[15-19]-OH to give
Z-[15-39]-OBu^t.

The protected pentacosapeptide ester thus obtained by either of
the two alternate routes was then hydrogenated to give the partially
protected pentacosapeptide ester, H-[15-39]-OBu^t. The nitro groups
of the Arg residues were reductively removed at this stage, too.

In view of their interest in preparing various ACTH peptides
with different chain lengths, they preferred to use the tetradeca-
peptide (positions 1-14) as the N-terminal unit [185, 186]. This
peptide was prepared by uniting three subunits, Boc-Ser-Tyr-Ser-
Met-NHNH$_2$, Z-Glu-(OBu^t)-His-Phe-NHNH$_2$, and H-Arg-Trp-Gly-Lys(Boc)-
Pro-Val-Gly-OEt.

The activation of Gly^{14} by DCC was reported to form the acyl
urea predominantly. Therefore the condensation between Boc-[1-14]-
OH and H-[15-39]-OBu^t was performed by DCC and pentachlorophenol.
The resulting fully protected nonatriacontapeptide was purified by
column chromatography on silica using the solvent ethyl acetate-
pyridine-water-acetic acid (60:20:6:5.5), and followed by treatment
with anhydrous TFA. The product, after conversion to the corresponding
acetate, was lyophilized (in vitro steroidogenetic activity, 78-153
U/mg).

In connection with the above synthesis of human ACTH, a number
of derivatives [187-190] and peptides with various chain lengths
were prepared:

Peptide	Activity, U/mg	Reference
ACTH-[1-15]-NH$_2$	0.2	186
ACTH-[1-16]-NH$_2$	1.4	186
ACTH-[1-17]-NH$_2$	10	186
ACTH-[1-18]-NH$_2$	27.5	186
ACTH-[1-23]-OH	69-115	184, 191
ACTH-[1-32]-OH	70-129	184, 191

E. Hydrogen Fluoride Procedure

Fujino, Hatanaka, and Nishimura synthesized a number of ACTH
peptides that carried protecting groups removable by hydrogen
fluoride. This method was introduced by Sakakibara and Shimonishi
in 1965 to peptide synthesis [192, 193]. This reagent cleaves not
only the Boc and Bu^t ester groups, but also the Z and Bzl ester
groups within 30 min at 0-20°. The nitro group attached at the
guanidino function of Arg is also removable by this reagent. It
will be realized that if this procedure can be applied at the final
stage of a synthesis, then elongation of a peptide chain containing
Lys(Z), Glu(OBzl), or Asp(OBzl) is possible. In this procedure, the
Boc group removable by TFA must play an exclusive role as the sole
α-amino protecting group. A Boc-amino acid is coupled with an
amino component containing the Z and Bzl ester groups, and then the
product is treated with TFA. The α-amino group becomes free, whereas

the other side-chain protecting groups remain unchanged. The presence
of a sulfur-containing amino acid does not interfere with this approach,
since catalytic hydrogenation is no longer necessary.

As was seen, the individual use of the Z and Boc groups in the
previous TFA procedure was completely reversed by this approach.
Therefore a large supply of Boc-amino acids is compulsory. To meet
this demand, Fujino and Hatanaka [194] introduced a new Boc-reagent,
tert-butyl pentachlorophenyl carbonate, which reacts with amino acids
in the presence of a calculated amount of sodium hydroxide at room
temperature to give Boc-amino acids in fairly good yield.

Reaction 4.   tert-Butyl Pentachlorophenyl carbonate,
Fujino and Hatanaka [194].

In addition, two new active ester procedures were examined.
In one, some simple ketoxime esters investigated by Losse et al.
[195] were found to be particularly useful for $Arg(NO_2)$, Thr, and
Ser [196]. The ketoxime esters (acetoxime, cyclopentanone oxime,
or cyclohexanone oxime) of these acyl-amino acids react with amines
in the presence of acid catalysts, such as acetic acid or formic
acid.

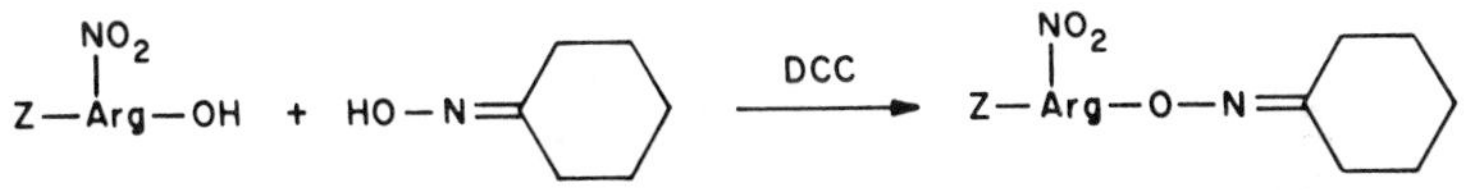

(5)

Reaction 5.   O-[Z-Arg(NO$_2$)]-cyclohexanone oxime,
Fujino and Nishimura [196].

The other procedure uses an ester-exchange reagent involving
pentachlorophenyl trichloroacetate or dichloroacetate [197]. By
the use of the compounds, the pentachlorophenyl ester of acyl-amino
acids or peptides can be prepared without DCC. Consequently the
troublesome acylurea formation as induced by DCC can be avoided
easily.

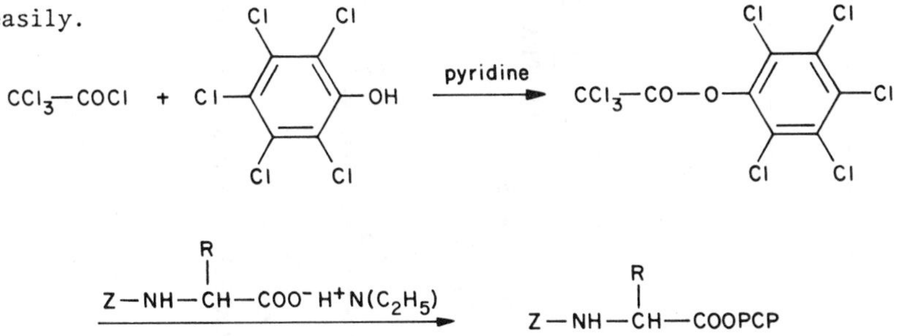

(6)

Reaction 6.   Pentachlorophenyl trichloroacetate (TCAOPCP),
Fujino and Hatanaka [197].

The synthesis of ACTH-[1-23]-NH$_2$ [198, 199] was accomplished
by applying these newly developed methodologies, as illustrated in
Scheme 8 a-g.

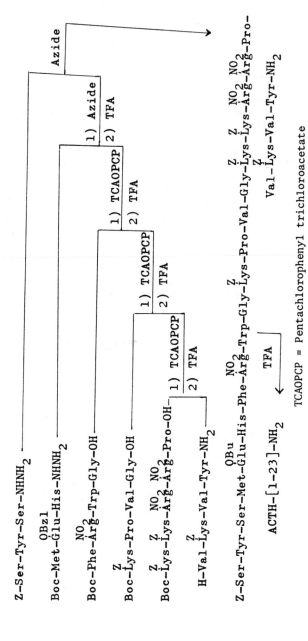

Scheme 8-a. Synthetic route to ACTH-[1-23]-NH$_2$
(Fujino et al. [198, 199]).

TCAOPCP = Pentachlorophenyl trichloroacetate

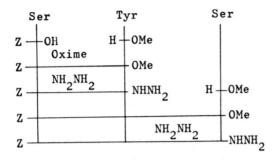

Scheme 8-b. Positions 1-3.

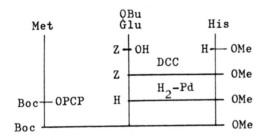

Scheme 8-c. Positions 4-6.

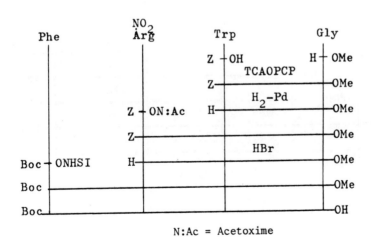

N:Ac = Acetoxime

Scheme 8-d. Positions 7-10.

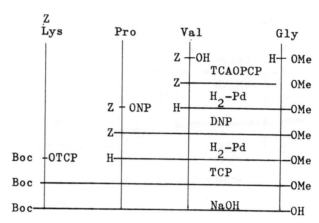

TCP = Trichlorophenyl ester

Scheme 8-e. Positions 11-14.

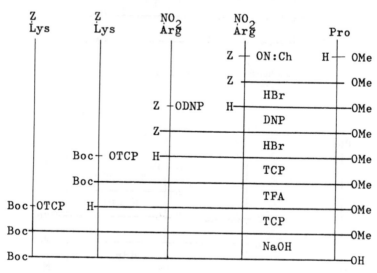

ON:Ch = Cyclohexanonoxime ester

Scheme 8-f. Positions 15-19.

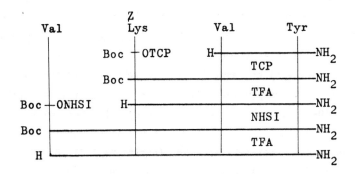

Scheme 8-g. Positions 20-23.

The synthetic outline for the C-terminal tetrapeptide amide,
H-Val-Lys(Z)-Val-Tyr-NH$_2$ (positions 20-23), is shown in Scheme 8-g.
Starting from H-Tyr-NH$_2$, this tetrapeptide was prepared in a stepwise fashion, in which Boc-Lys(Z)-OH was incorporated into the
peptide chain by the trichlorophenyl ester method. The product,
after treatment with TFA, was subsequently condensed with Boc-Val-OH
by means of N-hydroxysuccinimide. Treatment with TFA gave the acid
salt of H-[20-23]-NH$_2$, which was then converted into the free base
as fine crystals [200].

As illustrated in Scheme 8-f, the protected pentapeptide,
Boc-Lys(Z)-Lys(Z)-Arg(NO$_2$)-Arg(NO$_2$)-Pro-OH (positions 15-19),
was also prepared in a stepwise manner by the active ester method
[200]. First, the cyclohexanone oxime ester of Z-Arg(NO$_2$)-OH was
allowed to react with H-Pro-OMe acetate; the product (yield 90%)
was then treated with hydrogen bromide in acetic acid. The second
Z-Arg(NO$_2$)-OH was introduced by the dinitrophenyl ester method
(yield 70%), while the two Boc-Lys(Z)-OH residues were incorporated
through the trichlorophenyl ester procedure. The protected pentapeptide methyl ester was then saponified to give Boc-[15-19]-OH.

The tetrapeptide, Boc-Lys(Z)-Pro-Val-Gly-OH (positions 11-14), was synthesized in a stepwise fashion by the active ester procedure, as shown in Scheme 8-e. On comparison with the earlier routes to these ACTH derivatives, especially with reference to the use of the N^{ϵ}-Boc group in Lys-peptides, it will be noted that in Boc-[15-19]-OH and Boc-[10-14]-OH, the Z group was employed by these workers.

The N-terminal decapeptide (positions 1-10) was divided into three subunits, Z-Ser-Tyr-Ser-NHNH$_2$ (positions 1-3), Boc-Met-Glu (OBu^t)-His-NHNH$_2$ (positions 4-6), and Boc-Phe-Arg(NO$_2$)-Trp-Gly-OH (positions 7-10), as shown in Scheme 8-b, c, and d. Except for the N-terminal tripeptide, the other two subunits were protected by the Boc group.

These various subunits were assembled as illustrated in Scheme 8-a. Boc-[15-19]-OH was converted into the corresponding penta-chlorophenyl ester by TCAOPCP (yield 90%). This ester-exchange reaction proceeded at room temperature for 50 min. The active ester thus obtained was allowed to react with the C-terminal tetra-peptide amide, H-[20-23]-NH$_2$, at room temperature for 10 hr, and the product was isolated by precipitation from acetic acid with methanol (yield 76%). This procedure proved to be useful for purification of various protected peptides.

After removal of the Boc group from the protected nonapeptide amide by TFA, the resulting H-[15-23]-NH$_2$ was condensed with Boc-[11-14]-OH by the TCAOPCP procedure to give Boc[11-23]-NH$_2$ (yield 81%).

The similar TCAOPCP reaction was repeated to condense Boc-[7-10]-OH with H-[11-23]-NH$_2$. The protected heptadecapeptide amide, Boc-[7-23]-NH$_2$ (yield 83%), was treated with cold TFA containing 1% thioglycolic acid to remove the Boc group. Addition of a thiol compound proved to be effective in preventing destruction of the Trp residue. The resulting H-[7-23]-NH$_2$ was coupled with Boc-Met-Glu(OBu^t)-His-NHNH$_2$ via the azide procedure to give the protected eicosapeptide amide, Boc-[4-23]-NH$_2$ (yield 96%). By repeating a similar deblocking procedure, the resulting H-[4-23]-NH$_2$ was con-

densed with Z-Ser-Tyr-Ser-NHNH$_2$ by the azide procedure to give the
fully protected tricosapeptide amide, Z-[1-23]-NH$_2$, which was
purified by column chromatography on silica with ethyl acetate-
pyridine-acetic acid-water (60:20:6:10) (yield 86%).

The final deblocking of all protecting groups (Z, Bu^t, and
NO$_2$) was performed by hydrogen fluoride at 0° for 60 min. In
addition to anisole and thioglycolic acid, Met and Trp were added
to the reaction mixture to prevent undesirable alkylation reactions.
The resulting peptide was converted into the corresponding acetate
on Amberlite IRA-400 and then submitted to purification by column
chromatography on CM-cellulose. Gradient elution with ammonium
acetate separated the analytically pure H-[1-23]-NH$_2$ (yield 50%)
with an in vivo steroidogenetic potency of about 80 U/mg.

By using these newly developed procedures in peptide chemistry,
Fujino, Hatanaka, and Nishimura prepared a number of ACTH peptides
that were mainly modified around the N-terminal portion:

Peptides	Activity U/mg	Reference
ACTH-[1-24]-OH	90	201
ACTH-[4-23]-NH$_2$	15-20	202
ACTH-[5-23]-NH$_2$	1	202
ACTH-[6-24]-OH	0.1	202
ACTH-[7-23]-NH$_2$	inactive	202
β-Ala^1-ACTH-[1-23]-NH$_2$	90	199
β-Ala^1-ACTH-[1-24]-OH	100-160	201
γ-Aminobutylic acid^1-ACTH-[1-24]-OH	100-110	201
Sarcosine^1- ACTH-[1-24]-OH	100-160	201
Pro^1- ACTH-[1-24]-OH	50-65	201
Lys^1- ACTH-[1-24]-OH	30-50	201
Leu^4-ACTH-[1-24]-OH	55-85	203
Ile^4-ACTH-[1-24]-OH	55-85	203
Leu^7-ACTH-[1-24]-OH	15-25	204
N-Methyl-Trp^9-ACTH-[1-24]-OH	0.05	204

F.   Solid-Phase Procedure

A solid-phase synthesis of ACTH-[4-10]-OH, H-Met-Glu-His-Phe-Arg-Trp-Gly-OH, was reported by Blake and Li in 1968 [205]. From the protected heptapeptide resin (1.07 g, 0.24 mmole of peptide), the free heptapeptide (42 mg) was obtained after purification by countercurrent distribution and column chromatography on CM-cellulose. Later, Bayer et al. [206] briefly mentioned the solid-phase synthesis of ACTH-[1-20]-OH; however, no further information is available at this time. This method, as originally introduced by Merrifield in 1963 [207], can be classified as an application of an insoluble benzyl ester to a peptide synthesis.

A number of solid-phase preparations of biologically active peptides with relatively high molecular weights have appeared in the literature: insulin by Marglin and Merrifield (activity 8%) [208]; apoferredoxin by Bayer et al. (no biological activity) [209]; cytochrome-c-like compound by Sano and Kurihara (activity 2% in the presence of His) [210]; basic trypsin inhibitor by Noda et al. (activity 39%) [211]; ribonuclease A by Gutte and Merrifield [activity 13% (after tryptic treatment, the activity was increased to 78%)] [212, 213]; and human growth hormone by Li and Yamashiro [activity 10% (after this synthesis, the structure was revised)] [214]. The low activity of these materials produced by solid-phase synthesis is caused by an inseparable mixture of failure sequences accumulated on the resin [215]. These situations result from the incompleteness of α-amino deblocking [216] and coupling reactions [206], which are, in general, a feature of heterogeneous reactions.

Yet, solid-phase synthesis has some advantageous properties, since the excess carboxyl component and the reagent can be removed simply by washing the resin with appropriate solvents. It should be realized that during the course of one-at-a-time addition procedures, any byproducts or unreacted amino components attached to the resin cannot be removed by washing procedures, and they will have another chance to react at each of the later steps.

If well-characterized peptides with certain chain lengths can
be condensed on the resin, then such an approach reduces the number
of coupling reactions required to construct the desired peptide
chain. Even if quantitative coupling of succeeding peptides cannot
be achieved, the properties and size of the resulting byproducts
may be different from those of the desired compound. Therefore,
in theory, the purification of the desired product at the end of
the synthesis will not be difficult. Moreover, the unique properties of the insoluble benzyl ester are maintained by permitting
the purification to be done by the various washing procedures at
each step of the fragment condensation.

The principle of peptide fragment condensation on polymer
support is illustrated in Scheme 9-a. Here the authors have
examined the possibility of applying the principle of peptide
fragment condensation to the synthesis of ACTH peptides.

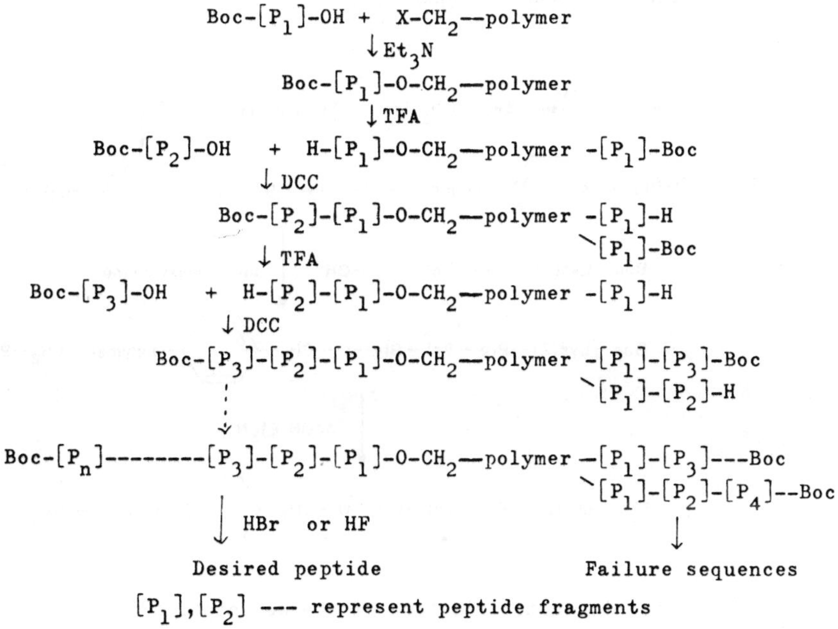

$$Boc\text{-}[P_1]\text{-}OH \ + \ X\text{-}CH_2\text{---}polymer$$
$$\downarrow Et_3N$$
$$Boc\text{-}[P_1]\text{-}O\text{-}CH_2\text{---}polymer$$
$$\downarrow TFA$$
$$Boc\text{-}[P_2]\text{-}OH \ + \ H\text{-}[P_1]\text{-}O\text{-}CH_2\text{---}polymer \ \ \text{-}[P_1]\text{-}Boc$$
$$\downarrow DCC$$
$$Boc\text{-}[P_2]\text{-}[P_1]\text{-}O\text{-}CH_2\text{---}polymer \ \ \text{-}[P_1]\text{-}H$$
$$\downarrow TFA \qquad\qquad\qquad\qquad \diagdown[P_1]\text{-}Boc$$
$$Boc\text{-}[P_3]\text{-}OH \ + \ H\text{-}[P_2]\text{-}[P_1]\text{-}O\text{-}CH_2\text{---}polymer \ \ \text{-}[P_1]\text{-}H$$
$$\downarrow DCC$$
$$Boc\text{-}[P_3]\text{-}[P_2]\text{-}[P_1]\text{-}O\text{-}CH_2\text{---}polymer \ \ \text{-}[P_1]\text{-}[P_3]\text{-}Boc$$
$$\vdots \qquad\qquad\qquad\qquad\qquad\qquad \diagdown[P_1]\text{-}[P_2]\text{-}H$$
$$Boc\text{-}[P_n]\text{------}[P_3]\text{-}[P_2]\text{-}[P_1]\text{-}O\text{-}CH_2\text{---}polymer \ \ \text{---}[P_1]\text{-}[P_3]\text{---}Boc$$
$$\diagdown[P_1]\text{-}[P_2]\text{-}[P_4]\text{--}Boc$$
$$\downarrow HBr \ \ or \ \ HF \qquad\qquad\qquad\qquad\qquad \downarrow$$

Desired peptide   Failure sequences

$[P_1], [P_2]$ --- represent peptide fragments

Scheme 9-a. Principle of peptide fragment
condensation on polymer support.

First, suitable conditions were examined for esterifying a model peptide onto the copolymer of styrene and 2% divinylbenzene. Originally, esterification of a Boc-amino acid onto the chloromethylated copolymer required heating for about 50 hr in the presence of triethylamine [207]. This condition does not seem to be suitable for the esterification of an acylpeptide with the resin.

It was found that as a model experiment, a fairly good esterification (yield more than 50%) of the tetrapeptide, Boc-Lys(Z)-Pro-Val-Gly-OH (1 equimole), was achieved by using bromomethylated resin (7 equimoles) and dicyclohexylamine at room temperature [217]. This resin can be prepared by bromomethylation of the resin with bromomethyl methyl ether in the presence of stannic bromide, according to the procedure developed for preparation of the chloromethylated resin [207]. Or, according to Tilak [218], it can be prepared by treatment of the acetoxymethylated resin with hydrogen bromide, as shown in Scheme 9-b.

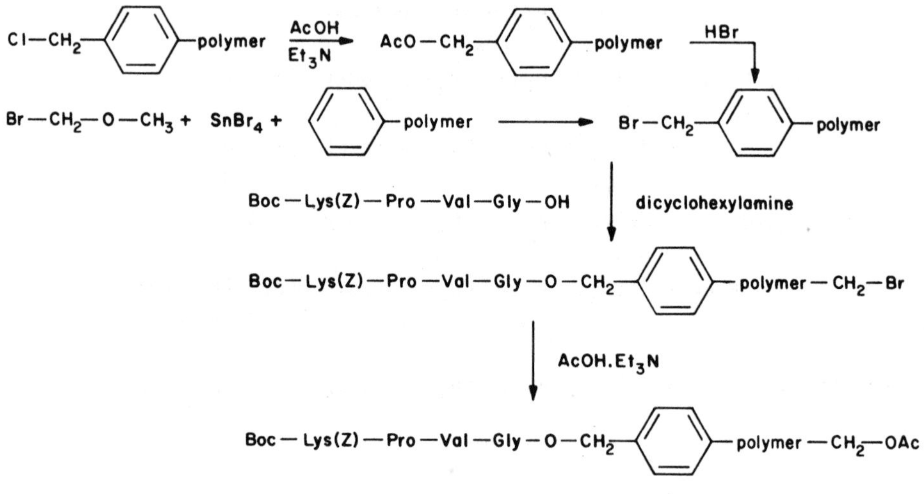

Scheme 9-b.   Esterification of the acylpeptide
onto the bromomethylated resin
(Yajima and Kawatani [217]).

When the chloromethylated resin was employed under identical conditions, the esterification yield was less than 1%. Attempts to esterify this peptide onto the hydroxymethylated resin with either DCC or carbonyldiimidazole were unsuccessful. The above esterification can also be performed by using the dicyclohexylamine salt of acylpeptides and the bromomethylated resin. It is recommended that one treat the esterified resin with triethylammonium acetate prior to the next coupling reaction in order to convert the remaining bromine atom to the corresponding acetoxy residue.

The peptide resin was then mixed with 1 N hydrochloric acid in dioxane or, preferably, 50% TFA in a mixture of dimethylformamide and methylene chloride to remove the Boc group. The resulting peptide resin was neutralized with triethylamine, and the free peptide resin was allowed to condense with Z-Glu(OBzl)-His-Phe-Arg(NO$_2$)-Trp-Gly-OH by means of DCC. At least three or more equivalents of acylpeptide had to be used to insure a quantitative coupling. The resin, after washing with dimethylformamide, was reacted with anhydrous hydrogen fluoride to cleave the ester bond from the resin [219] and to remove all protecting groups from the growing peptide chain. Anisole and Trp were added to prevent any possible alkylation reactions during this deblocking procedure. The product, H-Glu-His-Pre-Arg-Trp-Gly-Lys-Pro-Val-Gly-OH, ACTH-[5-14]-OH, was obtained as a single peak when purified by column chromatography on CM-cellulose. The yield in this model experiment was 40%, based on the tetrapeptide attached on the resin. Some improvements in the cleavage step may be expected if other agents are used.

On the basis of the amino component attached on the resin, at least three or more equivalents of the carboxyl component and the coupling reagent were employed in order to achieve a quantitative coupling reaction. However, despite the use of such a large excess of reagents, the reactivity of the coupling reagent, as pointed out by Omenn and Anfinsen [220], seems to be dependent on the molecular size and the nature of the terminal amino acid residues involved in

the coupling reaction. Further, Omenn and Anfinsen mentioned that
the N-terminal β-benzyl Asp residue offered special difficulty in
a solid-phase coupling. Thus, it became necessary to examine the
conditions required for a quantitative coupling between pMZ-Val-
Lys(Z)-Val-Tyr-Pro-OH (positions 19-24) and an insoluble benzyl
ester of H-Asp(OBzl)-Gly (positions 25-26), which was prepared
according to the procedure described above [221].

The reactivity of various coupling reagents was examined.
Satisfactory results were not obtained with DCC, DCC plus N-
hydroxysuccinimide, DCC plus pentachlorophenol, or N-ethyl-5-
phenylisoxazolium-3'-sulfonate. However, it is noteworthy that
N-ethoxycarbonyl-2-ethoxy-1,2-dihydroquinoline (EEDQ) gave 58%
coupling when 2 moles of pMZ-pentapeptide were employed, and nearly
90% coupling was achieved at 4 moles. Belleau and Málek [222] proposed that the amide-forming reaction by this reagent proceeds
through a mixed anhydride intermediate. From a practical standpoint,
if one needs a large excess of the carboxyl component for satisfactory
coupling reactions, then recovery of the starting material may be a
matter of importance. The fact that EEDQ meets such a demand seems
to be another superior property of this reagent in terms of solid-
phase synthesis.

Based on these two model experiments, a total synthesis of
bovine ACTH was undertaken according to Scheme 9-c.

A total of six peptide fragments with either Gly or Pro as
the terminus amino acids were synthesized in order to achieve a
racemization-free condensation on the polymer support [223].

The p-methoxybenzyloxycarbonyl group (pMZ) [224], removable by
TFA, was used for α-amino protection of these fragments. A newly
developed pMZ reagent, p-methoxybenzyl-8-quinolyl carbonate, served
to prepare these amino acid derivatives [225].

The C-terminal tripeptide, pMZ-Leu-Glu(OBzl)-Phe-OH, was prepared by the active ester method in a stepwise manner. For the
synthesis of the decapeptide, pMZ-Glu(OBzl)-Ala-Glu(OBzl)-Asp(OBzl)-
Ser-Ala-Gln-Ala-Phe-Pro-OH (positions 27-36), Z-Gln-Ala-NHNH$_2$ and

Scheme 9-c. Synthetic route to bovine ACTH
(Yajima, Kawatani, and Tamura [223]).

(7)

Reaction 7.   p-Methoxybenzyl-8-quinolyl carbonate
(Yajima, Tamura, and Kiso [225]).

Z-Ser-Ala-NHNH$_2$ were added consecutively via the azide procedure
to H-Phe-Pro-OH. Hydrogenation of the product gave the hexapeptide,
H-Ser-Ala-Gln-Ala-Phe-Pro-OH, to which the p-nitrophenyl esters of
pMZ-Asp(OBzl)-OH, pMZ-Glu(OBzl)-OH, and pMZ-Ala-OH were introduced
according to the previous sequence. The other subunits are known
peptides used previously by other investigators, except for the
α-amino masking group.

Esterification of pMZ-Leu-Glu(OBzl)-Phe-OH onto the bromomethylated resin was performed according to the earlier esterification
procedure [217]. The pMZ-tripeptide resin was then treated with
50% TFA in methylene chloride and the resin was washed with
triethylamine. EEDQ served to condense the pMZ-decapaptide prepared above to the tripeptide resin. This cycle was repeated
until the N-terminal decapeptide was condensed on the polymer
support. Amino acid ratio analysis in the acid hydrolysates of
peptide resins at each coupling stage was very close to those
expected by theory. The results indicate that peptide fragment
condensation on a polymer support can be applied to the synthesis
of relatively large peptides. Details of the synthesis of bovine
ACTH (revised) will be published in the future.

In summary, various procedures employed by independent investigators for the synthesis of ACTH-active peptides have been reviewed
and discussed in detail. During the past decade, the advancement in

synthetic methodology, the development of purification techniques, and the refinement in sensitive analytical tools has led to many improved synthetic procedures for ACTH peptides. Recently, the synthesis of ^{14}C-labeled polypeptides has been recorded, too. ^{14}C-Gly^{10}-ACTH-[1-10]-OH was prepared by Medzihradszky et al. [226], and ^{14}C-Phe^8-ACTH-[1-20]-NH$_2$ by Hofmann et al. [125]. The latter was used in a binding test with the adrenal cortex. From the use of these labeled peptides, studies on the mode of action of ACTH can be expected in the next few years.

V. STRUCTURE-FUNCTION OF ACTH PEPTIDES

The linear arrangement of amino acid residues in various peptide hormones looks like a cipher, from which one should be able to easily pick only the letters necessary to understand the biologically important or essential amino acid residues. Yet, the right key to deciphering such a password is still not in our hands.

Data obtained from the modification of natural or synthetic analogs enable us to a certain degree to look inside the ACTH molecule, especially with those compounds that elicit a characteristic physiologic response. However, our efforts to elucidate the structural requirements for ACTH activity suffer from serious limitations; two such restrictions are the methodology involving these biologic events and the difficulties of hormonal assay. Biologic evaluation still involves an assay with intact animals or in vitro systems that possess a high degree of organization, plus various factors involved in moving from the site of administration to ultimate receptor.

Keeping such limitations of hormone assay and various methodologies in mind, the accumulated data concerning structure-function relationships in ACTH can be summarized in the following manner: (a) chain length and activity; (b) synthetic analog and activity; (c) stereoisomers and activity; and (d) deschain peptides and activity.

A. Chain Length and Activity

Beginning with observations made by Bell et al. [54], synthetic efforts have concentrated on finding the shortest possible peptide with full biologic activity. Synthetic peptides prepared by various groups are listed in order of chain length, as shown in Table 2. Fortunately, peptides with chain lengths from 13 to 24 are available for comparison, except for 21 and 22. Some discrepancy in their potencies is inevitable, since they were assayed in different laboratories.

It can be seen from the data in Table 2 that the peptide chain can be shortened from 24 to around 20 residues (See XVIII, Table 2) without significant loss of its activity. When activities of NH_2-terminated peptides are compared, elimination of Val^{20} (XVI) and Pro^{19} (XII, XIII, XIV) causes some inactivation. With the elimination of Arg^{18} (IX, X) and Arg^{17} (VI), immense loss of activity occurs, decreasing to a level of 1/100 of the original. Further elimination of Lys^{16} (IV) and Lys^{15} (III or II) brings down the activity to 1/1000 of the natural ACTH activity. Such a trend of decreasing activity can be seen when the activities of various peptide amides are plotted against their chain length, as shown in Fig. 2.

Li et al. [97-99] and Otsuka et al. [168, 177] have been interested in the effect of the C-terminal carboxyl group in hexadeca-, octadeca-, and nonadecapeptides. They noticed that the amide-terminated peptides exhibited much higher activity than the corresponding peptides with the free carboxyl group. Therefore, the four basic net charges in -Lys-Lys-Arg-Arg- sequence around the C-terminal portion of these peptides seem to be of great im-

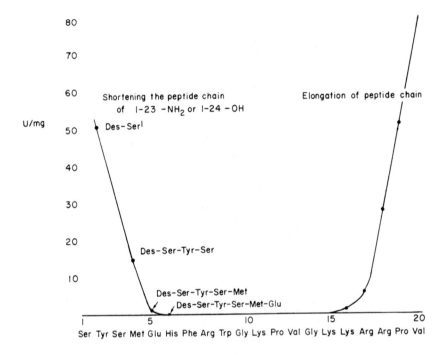

FIG. 2. Effect caused by elongation or shortening of the peptide chain.

portance in exerting a relatively high level of ACTH activity. In addition, the resistance of the amide peptides to the action of carboxypeptidase in vivo may participate in causing this observable difference.

The shortest peptide with ACTH activity known so far is the tridecapeptide amide, ACTH-[1-13]-NH_2 (II), and recently the ACTH activity of α-MSH was reexamined by Nakamura and Tanaka [37] by the adrenal cell suspension method. From these observations, it was concluded that the structural element that is essential for ACTH activity must reside within the first 13 amino acid residues, while the other element involving activity (the so-called binding site) must reside within sequences 14 to 20 or, more probably, positions 15 to 20.

TABLE 2

Relationship between Chain Length and Activity

Sequence (positions 1–24):
1 Ser – Tyr – Ser – Met – Glu 5 – His – Phe – Arg – Trp – Gly 10 – Lys – Pro – Val – Gly – Lys 15 – Lys – Arg – Arg – Pro – Val 20 – Lys – Val – Tyr – Pro

	Chain length	ACTH (U/mg) Ascorbic acid depletion	Steroidogenesis in vivo	Steroidogenesis in vitro	MSH (U/g)	Reference
I	Ac– 13─NH$_2$			0.0043	2.2×10^{10}	37
II	13─NH$_2$	0.1	0.1		1.9×10^{9}	109
III	14─OH					176
IV	15─NH$_2$	0.2				186
V	16─OH	0.1	0.1		3.7×10^{8}	118
VI	16─NH$_2$	1.4				186
VII	16─OMe	1		5–10		144
VIII	17─OH		5.4	5.2	2×10^{8}	98
IX	17─NH$_2$		33	42	4.2×10^{8}	99
X	17─NH$_2$	10				186
XI	18─OH	6.8	17.07	9.35		168
XII	18─NH$_2$	40	39.6	11.5		168
XIII	18─NH$_2$		133	33	4.5×10^{8}	99
XIV	18─NH$_2$	27.5				186

110

Compound	Structure					Ref.
XV	19—OH		47	35	1.4×10^7	96
XVI	19—NH_2		147	53	1.1×10^8	99
XVII	20—NH_2	118	83	2–3		120
XVIII	20—OMe				2.0×10^8	84
XIX	23—OH	116	91			111
XX	23—NH_2	100	80			160
XXI	23—NH_2	106				199
XXII	24—OH		90			136
XXIII	24—OH					201
XXIV	Beef 26—OH		130	93	4.8×10^8	101
XXV	Human 27—OH		356		6.5×10^8	175
XXVI	Human 28—OH	92		91		190
XXVII	Human 32—OH	104		88		190
XXVIII	Pig 39—OH	110				143
XXIX	Human 39—OH	109		107		184
XXX	23—NH_2	51	20			166
XXXI	23—NH_2		1			202
XXXII	23—NH_2		0.1			202
XXXIII	24—OH		0			202
XXXIV	23—NH_2					202

111

Concerning earlier observations around the N-terminal portion
of ACTH, it was originally believed that the Ser[1] residue was
essential for biologic activity. This view originated with White
[59], since he reported that inactivation of ACTH occurred parallel
to the elimination of Ser[1] and Tyr[2] when ACTH is exposed to leucine
aminopeptidase. Later, elimination of the Ser[1] residue from the
N-terminus of the natural ACTH was carried out by Dixon and Moret
[227], who treated ACTH with periodic acid, and then with
4-methylphenylene-1,2-diamine. This modified Des-Ser[1] compound
was reported to possess, in contrast to White's observation, a
relatively high ACTH activity. These conflicting views were
unequivocally resolved when Geiger et al. [166] synthesized
Des-Ser[1]-ACTH-[2-23]-NH$_2$ (XXX), which retains an activity of
51 U/mg.

Some Russian authors [228] have cleaved amino acid residues
from the N-terminus of ACTH by exposing the peptide to ultra-
sonic waves. They stated complete inactivation occurred at the
stage where the Phe residue in position 7 was eliminated from the
hormone. In connection with this statement, we may recall that
Pickering et al. [78] isolated a pig-type ACTH from sheep pituitary that does not have the Ser-Tyr-Ser-Met-Glu-His sequence. No
biologic data were assigned to this natural product.

In order to examine systematically the effects accompanying
the further shortening of the peptide chain from the N-terminus,
Fujino et al. [202] prepared peptides terminating at 23 or 24, but
lacking Ser-Tyr-Ser (XXXI), Ser-Tyr-Ser-Met (XXXII), and Ser-Tyr-
Ser-Met-Glu (XXXIII). When Ser-Tyr-Ser-Met-Glu-His was deleted
from the N-terminus (XXXIV), complete inactivation was observed.
These results predict that the heterogeneous ACTH mentioned above
from sheep pituitary would have no ACTH activity.

The effects caused by elimination of the N-terminal portion
from the tricosa- or tetracosapeptide are shown in Fig. 2. This
type of systematic synthesis seems to indicate that the His
residue at position 6 belongs to the essential part of the mole-

cule, since this amino acid is located within the essential octapeptide fragment (positions 6-13). The amino acid residue or
residues that play an essential role for ACTH activity all reside
within this octapeptide sequence, His-Phe-Arg-Trp-Gly-Lys-Pro-Val.
Further sensitive assays may hopefully pinpoint the shortest peptide
chain length that endows full ACTH activity.

The binding site in hormone molecules is an idea that seeks to
explain the interaction between the target organ and the hormones
that are circulating in the plasma. If one assumes strong binding
is established, then the greater will be the activity. This concept is also helpful in explaining the phenomena of increasing activity accompanied by elongation of the peptide chain and decreasing
activity accompanied by shortening the peptide chain. From this
point of view, the binding sites of the ACTH molecule are considered to be located within the N-terminal Ser-Tyr-Ser-Met-Glu and
between residues 14 and 20.

B. Synthetic Analogs and Activity

In order to examine the role of individual amino acid residues
on the one hand, and to prepare long-acting or potentiated peptides
on the other, a number of analogous peptides were synthesized, as
listed in Table 3.

From the latter aspect, amide protection of the C-terminus will
be effective in preventing the attack of carboxypeptidase. Along
these lines, Li et al. [100] replaced the C-terminal Pro[19] with
prolinol in the nonadecapeptide (See I, Table 3). Indeed, a
remarkable increase in the extravascular potency of this peptide
was noted.

Concerning the structure-function relationships between the N-
terminal portion of ACTH, some misleading observations have come
from modification of natural ACTH. As mentioned earlier, it was
thought that inactivation of ACTH by leucine aminopeptidase was

TABLE 3

Relationship between Synthetic Analogs and Activity

| Synthetic analogs | ACTH (U/mg) | | | MSH U/g (μM) | Reference |
	Ascorbic acid depletion	Steroidogenesis in vivo	in vitro		
I Prolinol[19]-[1-19]-OH			338 U/μM	6.8×10^8	100
II Gly[1]-[1-14]-OH	week				176
III Gly[1]-[1-18]-NH$_2$	26	151	27	6.7×10^8	177
IV Gly[1]-[1-23]-NH$_2$	104				166
V β-Ala[1]-[1-18]-NH$_2$	135	285	237	9.6×10^9	179
VI β-Ala[1]-[1-23]-NH$_2$	478				163
VII β-Ala[1]-[1-23]-NH$_2$		90			197
VIII β-Ala[1]-[1-24]-OH		160			201
IX Pro[1]-[1-23]-NH$_2$	48				163
X Pro[1]-[1-24]-OH		65			201
XI Ibu[1]-[1-18]-NH$_2$	266	220	83	7.9×10^9	180
XII Ibu[1]-[1-24]-OMe		236			176
XIII Ibu[1]-[1-27]-OH		325	302		176
XIV Abu[1]-[1-24]-OH		110			201
XV Sar[1]-[1-24]-OH		160			201
XVI Lys[1]-[1-24]-OH		50			201
XVII Phe[2]-[1-23]-NH$_2$	66				166
XVIII Ala[3]-[1-23]-NH$_2$	50				166
XIX Abu -Gln[5]-[1-20]-NH$_2$	48.4			1.6×10^7	122

No.	Peptide					Ref.
XXI	Leu^4-[1-24]-OH		85			203
XXII	Ile^4-[1-24]-OH		85			203
XXIII	Gln^5-[1-19]-OH	20-50				134
XXIV	Gln^5-Pyrazolyl-alanine^6-[1-20]-NH$_2$	50	60		8.4×10^7	124
XXV	Leu^7-[1-24]-OH		25			204
XXVI	Lys^8-[1-17]-NH$_2$			1.0	1×10^7	102
XXVII	Gln^5-Phe^9-[1-20]-NH$_2$	2		2	107	124
XXVIII	N-CH$_3$-Trp^9-[1-24]-OH		0.05			204
XIX	Gly^1-Gly^{11}-[1-18]-NH$_2$		1.1			178
XXX	Gly^{13}-[1-17]-NH$_2$			2.8/μmole	(8×10^4)	103
XXXI	Gly^1-Gly^{11}-Gly^{13}-[1-18]-NH$_2$		0.1	0.003		178
XXXII	Gly^{11}-Gly^{13}-[1-17]-NH$_2$			0.5/μmole	(6×10^3)	103
XXXIII	Gly^{12}-Gly^{13}-[1-17]-NH$_2$			1.0/μmole	(4×10^3)	103
XXXIV	Gly^{11}-Gly^{12}-Gly^{13}-[1-19]-OH			0.3/μmole	(6×10^3)	103
XXXV	Orn^{17}-Orn^{18}-[1-24]-OH	100				145
XXXVI	Nle^4-Lys^{17}-Lys^{18}-Val^{25}-[1-25]-NH$_2$	100				87
XXXVII	β-Ala-Orn^{15}-[1-18]-NH$_2$		170	50	7.4×10^9	178
XXXVIII	Ibu^1-Orn^{15}-[1-18]-NH$_2$	249	305	137	7.4×10^9	178
XXXIX	Ac^1-Gln^5-Lys^{11}-$\overline{\text{F}}$-Lys^{15}-Lys^{16}-[1-16]-NH$_2$	0	0		2.0×10^9	118
XL	Ac^1-Gln^5-Lys^{11}-$\overline{\text{F}}$-Lys^{15}-Lys^{16}-[1-20]-NH$_2$	0	0		4.2×10^8	120
XLI	Ac^1-Gln^5-Lys^{11}-$\overline{\text{F}}$-Lys^{15}-Lys^{16}-Lys^{21}-[1-23]-NH$_2$	0.05			2.0×10^8	111
XLII	Nle^4-Val^{25}-[1-25]-NH$_2$	100				87
XLIII	Nva^4-Val^{25}-[1-25]-NH$_2$	100				87
XLIV	β-Ala^1-Leu^4-[1-23]-NH$_2$	73				163

followed by loss of the Ser[1] residue [59]. It was also stated
that acetylation [229], or periodic acid oxidation (conversion
of Ser to glycolyl), resulted in complete loss of activity [230].
However, when Dixon and Weitkamp [231] treated the periodic acid-
oxidized ACTH with glutamine in the presence of copper ion, the
resulting peptide, presumably Gly[1]-ACTH, recovered almost full
activity. This data led to the conclusion that the free amino
group, not the hydroxyl group of Ser, is essential for biologic
activity [231, 232]. Indeed, the synthetic peptides, Gly[1]-ACTH-
[1-23]-NH_2 (IV) and Gly[1]-ACTH-[1-18]-NH_2 (III), are both active as
the corresponding tricosa- and octadecapeptide amides respectively.
Furthermore, it was found that Pro[1] peptides (IX, X) containing the
secondary amine still retain about 48-65% of the original activity.
These activities are nearly equivalent to ACTH-[2-23]-NH_2 [166],
which does not possess the Ser[1] residue. Analogous synthesis has
thus eliminated Ser[1] as an essential functional unit in ACTH activity.

The substitution of Ser[1] by other amino acids (β-Ala, isobutylic
acid, γ-aminobutylic acid, or sarcosine) was undertaken, since these
residues seem to resist the action of leucine aminopeptidase. Most
of these synthetic peptides, especially β-Ala peptides (VI, VII,
VIII), showed significantly high apparent activity. Fujino [233]
added one Leu, two Leu, and three Leu at the N-terminal tetracosapeptide. Although Leu is known to be most susceptible to
leucine aminopeptidase, the addition of these extra units caused
considerable inactivation. As a result, the response slowly
dropped to the level of 50%, 10%, and 5% of the original activity
of the tetracosapeptide. Further examples showed that substitution
of Tyr[2] by Phe (XVII), and Ser[3] by Ala (XVIII), reduced the activity
to about 40-50%.

As for Met[4], it has been well known that treatment of ACTH with
hydrogen peroxide converts Met[4] to the corresponding sulfoxide, and
that this decreases the adrenal stimulating activity considerably
[75, 234-237]. However, it is interesting to note that the
synthetic eicosapeptide analogs, in which Met[4] was replaced by

α-aminobutyric acid, exhibited high ACTH activity (48 U/mg). This
result seems to indicate that the hydrophobic nature of Met, but
not its oxidation-reduction nature, plays a role in a binding site
of the hormone. It seems that in addition to the basic charges,
the hydrophobic quality of Met is also participating in the binding
mechanism of ACTH to the receptor.

Indeed, the Leu[4] (XX, XXI) and Ile[4] (XXIII) analogs are also
very active peptides. As mentioned above in this section on
synthetics, most of the synthetic difficulty associated with ACTH
peptides comes from the presence of the Met residue. This particular amino acid restricts the use of catalytic removal of the
Z-protecting group. In addition, during purification procedures,
partial oxidation of the Met takes place, causing considerable
inactivation. This is why Hofmann et al. [110, 111] and other
investigators used thiol reagents at the final purification step.
Substitution of Met by other non-sulfur-containing amino acids will
easily open many synthetic routes to other ACTH active peptides.

Glu[5] can be substituted by Gln without altering its activity,
as shown in the synthesis of Gln[5]-[1-19]-OH (XXIII).

Previously it had been observed that the shortening of the
peptide chain beyond His[6] from the N-terminus caused complete
loss of the activity [202]. Therefore, replacement of amino acid
residues at the His residue seems interesting. The following
analogs are found in the literature: His[6] to 3-pyrazolylalanine,
Pyr[6]-Gln[5]-[1-20]-NH$_2$ (48.4 U/mg) (XXIV); Phe[7] to Leu, Leu[7]-[1-24]-OH
(25 U/mg) (XXV); Arg[8] to Lys, Lys[8]- [1-18]-NH$_2$ (1 U/mg) (XXVI);
Trp[9] to Phe, Phe[9]-Gln[5]-[1-20]-NH$_2$ (2 U/mg) (XXVII); and Trp[9] to
N-methyl-Trp, N-methyl-Trp[9]-[1-24]-OH (0.05 U/mg) (XXVIII). Among
these, substitution of Arg[8] and Trp[9], although dependent on the
nature of the substituents, does seem to produce a substantial depres-
sant effect on activity. Moreover, the very low activity of the
N-methyl-Trp analogs (XXVIII) is noteworthy. No significant
difference was noted between the ORD spectrum on this compound
and the parent tetracosapeptide, indicating that the configuration

of the tetracosapeptide seemed to suffer no fundamental change by
this substitution [233].

Between positions 11 and 13, only Gly-substituted analogs are
known. Considerable loss of activity was noted in the substitution
of Lys^{11} by Gly (XXIX) and even Val^{13} by Gly (XXX, XXXI, XXXII,
XXXIII, and XXXIV).

As far as substitution by basic amino acids goes, no significant
change in activity was seen in the analogs at positions 15 to 18, as
found in the conversion of Arg^{17} and Arg^{18} to Orn (XXXV, XXXVI).
Lys^{15}-substituted analogs, β-Ala-Orn^{15}-[1-18]-NH_2 (XXXVII) and
Ibu^1-Orn^{15}-[1-18]-NH_2 (XXXVIII), are all quite active. Previously
Hofmann et al. [110, 111, 118-120] treated the N^{ϵ}-formyl-Lys derivatives of the eicosa- (XL) and tricosapeptide (XLI) with dilute
hydrochloric acid. This treatment removed the formyl groups, as
well as the N-acetyl group, from these peptides and converted the
essentially inactive compounds into fully active compounds. This
strongly suggested that the free basic charges at Lys^{11}, Lys^{15},
and Lys^{16} are more important than Lys^{21}.

The data available in Table 3 seem to indicate that substitution
occurring at positions 8 and 9 considerably reduces ACTH activity.
It seems that the Arg-Trp residue in the ACTH molecule contri-
butes greatly to the intrinsic activity.

It should be noted, however, that even if one or two amino acid
residues were substituted by other amino acids, synthetic fragments
with certain chain lengths still retain some activity. This information suggests that the characteristic that defines the specific
activity of ACTH is not represented by a single amino acid residue
bearing a special functional group, but rather that it is the
result of the specific arrangement of a few amino acid residues.
In other words, the cause of ACTH activity should not be sought in
terms of a pinpoint in the molecule, as in the so-called active
site represented by one amino acid residue; rather, it should be
looked for as a special circumstance produced by a combination of
a few amino acid residues.

The smallest active fragment known so far, found in hormone
research, is a tetrapeptide or a tripeptide amide, for example,
His-Phe-Arg-Trp [238] in MSH and TRF, pyroGlu-His-Pro-NH$_2$ [239].
These imply that a short, biologically active core may be generated
by at least a tetrapeptide or a tripeptide amide [103, 233]. A
single nucleotide has no meaning in itself, since three nucleotides
specify one amino acid. In a similar manner, it may be possible to
assume that a combination of at least four amino acids can form
one word, and a grouping of these words in a molecule could define
the specific nature of hormone molecules. One might predict that
the active core of ACTH is His-Phe-Arg-Trp, which is identical with
MSH, but the small molecule of MSH has no other core words to exert
full ACTH activity.

At present, we do not have the right tools to interpret the
specific activity that is produced by various combinations of
amino acid residues, even if only a small molecule is involved in
the response. Any physicochemical data that have a close correlation with activity may or may not be useful, depending on how
the conformation of the molecule changes with the environment.

C. Stereoisomers and Activity

It is well documented that heating pituitary extracts with
sodium hydroxide causes complete destruction of the adrenal
ascorbic acid-depleting and steroidogenetic activities [3, 240,
241]. Pickering and Li [242] treated sheep ACTH with 0.1 N sodium
hydroxide and found that this treatment caused a complete loss of
both adrenal-stimulating and lipolytic activities. After extensive
physical and chemical studies, it was concluded that a limited
cleavage of the second bond in the hormone occurred and a number
of constituent amino acid residues within the molecule were
racemized. However, it is interesting to note that the same
treatment not only potentiated, but also prolonged the intrinsic

melanocyte-stimulating potency of ACTH [243]. A similar treatment
of α-MSH [244] and β-MSH [245] also caused racemization of constituent amino acids, and these racemized MSHs exerted similar
phenomena in terms of MSH activity.

In 1965, Yajima and Kubo [246] made the first attempt to
examine the biologic activity of an optical antipode of an active
fragment of MSH. The synthetic D-His-D-Phe-D-Arg-D-Trp-Gly-OH has
no MSH activity. However, D-Phe-pentapeptide, His-D-Phe-Arg-Trp-
Gly-OH and D-Trp pentapeptide, His-Phe-Arg-D-Trp-Gly-OH [247, 248]
exhibited considerable potentiated MSH activity, which indicated
that the receptor site of MSH does not exert a high degree of
stereospecificity. The D-Phe pentapeptide was synthesized by
Schnabel and Li [249] and this compound also exhibited a much higher
MSH activity, as compared with the all-L-pentapeptide.

Later, two stereoisomers of ACTH peptides were synthesized
by Kappeler et al. [146], as shown in Table 4: D-Ser[1]-D-Tyr[2]-D-
Ser[3]-D-Met[4]-ACTH-[1-24]-OH (I) and D-Glu[5]-D-His[6]-D-Phe[7]-D-Arg[8]-ACTH-
[1-24]-OH (II). It was reported that the former is nearly as active
as natural ACTH, while the latter is completely inactive. These
results indicate the absolute necessity for the L-configuration in
the constituent amino acids at residues 5 to 8 in order that the
molecule exert some ACTH activity. In connection with the earlier
observations on the relationship between chain length activity and
synthetic analog activity, these studies on stereoisomers also
suggest that the active core of ACTH is the tetrapeptide, His-Phe-
Arg-Trp, which is identical with that of MSH. But the studies
differ from each other in determining the requirements for stereospecificity in their receptor sites.

Information pertaining to the stereochemical nature of the
entire ACTH molecule is limited [250, 251]. Squire and Bewley
[252] measured the ORD of sheep ACTH and found an apparent helical
content of 11-15% at pH 8.1. It was suggested that the helical
content at this pH value probably involves the first 11 residues
from the N-terminus. Recently, Ikeda et al. [253] examined the

TABLE 4

Relationship between Stereoisomers and Activity

| | | ACTH (U/mg) | | | |
	Stereoisomers	Ascorbic acid depletion	Steroidogenesis in vivo	in vitro	Reference
I	D-Ser^1-D-Tyr^2-D-Ser^3-D-Met^4-[1-24]-OH	100			146
II	D-Glu^5-D-His^6-D-Phe^7-D-Arg^8-[1-24]-OH	0.01		0	146
III	D-Ser^1 -[1-24]-OH	750		300	146
IV	D-Ser^1 -[1-13]-NH$_2$		0.1		147
V	D-Ser^1 -[1-16]-NH$_2$		1.4		147
VI	D-Ser^1 -[1-19]-NH$_2$		72.4		147
VII	D-Ser^1 -[1-39]-OH		140		147
VIII	D-Ser^1-D-Tyr^2-[1-24]-OH	100			146
IX	D-Ala^1-[1-24]-OH		300		146
X	D-Ser^1-Nva^{17}-Nva^{18}-[1-19]-NH$_2$	1/2 of 19			148
XI	D-Ser^1-Nle^{17}-Nle^{18}-[1-19]-NH$_2$	1/2 of 19			148
XII	D-Ser^1-Nle^4-Val^{25}-[1-25]-NH$_2$	625			86
XIII	Nle^4-Lys^{17}Lys^{18}-D-Val^{25}-[1-25]-NH$_2$	100			87
XIV	β-Ala^1-D-Phe^7-Orn^{15}-[1-18]-NH$_2$		6.1	9.6	178
	D-Ser^1-Lys^{17}-Lys^{18}-[1-18]-NH$_2$	prolonged	active		149

CD of α-MSH and concluded that the α-MSH molecule is entirely a random structure. Since α-MSH occupies the first 13 amino acid sequences of ACTH, the possibility that the helical conformation in the ACTH molecule does not lie around the N-terminal portion cannot be excluded.

Most of the synthetic stereoisomers listed in Table 4 are peptides prepared in terms of protecting fragments from the attack of leucine aminopeptidase. The D-Ser[1] analogs are all very active as compared with the corresponding all-L-peptides. The results indicate definitely that the L-configuration of the N-terminal Ser is not essential for ACTH activity.

Further studies seem to be required in order for us to understand the relationship between the L-configuration of various constituent amino acid residues in biologically active peptides and their characteristic physiologic functions.

D. Des-Chain Peptides and Activity

In order to examine whether the tetrapeptide, Lys-Pro-Val-Gly (positions 11-14), is necessary for ACTH activity or is merely a bridge or spacing factor between the basic core, Lys-Lys-Arg-Arg- (positions 15-18), and the N-terminal decapeptide, Li et al. [104, 105] synthesized Des-Lys[11]-Pro[12]-Val[13]-Gly[14]-pentadecapeptide, ACTH-[1-10]-[15-19]-OH (I), as shown in Table 5. They found that this peptide exhibited an adrenal-stimulating activity of less than 1 U/mg. However, it is a potent lipolytic agent, as is natural ACTH. The Lys-Pro-Val- sequence in the heptadecapeptide was replaced by Gly [103], as previously shown in Table 3. The results indicated the dual importance of the tetrapeptide sequence at positions 11 to 14 of the parent nonadecapeptide (a) in terms of the effect on the distance between the basic core at positions 15 to 18 and the N-terminal decapeptide and (b) in terms of the effect on the specific character of the amino acids at positions

TABLE 5

Relationship between Des-Chain Peptides and Activity

	ACTH (U/mg)			
Des-chain Peptides	Ascorbic acid depletion	Steroidogenesis		Reference
		in vivo	in vitro	
I (Des-Lys^{11}-Pro^{12}-Val^{13}-Gly^{14})-[1-10]-[15-19]-OH		0.006	0.4	105
II (Des-Lys^{11}-Gly^{14})-Gly^1-[1-10]-[12-13]-[15-18]-NH$_2$		0.06	0.05	178
III (Des-Lys^{15})-Gly^1-[1-14]-[16-18]-NH$_2$		2.7	2.1	182
IV (Des-Lys^{15}Lys^{16})-Gly^1-[1-14]-[17-18]-OH		0.124	0.13	181

11 to 13. The intact Lys-Pro-Val sequence is thus necessary for
full ACTH activity.

As demonstrated by Otsuka et al [178, 181, 182], the sub-
traction of Lys^{11} (II), Lys^{15} (III), and Lys^{15}-Lys^{16}(IV) from the
Gly^1-octadecapeptide caused considerable loss of activity.

It should be noted that the subtraction of one or two letters
from a cipher will result in the complete misreading of its
intrinsic meaning. The low activity of these des-chain peptides
implies that such subtraction did not take place in an essential
part of the ACTH molecule.

In summarizing the data concerning the structure-function of
ACTH, a tentative structural feature of ACTH can be drawn, as shown
in Fig. 3. This gives us the impression that the ACTH molecule
has a surprisingly systematic structure, in which the steroidogenetically active portion and the immunologically important section
are divided into two areas. The active core of ACTH is located
near the N-terminal portion in the entire molecule, but it is
still protected by a few amino acid residues from the attack of
various proteolytic enzymes. It would be interesting to find out
whether such a systematic biologic feature is shared by other
large, peptide hormones.

VI. ACTH RECEPTOR

The relationship between a hormone molecule and a receptor
could be like a mirror image. Although the chemical structures
of a number of peptide hormones are known, and synthetic studies
have uncovered much information in this area, the true chemical
nature of the receptor has hitherto remained obscure.

Recently, Pastan and Lefkowitz et al. [254-260] undertook a
series of investigations on the early mode of action of ACTH.

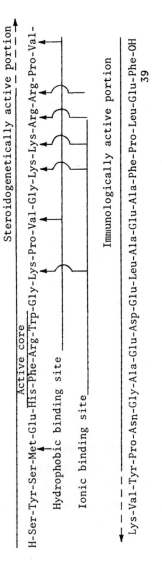

FIG. 3. Tentative structure-function of ACTH.

They summarized the current concepts as follows: adrenal steroidogenesis, as stimulated by ACTH, seems to proceed by (a) the binding of ACTH to a receptor [254]; (b) activation of adrenal adenyl cyclase by ACTH [261, 255, 256]; (c) the formation of cyclic AMP from ATP [262]; and (d) the conversion of cholesterol to corticosteroids [263].

Taunton et al. [254] observed that trypsin-treated adrenal slices lost steroid production ability. He suggested that the first step in ACTH stimulation of the adrenal is the binding of the hormone to the receptor on the outside of the target tissue. Further confirmative evidence along these lines was offered by Schimmer et al. [264]. ACTH or synthetic eicosapeptide was diazotized to p-aminobenzoyl-cellulose, and when such a large, insoluble ACTH-cellulose was incubated in adrenal cell cultures, steroid production was observed proportionally to the ACTH concentration on the resin. Therefore it is evident that ACTH stimulates the surface of adrenal cells without entering the target tissues.

It has been known that elevation of cyclic $3',5'$-AMP concentration occurs within 1 min after addition of ACTH, but before a rise in steroid production [265]. In the presence of fluoride, ACTH elevates the cyclic $3',5'$-AMP concentration by stimulating adenyl cyclase [255, 256]. Thus it seems even more likely that the membrane-bound enzyme adenyl cyclase may be the actual ACTH binding site.

Efforts are now being concentrated on solubilizing this enzyme [257]. Adenyl cyclase from an ACTH-sensitive mouse adrenal tumor was reduced to a molecular weight of 3 to 7 million by treatment of a particulate enzyme in a French pressure cell, a Nossal shaker, or by sonication in the presence of both phosphatidyl compounds and fluoride. Using this solubilized enzyme preparation in the absence of calcium ion, it was found that the ACTH activation of adenyl cyclase is totally inhibited, but activation by fluoride is not [258]. This soluble extract of mouse adrenal tumor cells binds [^{125}I]-ACTH and the binding was correlated with stimulation

of the ACTH-sensitive adenyl cyclase. ACTH preparations treated
with sodium hydroxide, trypsin, leucineaminopeptidase, or oxidation
with periodate largely destroyed both binding and cyclase activation. This evidence led Lefkowitz et al. [260] to suggest a
radioreceptor assay of ACTH, since unlabeled ACTH at 1 pg/ml
significantly displaced labeled ACTH from these receptors.

Hofmann et al. [266] isolated from homogenated beef adrenal
slices a particulate fraction that binds synthetic [^{14}C-Phe]-Gln^5-
ACTH-[1-20]-NH$_2$. This fraction failed to bind [^{14}C-Phe]-ACTH-
[1-10]-OH. The former peptide is active, but the latter is inactive.
The binding of the eicosapeptide with this fraction was displaced
by the nonradioactive analogs or by fragments of the eicosapeptide.
Hofmann et al. concluded that charge-charge interaction plays a
major role in the attachment of ACTH to the particulate fraction,
and the sequence Lys-Lys-Arg-Arg (positions 15-18) proved to be
particularly significant.

Similarly prepared particulates from beef kidney, liver, or
adrenal medula exhibited little affinity towards the radioactive
peptide. This suggested that the above particulate indeed contains
ACTH receptor.

The concept of binding between a hormone and a receptor has
thus been experimentally substantiated. Additional studies to
elucidate the chemical nature of such particulate fractions, as
well as adenyl cyclase, are awaited.

VII. ACTH RELEASING FACTOR

As stated earlier, recent developments on the radioimmunoassay
of ACTH have made it possible to determine diurnal variation in
ACTH secretion in both normal subjects and patients with various
endocrine disorders. These developments allow us to investigate and

to clarify the various factors that control ACTH secretion [267]:
(a) a regulation mechanism as controlled by the hypothalamus;
(b) a negative feedback mechanism through adrenal cortex stimulation; and (c) promotion of secretion by stress.

During the period 1959 to 1964, Schally and Guillemin [268-271] reported the presence of various factors in the hypothalamus that stimulate the secretion of ACTH; the so-called corticotropin releasing factors and three factors, α_1, α_2, and β-CRF, were partially purified. It was stated that α_2-CRF is a compound resembling α-MSH, while β-CRF is a vasopressin-like compound.

Doepfner et al. [272] assayed His-Ser-vasopressin and acetyl-Ser-Tyr-Ser-vasopressin, but the results showed considerable variation. It was suggested the pressor and the CRF activity of peptides related to the neurophypophysial hormones need not be strictly correlated. Some of the ACTH peptide fragments, for example, Met-Glu-His-Phe-Arg-Trp-Gly, were reported to possess some CRF activity [273]. However, the CRF activity of such a partial sequence of ACTH could not be confirmed by other authors. Pure α-MSH and its derivatives possess no CRF activity at all [274].

De Garilhe et al. [275-279] isolated from pig posterior lobe a peptide with CRF activity. This peptide is very similar to α_1-CRF and contains Thr, Ala, Leu, and Asp, as well as the amino acids of α-MSH. In addition to this fraction, they isolated Leu-Arg-Leu and other uncharacterized neutral and basic peptides. CRF must be a small peptide or peptide derivative, since it is produced by nerve cells. The elucidation of its chemical nature is important in attempting to establish a correlation between hormones and nerve systems.

Thus the chemical nature of ACTH is gradually being clarified by a dual approach, one from the receptor site and the other from the control mechanism. In summary, one of the greatest achievements in the field of endocrine chemistry has been the amount of information provided by the many indefatigable studies devoted to ACTH.

Note added in proof: In 1972, Sieber et al. (footnote of Fig. 1)
synthesized human ACTH by essentially the same method as described
by Schwyzer and Sieber [142, 143]. Kisfaludy et al. [280] also
achieved the total synthesis of human ACTH using the pentafluoro-
phenyl active ester procedure. This synthesis was also reported by
Nishimura and Fujino [281] who applied the HF procedure. Under ideal
controlled conditions Yamashiro and Li [282] performed successfully
the solid phase synthesis of human ACTH.

REFERENCES

1. E. B. Astwood, M. S. Raben, and R. W. Payne, Recent Progr.
    Hormone Res., 7, 1 (1952).
2. E. E. Hays and W. F. White, Recent Progr. Hormone Res., 10,
    265 (1954).
3. F. L. Engel, Vitamins Hormones, 19, 189 (1961).
4. G. W. Liddle, D. Island, and C. K. Meador, Recent Progr.
    Hormone Res., 18, 125 (1962).
5. J. G. Phillips and D. Bellamny, in Comparative Endocrinology
    (U.S. von Euler and H. Heller, eds.), Vol. 1, Academic, New
    York, 1963, p. 208.
6. H. E. Lebovitz and F. L. Engel, Metab. Clin. Exptl., 13, 1230
    (1964).
7. K. Hofmann and H. Yajima, Recent Progr. Hormone Res., 18, 41
    (1962).
8. K. Hofmann, Ann. Rev. Biochem., 31, 213 (1962).
9. K. Hofmann, Pure Appl. Chem., 6, 245 (1963).
10. K. Hofmann and P. G. Katsoyannis, in The Proteins (H. Neurath,
    ed.), Vol. 1, Academic, New York, 1963, p. 53.
11. C. H. Li, Advan. Protein Chem., 11, 101 (1956).
12. C. H. Li, Lab. Invest., 8, 574 (1959).
13. C. H. Li, Surv. Biol. Progr., 4, 93 (1962).
14. C. H. Li, Recent Progr. Hormone Res., 18, 1 (1962).

15. R. Schwyzer, Ergeb. Physiol. Biol. Chem. Exptl. Pharmakol., 53, 41 (1963).

16. R. Schwyzer, Ann. Rev. Biochem., 33, 259 (1964).

17. R. Schwyzer, Z. Naturwisch., 53, 189 (1966).

18. E. Schröder and K. Lübke, in The Peptides, Vol. 2, Academic, New York, 1965, p. 194.

19. J. E. White and F. L. Engel, J. Clin. Invest., 37, 1556 (1958).

20. C. H. Hollenberg, M. S. Raben, and E. B. Astwood, Endocrinology, 68, 589 (1961).

21. M. S. Raben, R. Landolt, F. A. Smith, K. Hofmann, and H. Yajima, Nature, 189, 681 (1961).

22. A. Tanaka, B. T. Pickering, and C. H. Li, Arch. Biochem. Biophys., 99, 294 (1962).

23. A. Tanaka, K. Kubo, and H. Yajima, Endocrinol. Japon., 16, 647 (1969).

24. J. I. Harris and A. B. Lerner, Nature, 179, 1346 (1957).

25. J. D. Fisher, in Methods in Hormone Research Vol. 2 (R. I. Dorfman, ed.), Academic, New York, 1962, p. 641.

26. M. A. Sayers, G. Sayers, and L. A. Woodbury, Endocrinology, 42, 347 (1948).

27. The U. S. Pharmacopeia, XVII, 147 (1965).

28. R. Guillemin, G. W. Clayton, J. D. Smith, and H. S. Lipscomb, Endocrinology, 63, 349 (1958).

29. M. Saffran and A. V. Schally, Endocrinology, 56, 523 (1955).

30. P. A. Desaulles, P. Barthe, B. Schär, and M. Staehelin, Acta Endocrinol., 51, 609 (1966).

31. R. Schwyzer, P. Schiller, S. Seelig, and G. Sayers, FEBS Letters, 19, 229, 232 (1971).

32. P. W. C. Kloppenborg, D. P. Island, G. W. Liddle, A. M. Michelakis, and W. E. Nicholson, Endocrinology, 82, 1053 (1968).

33. I. D. K. Halkerston and M. Feinstein, Fed. Am. Soc. Exptl. Biol. Proc., 27, 626 (1968).

34. R. L. Swallow and G. Sayers, Proc. Soc. Exptl. Biol. Med., 131, 1 (1969).

35. R. Haning, S. A. S. Tait, and J. F. Tait, Endocrinology, 87, 1147 (1970).

36. G. Sayers, R. L. Swallow, and N. D. Giordano, Endocrinology, 88, 1063 (1971).

37. M. Nakamura and A. Tanaka, Endocrinol. Japon, 18, 289 (1971).

38. H. S. Lipscomb and D. H. Nelson, Endocrinology, 71, 63 (1962).

39. J. P. Felber, Experientia, 19, 227 (1963).

40. S. A. Berson and R. S. Yalow, J. Clin. Invest., 38, 2017 (1959).

41. R. S. Yalow and S. A. Berson, Nature, 212, 357 (1959).

42. J. Fishman, Proc. Soc. Exptl. Biol. Med., 102, 446 (1959).

43. H. Matsuyama, R. B. Mims, W. Ruhman, and D. H. Nelson, Endocrinology, 88, 692 (1971).

44. S. A. Berson and R. S. Yalow, New Engl. J. Med., 277, 640 (1967).

45. H. Imura, L. L. Sparks, G. M. Grodsky, and P. H. Forsham, J. Clin. Endocrinol. Metab., 25, 1361 (1965).

46. W. R. Lyons, Proc. Soc. Exptl. Biol. Med., 35, 645 (1937).

47. C. H. Li, H. M. Evans, and M. E. Simpson, J. Biol. Chem., 149, 413 (1943).

48. G. Sayers, A. White, and C. N. H. Long, J. Biol. Chem., 149, 425 (1943).

49. E. B. Astwood, M. S. Raben, R. W. Payne, and A. B. Grady, J. Am. Chem. Soc., 73, 2969 (1951).

50. R. W. Payne, M. S. Raben, and E. B. Astwood, J. Biol. Chem., 187, 719 (1950).

51. P. H. Bell, J. Am. Chem. Soc., 76, 5565 (1954).

52. R. G. Shepherd, K. S. Howard, P. H. Bell, A. R. Cacciola, R. G. Child, M. C. Davies, J. P. English, B. M. Finn, J. H. Meisenhelder, A. W. Moyer, and van der Scheer, J. Am. Chem. Soc., 78, 5051 (1956).

53. R. Guillemin, Endocrinology, 66, 819 (1960).

54. P. H. Bell, K. S. Howard, R. G. Shepherd, B. M. Finn, and J. H. Meisenhelder, J. Am. Chem. Soc., 78, 5059 (1956).

55. R. G. Shepherd, S. D. Wilson, K. S. Howard, P. H. Bell,

   D. S. Davis, S. B. Davis, E. A. Eigner, and N. E. Shakespear, J. Am. Chem. Soc., 78, 5067 (1956).

56. W. F. White, J. Am. Chem. Soc., 75, 503 (1953).

57. W. F. White and W. L. Fierce, J. Am. Chem. Soc., 75, 245 (1953).

58. W. A. Landmann, M. P. Drake, and W. F. White, J. Am. Chem. Soc., 75, 4370 (1953).

59. W. F. White, J. Am. Chem. Soc., 75, 4877 (1953).

60. W. F. White and W. A. Landmann, J. Am. Chem. Soc., 76, 4193 (1954).

61. W. F. White, J. Am. Chem. Soc., 76, 4194 (1954).

62. W. F. White and W. A. Landmann, J. Am. Chem. Soc., 77, 771 (1955).

63. W. F. White and W. A. Landmann, J. Am. Chem. Soc., 77, 1711 (1955).

64. J. I. Harris, Brit. Med. Bull., 16, 189 (1960).

65. A. L. Levy, I. I. Geschwind, and C. H. Li, J. Biol. Chem., 231, 187 (1955).

66. C. H. Li, I. I. Geschwind, A. L. Levy, J. I. Harris, J. S. Dixon, N. G. Pon, and J. O. Porath, Nature, 173 251 (1954).

67. C. H. Li, I. I. Geschwind, J. S. Dixon, A. L. Levy, and J. I. Harris, J. Biol. Chem., 213, 171 (1955).

68. J. I. Harris and C. H. Li, J. Am. Chem. Soc., 76, 3607 (1954).

69. C. H. Li, I. I. Geschwind, R. D. Cole, I. D. Raacke, J. I. Harris, and J. S. Dixon, Nature, 176, 687 (1955).

70. J. Leonis, C. H. Li, and D. Chung, J. Am. Chem. Soc., 81, 419 (1959).

71. C. H. Li, J. S. Dixon, and D. Chung, J. Am. Chem. Soc., 80, 2587 (1958).

72. C. H. Li, J. S. Dixon, and D. Chung, Biochim. Biophys. Acta, 46, 324 (1961).

73. T. H. Lee, A. B. Lerner, and V. B. Janusch, J. Am. Chem. Soc., 81, 6084 (1959).

74. T. H. Lee, A. B. Lerner, and V. B. Janusch, J. Biol. Chem., 236, 2970 (1961).

75. H. B. F. Dixon and M. P. Stack-Dunne, Biochem. J., 61, 483 (1955).

76. R. J. Barrett, H. Friesen, and E. B. Astwood, J. Biol. Chem., 237, 432 (1962).

77. S. Lande, M. Sribney, R. K. McDonald, and W. Baxst, Biochim. Biophys. Acta, 154, 429 (1968).

78. B. T. Pickering, R. N. Andersen, P. Lohmar, Y. Birk, and C. H. Li, Biochim. Biophys. Acta, 74, 763 (1963).

79. W. F. White and R. L. Peters, J. Am. Chem. Soc., 78, 4181 (1956).

80. C. H. Li, Science, 129, 969 (1959).

81. J. B. Lesh, J. D. Fisher, I. M. Bunding, J. J. Kocsis, L. J. Walaszek, W. F. White, and E. E. Hays, Science, 112, 43 (1950).

82. C. H. Li and K. O. Pedersen, Arkiv. Kemi., 1, 533 (1950).

83. C. H. Li, A. Tiselius, K. O. Pedersen, L. Hagdahl, and H. Carstensen, J. Biol. Chem., 190, 317 (1951).

84. R. A. Boissonnas, St. Guttmann, J. P. Waller, and P. A. Jaquenoud, Experientia, 12, 446 (1956).

85. R. A. Boissonnas, St. Guttmann, P. A. Jaquenoud, Ed. Sandrin, and J. P. Waller, Helv. Chim. Acta, 44, 123 (1961).

86. R. A. Boissonnas, St. Guttmann, and J. Pless, Experientia, 22, 526 (1966).

87. St. Guttmann, J. Pless, and R. A. Boissonnas, Acta Chim. Acad. Sci. Hung., 44, 23 (1965).

88. V. du Vigneaud, C. Ressler, J. M. Swan, C. W. Roberts, P. G. Katsoyannis, and S. Gordon, J. Am. Chem. Soc., 75, 4879 (1953).

89. P. G. Katsoyannis, D. T. Gish, and V. du Vigneaud, J. Am. Chem. Soc., 79, 4516 (1957).

90. R. Schwyzer and C. H. Li, Nature, 182, 1669 (1958).

91. E. Schnabel and C. H. Li, J. Am. Chem. Soc., 82, 4576 (1960).

92. J. Ramachandran and C. H. Li, J. Org. Chem., 27, 4006 (1962).

93. C. H. Li, J. Meienhofer, E. Schnabel, D. Chung, T. B. Lo, and J. Ramachandran, J. Am. Chem. Soc., 82, 5760 (1960).

94. C. H. Li, J. Meienhofer, E. Schnabel, D. Chung, T. B. Lo, and
     J. Ramachandran, J. Am. Chem. Soc., 83, 4449 (1961).

95. J. Meienhofer and C. H. Li, J. Am. Chem. Soc., 84, 2434 (1962).

96. C. H. Li, D. Chung, and J. Ramachandran, J. Am. Chem. Soc.,
     86, 2715 (1964).

97. C. H. Li, D. Chung, J. Ramachandran, and B. Gorup, J. Am. Chem.
     Soc., 84, 2460 (1962).

98. C. H. Li, J. Ramachandran, D. Chung, and B. Gorup, J. Am.
     Chem. Soc., 86, 2703 (1964).

99. J. Ramachandran, D. Chung, and C. H. Li, J. Am. Chem. Soc.,
     87, 2696 (1965).

100. W. Oelofsen and C. H. Li, J. Am. Chem. Soc., 88, 4254 (1966).

101. J. Ramachandran and C. H. Li, J. Am. Chem. Soc., 87, 2691 (1965).

102. D. Chung and C. H. Li, J. Am. Chem. Soc., 89, 4208 (1967).

103. J. Blake and C. H. Li, Biochim. Biophys. Acta, 147, 386 (1967).

104. C. H. Li, J. Ramachandran, and D. Chung, J. Am. Chem. Soc.,
     85, 1895 (1963).

105. C. H. Li, J. Ramachandran, and D. Chung, J. Am. Chem. Soc.,
     86, 2711 (1964).

106. M. Wilchek, S. Sarid, A. Patchornik, Biochim. Biophys. Acta,
     104, 616 (1965).

107. A. Patchornik, M. Wilchek, and S. Sarid, J. Am. Chem. Soc.,
     86, 1457 (1964).

108. W. F. Benisck, M. A. Raftery, and R. D. Cole, Biochemistry,
     6, 3780 (1967).

109. K. Hofmann and H. Yajima, J. Am. Chem. Soc., 83, 2289 (1961).

110. K. Hofmann, H. Yajima, N. Yanaihara, T. Y. Liu, and S. Lande,
     J. Am. Chem. Soc., 83, 487 (1961).

111. K. Hofmann, H. Yajima, T. Y. Liu, and N. Yanaihara, J. Am.
     Chem. Soc., 84, 4475 (1962).

112. K. Hofmann, E. Stutz, G. Spühler, H. Yajima, and E. T. Schwartz,
     J. Am. Chem. Soc., 82, 3727 (1960).

113. K. Hofmann, T. A. Thompson, H. Yajima, E. T. Schwartz, and H.
     Inouye, J. Am. Chem. Soc., 82, 3715 (1960).

114. K. Hofmann, H. Yajima, E. T. Schwartz, J. Am. Chem. Soc.,
     82, 3732 (1960).

115. K. Hofmann, H. Yajima, and E. T. Schwartz, Biochim. Biophys.
     Acta, 36, 252 (1959).

116. K. Hofmann and S. Lande, J. Am. Chem. Soc., 83, 2286 (1961).

117. K. Hofmann, T. Y. Liu, H. Yajima, N. Yanaihara, and S. Lande,
     J. Am. Chem. Soc., 83, 2294 (1961).

118. K. Hofmann, N. Yanaihara, S. Lande, and H. Yajima, J. Am.
     Chem. Soc., 84, 4470 (1962).

119. K. Hofmann, T. Y. Liu, H. Yajima, N. Yanaihara, C. Yanaihara,
     and J. L. Humes, J. Am. Chem. Soc., 84, 1054 (1962).

120. K. Hofmann, H. Yajima, T. Y. Liu, N. Yanaihara, C. Yanaihara,
     and J. L. Humes, J. Am. Chem. Soc., 84, 4481 (1962).

121. K. Hofmann, R. D. Wells, H. Yajima, and J. Rosenthaler, J. Am.
     Chem. Soc., 85, 1546 (1963).

122. K. Hofmann, J. Rosenthaler, R. D. Wells, and H. Yajima, J. Am.
     Chem. Soc., 86, 4991 (1964).

123. K. Hofmann, H. Bohn, and R. Andreatta, J. Am. Chem. Soc., 89,
     7126 (1967).

124. K. Hofmann, R. Andreatta, H. Bohn, and L. Moroder, J. Med.
     Chem., 13, 339 (1970).

125. L. Moroder and K. Hofmann, J. Med. Chem., 13, 839 (1970).

126. H. Yajima, K. Kawasaki, Y. Okada, H. Minami, K. Kubo, and
     I. Yamashita, Chem. Pharm. Bull. (Tokyo), 16, 919 (1968).

127. H. Yajima, Y. Okada, Y. Kinomura, and H. Minami, J. Am.
     Chem. Soc., 90, 527 (1968).

128. H. Yajima, Y. Okada, Y. Kinomura, N. Mizokami, and H.
     Kawatani, Chem. Pharm. Bull. (Tokyo), 17, 1237 (1969).

129. H. Yajima, K. Kawasaki, H. Minami, H. Kawatani, N. Mizokami,
     and Y. Okada, Biochim. Biophys. Acta, 175, 228 (1969).

130. H. Yajima, K. Kawasaki, H. Minami, H. Kawatani, N. Mizokami,
     Y. Kiso, and F. Tamura, Chem. Pharm. Bull. (Tokyo), 18, 1394
     (1970).

131. R. Geiger and W. Siedel, Chem. Ber., 101, 3386 (1968).

132. R. Geiger and W. Siedel, Chem. Ber., 102, 2487 (1969).

133. R. Schwyzer and W. Rittel, Helv. Chim. Acta, 44, 159 (1961).

134. R. Schwyzer, W. Rittel, H. Kappeler, and B. Iselin, Angew. Chem., 72, 915 (1960).

135. H. Kappeler and R. Schwyzer, Helv. Chim. Acta, 44, 1136 (1961).

136. R. Schwyzer and H. Kappeler, Helv. Chim. Acta, 46, 1550 (1963).

137. R. Schwyzer, B. Riniker, and H. Kappeler, Helv. Chim. Acta, 46, 1541 (1963).

138. W. Rittel, Helv. Chim. Acta, 45, 2465 (1962).

139. R. Schwyzer and H. Kappeler, Helv. Chim. Acta, 44, 1991 (1961).

140. B. Iselin and R. Schwyzer, Helv. Chim. Acta, 44, 169 (1961).

141. H. Kappeler, Helv. Chim. Acta, 44, 476 (1961).

142. R. Schwyzer and P. Sieber, Nature, 199, 172 (1963).

143. R. Schwyzer and P. Sieber, Helv. Chim. Acta, 49, 134 (1966).

144. R. Schwyer, W. Rittel, and A. A. Costopanagiotis, Helv. Chim. Acta, 45, 2473 (1962).

145. G. I. Tesser and R. Schwyzer, Helv. Chim. Acta, 49, 1013 (1966).

146. H. Kappeler, B. Riniker, W. Rittel, P. Desaulles, R. Maier, B. Schär, and M. Staehelin, in Peptides (H. C. Beyermann, A. van de Linde, and W. M. van der Brink, eds.), North-Holland Pub., Amsterdam, 1967, p. 214.

147. W. Rittel, in Advances in Experimental Medical Biology (N. Back, L. Martin, and R. Paoletti, eds.), Vol. 2, Plenum, New York, 1968, p. 35.

148. M. Brugger, P. Barthe, and P. A. Desaulles, Experientia, 26, 1050 (1970); Helv. Chim. Acta, 54, 1261 (1971).

149. B. Riniker and W. Rittel, Helv. Chim. Acta, 53, 513 (1970).

150. G. Losse, D. Zeidler, and T. Grieshaber, Ann. Chem., 715, 196 (1968).

151. L. Zervas, D. Borovas, and E. Gazis, J. Am. Chem. Soc., 85, 3660 (1963).

152. L. Zervas and C. Hamalidis, J. Am. Chem. Soc., 87, 99 (1965).

153. P. Sieber and B. Iselin, Helv. Chim. Acta, 51, 614 (1968).

154. P. Sieber and B. Iselin, Helv. Chim. Acta, 51, 622 (1968).

155. P. Sieber and B. Iselin, Helv. Chim. Acta, 52, 1525 (1969).

156. E. Wünsch, Z. Naturforsch., 22b, 1269 (1967).

157. E. Wünsch and G. Wendlberger, Chem. Ber., 101, 3659 (1968).

158. W. Rittel, M. Brugger, B. Kamber, B. Riniker, and P. Sieber, Helv. Chim. Acta, 51, 924 (1968).

159. St. Guttmann, J. Pless, Ed. Sandrin, P. A. Jaquenoud, H. Bossert, and H. Willems, Helv. Chim. Acta, 51, 1155 (1968).

160. R. Geiger, K. Sturm, and W. Siedel, Chem. Ber., 97, 1207 (1964).

161. K. Sturm, R. Geiger, and W. Siedel, Chem. Ber., 97, 1197 (1964).

162. R. Geiger, K. Sturm, and W. Siedel, Chem. Ber., 96, 1080 (1963); 97, 1207 (1964).

163. R. Geiger, H. G. Schröder, and W. Siedel, Ann. Chem., 726, 177 (1969).

164. W. König and R. Geiger, Chem. Ber., 103, 788 (1970).

165. W. König and R. Geiger, Chem. Ber., 103, 2024 (1970).

166. R. Geiger, K. Sturm, G. Vogel, and W. Siedel, Z. Naturforsch., 19b, 858 (1964).

167. R. Geiger, Angew. Chem., 83, 155 (1971).

168. H. Otsuka, K. Inouye, F. Shinozaki, and M. Kanayama, J. Biochem. (Tokyo), 58, 512 (1965).

169. H. Otsuka, K. Inouye, and Y. Jono, Bull. Chem. Soc. Japan, 37, 1471 (1964).

170. H. Otsuka, K. Inouye, F. Shinozaki, and M. Kanayama, Bull. Chem. Soc. Japan, 39, 882 (1966).

171. H. Otsuka, K. Inouye, F. Shinozaki, and M. Kanayama, Bull. Chem. Soc. Japan, 39, 1171 (1966).

172. K. Inouye and H. Otsuka, Bull. Chem. Soc. Japan, 34, 1 (1961).

173. K. Inouye, Bull. Chem. Soc. Japan, 38, 1148 (1965).

174. K. Inouye and H. Otsuka, J. Org. Chem., 27, 4236 (1962).

175. H. Otsuka, K. Watanabe, and K. Inouye, Bull. Chem. Soc. Japan, 43, 2278 (1970).

176. K. Inouye and H. Otsuka, private communication.

177. H. Otsuka, M. Shin, Y. Kinomura, and K. Inouye, Bull. Chem. Soc. Japan, 43, 196 (1970).

178. M. Shin, K. Inouye, and H. Otsuka, in Proc. 8th Symp. Peptide Chem. (T. Kaneko, ed.), 1971, p. 91.

179. K. Inouye, A. Tanaka, and H. Otsuka, Bull. Chem. Soc. Japan, 43, 1163 (1970).

180. K. Inouye, K. Watanabe, K. Namba, and H. Otsuka, Bull. Chem. Soc. Japan, 43, 3873 (1970).

181. H. Otsuka, K. Inouye, M. Kanayama, and F. Shinozaki, Bull. Chem. Soc. Japan, 38, 679 (1965).

182. H. Otsuka, K. Inouye, M. Kanayama, and F. Shinozaki, Bull. Chem. Soc. Japan, 38, 1563 (1965).

183. S. Bajusz, K. Medzihradszky, Z. Paulay, and Z. Láng, Acta Chim. Acad. Sci. Hung., 52, 335 (1967).

184. S. Bajusz, Z. Paulay, Z. Láng, K. Medzihradszky, and M. Löw, in Peptides (E. Bricas, ed.), North-Holland Pub., Amsterdam, 1968, p. 237.

185. S. Bajusz and K. Medzihradszky, Kem. Kozlem., 29, 369 (1968).

186. S. Bajusz and K. Medzihradszky, in Peptides (H. C. Beyerman, A. van de Linde, and W. M. van den Brink, eds.) North-Holland Pub., Amsterdam, 1967, p. 209.

187. K. Medzihradszky, V. Bruckner, M. Kajtár, M. Löw, S. Bajusz, and L. Kisfaludy, Acta Chim. Acad. Sci. Hung., 30, 105 (1962).

188. S. Bajusz, K. Lenard, L. Kisfaludy, K. Medzihradszky, and V. Bruckner, Acta Chim. Acad. Sci. Hung., 30, 239 (1962).

189. L. Kisfaludy, S. Dualszky, K. Medzihradszky, S. Bajusz, and V. Bruckner, Acta Chim. Acad. Sci. Hung., 30, 473 (1962).

190. S. Bajusz and T. Lázár, Acta Chim. Acad. Sci. Hung., 48, 111 (1966).

191. L. Kisfaludy and M. Löw, Acta Chim. Acad. Sci. Hung., 58, 231 (1968).

192. S. Sakakibara and Y. Shimonishi, Bull. Chem. Soc. Japan, 38, 1412 (1965).

193. S. Sakakibara, Y. Shimonishi, Y. Kishida, M. Okada, and H. Sugihara, Bull. Chem. Soc. Japan, 40, 2164 (1967).

194. M. Fujino and C. Hatanaka, Chem. Pharm. Bull. (Tokyo), 15, 2015 (1967).

195. G. Losse, A. Barth, and K. Schatz, Ann. Chem., 677, 185 (1964).

196. M. Fujino and O. Nishimura, Chem. Pharm. Bull. (Tokyo), 17, 1937 (1969).

197. M. Fujino and C. Hatanaka, Chem. Pharm. Bull. (Tokyo), 16, 929 (1968).

198. M. Fujino, C. Hatanaka, and O. Nishimura, Chem. Pharm. Bull. (Tokyo), 17, 2186 (1969).

199. M. Fujino, C. Hatanaka, and O. Nishimura, Chem. Pharm. Bull. (Tokyo), 18, 771 (1970).

200. M. Fujino, O. Nishimura, and C. Hatanaka, Chem. Pharm. Bull. (Tokyo), 17, 2135 (1969).

201. M. Fujino, C. Hatanaka, and O. Nishimura, Chem. Pharm. Bull. (Tokyo), 18, 1288 (1970).

202. M. Fujino, C. Hatanaka, and O. Nishimura, Chem. Pharm. Bull. (Tokyo), 19, 1066 (1971).

203. M. Fujino, O. Nishimura, and C. Hatanaka, Chem. Pharm. Bull. (Tokyo), 18, 1291 (1970).

204. M. Fujino, C. Hatanaka, O. Nishimura, and S. Shinagawa, Chem. Pharm. Bull. (Tokyo), 19, 1075 (1971).

205. J. Blake and C. H. Li, J. Am. Chem. Soc., 90, 5882 (1968).

206. E. Bayer, H. Eckstein, K. Hägele, W. A. König, W. Brüning, H. Hagenmaier, and W. Pan, J. Am. Chem. Soc., 92, 1735 (1970).

207. R. B. Merrifield, J. Am. Chem. Soc., 85, 2149 (1963).

208. A. Marglin and R. B. Merrifield, J. Am. Chem. Soc., 88, 5051 (1966).

209. E. Bayer, G. Jung, and H. Hagenmaier, Tetrahedron, 24, 4853 (1968).

210. S. Sano and M. Kurihara, Z. Physiol. Chem., 350, 1183 (1969).

211. K. Noda, S. Terada, N. Mitsuyasu, N. Yoshida, M. Waki, T. Kato, and N. Izumiya, J. Japan. Biochem. Soc., 42, 487 (1970).

212. B. Gutte and R. B. Merrifield, J. Am. Chem. Soc., 91, 501 (1969).

213. B. Gutte and R. B. Merrifield, J. Biol. Chem., 246, 1922 (1971).

214. C. H. Li and D. Yamashiro, J. Am. Chem. Soc., 92, 7608 (1970).

215. J. H. Jones, in Amino Acids, Peptides and Proteins, The
      Chemical Society, London, Vol. 1, 1969, p. 201; Vol. 2,
      1970, p. 159.

216. F. C. H. Chou, R. K. Chawla, R. F. Kibler, and K. Shapira,
      J. Am. Chem. Soc., 93, 267 (1971).

217. H. Yajima, H. Kawatani, and H. Watanabe, Chem. Pharm. Bull.
      (Tokyo), 18, 1333 (1970).

218. M. A. Tilak, Tetrahedron Letters, 6323 (1968).

219. J. Lenard and A. B. Robinson, J. Am. Chem. Soc., 89, 181 (1967).

220. G. S. Omenn and C. B. Anfinsen, J. Am. Chem. Soc., 90, 6571
      (1968).

221. H. Yajima and H. Kawatani, Chem. Pharm. Bull. (Tokyo), 19,
      1905 (1971).

222. B. Belleau and G. Malek, J. Am. Chem. Soc., 90, 1651 (1968).

223. H. Yajima, H. Kawatani, and F. Tamura, unpublished.

224. F. Weygand and K. Hunger, Chem. Ber., 95, 1 (1962).

225. H. Yajima, F. Tamura, and Y. Kiso, Chem. Pharm. Bull. (Tokyo),
      18, 2574 (1970).

226. K. Medzihradszky, H. S. Vargha, S. Fittkau, and I. Marquardt,
      Acta Chim. Acad. Sci. Hung., 65, 449 (1970).

227. H. B. F. Dixon and V. Moret, Biochem. J., 93, 250 (1964).

228. J. E. Elpiner and L. I. Stekolnikov, Dokl. Akad. Nauk. USSR,
      153, 710 (1963); through CA, 60, 8279 (1964).

229. J. P. Waller and H. B. F. Dixon, Biochem. J., 75, 320 (1960).

230. H. B. F. Dixon, Biochem. J., 62, 25 p (1956).

231. H. B. F. Dixon and L. R. Weitkamp, Biochem. J., 84, 462 (1962).

232. H. E. Lebovitz and F. L. Engel, Endocrinology, 73, 573 (1963).

233. M. Fujino, J. Takeda Res. Lab., 30, 358 (1971).

234. M. L. Dedman, T. H. Farmer, and C. J. O. R. Morris, Biochem. J.,
      59, xii (1955).

235. H. B. F. Dixon, Biochim. Biophys. Acta, 18, 599 (1955).

236. T. H. Farmer and C. J. O. R. Morris, Nature, 178, 1465 (1956).

237. M. L. Dedman, T. H. Farmer, and C. J. O. R. Morris, Biochem. J., 66, 166 (1957).

238. H. Otsuka and K. Inouye, Bull. Chem. Soc. Japan, 37, 289, 1465 (1964).

239. A. V. Schally, T. W. Redding, C. Y. Bowers and J. F. Barrett, J. Biol. Chem., 244, 4077 (1969).

240. F. W. Landgrebe and H. Waring, Quart, J. Exptl. Physiol., 31, 31 (1941).

241. W. O. Reinhardt, I. I. Geschwind, J. O. Porath, and C. H. Li, Proc. Soc. Exptl. Biol. Med., 80, 439 (1952).

242. B. T. Pickering and C. H. Li, Arch. Biochem. Biophys., 104, 119 (1964).

243. B. T. Pickering and C. H. Li, Biochim. Biophys. Acta, 62, 475 (1962).

244. T. H. Lee and V. B. Janusch, J. Biol. Chem., 238, 2012 (1963).

245. I. I. Geschwind and C. H. Li, Arch. Biochem. Biophys., 106, 200 (1964).

246. H. Yajima and K. Kubo, J. Am. Chem. Soc., 87, 2039 (1965).

247. K. Hano, M. Koida, H. Yajima, and T. Oshima, Biochim. Biophys. Acta, 115, 337 (1966).

248. H. Yajima and K. Kubo, Biochim. Biophys. Acta, 97, 596 (1965).

249. E. Schnabel and C. H. Li, J. Am. Chem. Soc., 82, 4576 (1960).

250. R. A. Brown, M. Davies, M. Englert, and H. R. Cox, J. Am. Chem. Soc., 78, 5077 (1956).

251. J. Léonis and C. H. Li, J. Am. Chem. Soc., 81, 415 (1959).

252. P. G. Squire and T. Bewley, Biochim. Biophys. Acta, 109, 234 (1965).

253. K. Ikeda, W. Urano, K. Hamaguchi, K. Kubo, and H. Yajima, Chem. Pharm. Bull. (Tokyo), 46, 1122 (1967).

254. O. D. Taunton, J. Roth, and I. Pastan, J. Clin. Invest., 46, 1122 (1967).

255. O. D. Taunton, J. Roth, and I. Pastan, Biochem. Biophys. Res. Commun., 29, 1 (1967).

256. O. D. Taunton, J. Roth, and I. Pastan, J. Biol. Chem., 244, 247 (1969).

257. I. Pastan, W. Pricer, and J. B. Mackie, Metab. Clin. Exptl., 19, 809 (1970).

258. R. J. Lefkowitz, J. Roth, and I. Pastan, Nature, 228, 864 (1970).

259. R. J. Lefkowitz, J. Roth, W. Pricer, and I. Pastan, Proc. Natl. Acad. Sci. U.S., 65, 745 (1970).

260. R. J. Lefkowitz, J. Roth, and I. Pastan, Science, 170, 633 (1970).

261. E. W. Sutherland, I. Oye, and R. W. Butcher, Recent Progr. Hormone Res., 21, 623 (1965).

262. E. W. Sutherland and T. W. Rall, J. Biol. Chem., 232, 1077 (1958).

263. R. C. Haynes and L. Berthet, J. Biol. Chem., 225, 115 (1957).

264. B. P. Schimmer, K. Ueda, and G. H. Sato, Biochem. Biophys. Res. Commun., 32, 806 (1968).

265. D. G. G. Smith, R. W. Butcher, R. L. Ney, and E. W. Sutherland, Clin. Res., 15, 259 (1967).

266. K. Hofmann, W. Wingender, and F. M. Finn, Proc. Natl. Acad. Sci. U.S., 67, 829 (1970).

267. G. Sayers, E. S. Redgate, and P. C. Royce, Ann. Rev. Physiol., 20, 243 (1958).

268. A. V. Schally, R. N. Andersen, H. S. Lipscomb, J. M. Long, and R. Guillemin, Nature, 188, 1192 (1960).

269. R. Guillemin, A. V. Schally, R. N. Andersen, H. S. Lipscomb, and J. Long, Compt. Rend., 250, 4462 (1960).

270. A. V. Schally, H. S. Lipscomb, and R. Guillemin, Endocrinology, 71, 164 (1962).

271. A. V. Schally and C. Y. Bowers, Meta. Clin. Exptl., 13, 1190 (1964).

272. W. Doepfner, E. Stürmer, and B. Berde, Endocrinology, 72, 897 (1963).

273. H. Kappeler and R. Schwyzer, Helv. Chim. Acta, 43, 1453 (1960).

274. M. P. de Garilhe, C. Gros, Y. Lozaćh, and B. Garnuchot, Experientia, 18, 92 (1962).

275. C. Gros and M. P. de Garilhe, Compt. Rend., 249, 2234 (1959).

276. M. P. de Garilhe, C. Gros, J. Chauvet, C. Fromageot, C. M. Voloss, and J. Benoit, Biochim. Biophys. Acta, 29, 603 (1958).

277. M. P. de Garilhe, C. Gros, J. Porath, and E. B. Lindner, Experientia, 16, 414 (1960).

278. M. P. de Garilhe and C. Gros, Arch. Exptl. Pathol. Pharmakol., 245, 184 (1963).

279. C. Gros, M. P. de Garilhe, A. A. Costopanagiotis, and R. Schwyzer, Helv. Chim. Acta, 44, 2042 (1961).

280. Kisfaludy et al., Chemistry and Biology of Peptides (J. Meienhofer, ed.), Ann Arbor Science Publishers, Michigan, 1972, p. 297.

281. Nishimura and Fujino, Tenth Symposium on Peptide Chemistry, Protein Research Foundation, Osaka, Japan, 1972, p. 154.

282. D. Yamashiro and C. H. Li, J. Am. Chem. Soc., 95, 1310 (1973).

CHAPTER 3
REACTION OF SMALL HETERO-
CYCLIC COMPOUNDS WITH AMINO ACIDS

K. Jankowski

Université de Moncton
Moncton, New Brunswick, Canada

I. INTRODUCTION.   146

    A. General Considerations.   146

    B. Nucleophilic Ring-Opening Reactions   147

    C. Electrophilic Ring-Opening Reactions.   153

    D. Extrusion and Rearrangement Reactions   153

    E. Amino Acids as Opening Reagents   154

    F. Scope and Application of the Reactions.   160

II. REACTION OF THREE-MEMBERED RINGS WITH AMINO ACIDS . . .   163

    A. Reactions of Epoxides   163

    B. Reactions of Episulfides.   174

    C. Reactions of Aziridines   176

    D. Reactions of Seleniranes.   178

    E. Reactions of Other Three-Membered Rings   180

    F. Future Applications   180

III. REACTION OF FOUR-MEMBERED RINGS WITH AMINO ACIDS. . . .   181

    A. Introduction.   181

    B. Openings of Carbonyl-Containing Four-Membered
        Rings   181

    C. Openings of Other Four-Membered Rings   184

D. Polymerization. 185

E. Conclusion. 186

IV. ILLUSTRATIVE PROCEDURES 187

A. Reaction of 3β-Hydroxy-16α,17α-epoxy-5-pregnen-
20-one with Ethyl Glycinate 187

B. Reaction of Styrene Oxide with Ethyl Glycinate. . . 188

C. Reaction of Epoxyethylene with Ethyl Esters of
Amino Acids 188

D. Reaction of Epithiocyclohexane with Ethyl
Sarcosinate 189

E. Reaction of Propylenimine with Sodium Glycinate . . 189

F. Reaction of Episelenate with Ethyl Sarcosinate. . . 190

G. Reaction of β-Lactams with Ethyl Sarcosinate. . . . 190

H. Reaction of β-Lactone with Ethyl Phenylglycinate. . . 190

I. Reaction of Methyl Cyclopropane with Ethyl
N-Bromoglycinate. 191

V. TABULAR SURVEY OF THE REACTIONS 191

REFERENCES. 200

I. INTRODUCTION

A. General Considerations

Many substances are the products of the reactions of mono-
heteroatomic three- or four-membered rings. The exceptional
susceptibility of these so-called small rings to cleavage can
be explained by the release of strain energy. The three-member
rings are more reactive than their four-member homologs because
of a bigger ring strain. The majority of these compounds are
useful as starting materials in simple synthesis, since they give
more functionalized products.

Generally, three different types of opening reactions are

observed in practice. First, the most common reaction in the case
of three-membered rings and with amino acids, is nucleophilic attack;
second, the principal result with N-blocked amino acids on four-
membered rings is an electrophilic opening; and finally, a rearrangement of a three- or four-membered ring occurs in the presence of an
amino acid. The last category also covers the extrusion reaction observed for all selenium derivatives and some other rings. In
summary, the opening of three- or four-membered rings by an amino
acid derivative is the most common result. However, the reaction
of many small cyclic compounds with simple amino acids is almost
unknown at this time.

B. Nucleophilic Ring-Opening Reactions

Small heterocyclic compounds (three- or four-membered [1]) react
well with amines [2] or carboxylic acids [3]. The yields of the
opened products vary with the nature of the reagents and the common
physical parameters (temperature, solvent, time, catalysts, etc.).
The most influential of these parameters is the solvent [4], for
without such a fluid the reaction gives only (mostly) decomposition,
rearrangement, or polymerization (dimerization) products [5]. At
the same time, the presence of solvent (especially a polar medium)
completes the necessary conditions for a push-pull mechanism in the
case of nucleophilic ring openings.

It is difficult to compare the chemistry of epoxides and episulfides on the one hand, to aziridines and seleniranes, on the other.
The reactivity of the amino acids--or the carboxylate forms of
amino acids--to different heterocyclic rings decreases in the
following order: X = NH, NR, O, S, Se, SO_2, CH_2. The carboxylate
forms are generally the metallic salts of the amino acids. The
reactivity of the aziridines, which have a basic character to amino
acids possessing a NHR group, seems to be greater than with the
corresponding epoxides and episulfides. At this point, our obser-

vations do not agree with some general comments by other authors
[6, 7]. However, the opening of aziridines by basic amino acids
seems to follow the general nucleophilic route. It is also hard
to classify heterocyclic rings that possess a carbonyl or hydroxyl
group alpha to the heteroatom. For example, rings with a carbonyl
group react as lactones, amides, etc., because of the changing
nature of the reactivity site.

A nucleophilic reagent attacks an unsymmetrical heterocycle
simultaneously on both the most substituted and less substituted
carbon atoms. As a result, two series of reaction products are
observed in most cases. According to Krassusky [8], attack on
the less substituted carbon gives a normal product, while the more
substituted carbon yields an abnormal series of products (see
Reaction 1).

$$Nu$$

$$ABNORMAL \diagup \quad \diagdown NORMAL$$

$$R-CH-CH_2$$

$$X$$

$$X = (CH_2)_n Y \quad Y = O, N, S, Se \quad n = 0, 1$$

$$\underline{(1)}$$

Two important steric factors must be mentioned in explaining
the direction of the openings. The nucleophilic reagent (of the
amino acid ester type) can easily attack the small ring if the
amino group is not substituted by a large, bulky group. This
general consideration is seen for the glycine series, where the
product yield decreases in the order glycine < sarcosine < N-
phenylglycine. At the same time, an unsymmetrically substituted
small ring presents two different sites for attack, and one of
them will be more open to the incoming nucleophilic reagent. In

this case, some regularity is observed if the same type of substituents are compared.

The steric factor is less important when the small ring is
substituted by an electron-releasing substituent.   A general
mechanism scheme is shown in Reaction 2 [9].

$$
R'-CH\underset{\displaystyle \diagdown_{\!\!X}\diagup}{-\!\!-}CH-R'' \quad \xrightarrow[E^{\oplus}]{:Nu} \quad
\left[
\begin{array}{c}
:Nu \\
\parallel \\
R'-CH\!\!-\!\!-\!\!CH-R'' \\
\diagdown_{\!\!X}\diagup \\
\parallel \\
E \\
\text{or/and} \\
:Nu \\
\parallel \\
R'-CH\!\!-\!\!-\!\!CH-R'' \\
\diagdown_{\!\!X} \\
\parallel \\
E
\end{array}
\right]^{\oplus}
$$

<u>(2)</u>

The $E^{\ominus}$ factor is generally a protic solvent that stabilizes
the reaction transition state.   Resonance is of the greatest importance, especially in the case of aromatic substituents on the
small rings.   The electronic distribution at this point controls
the ultimate direction of the opening of the ring.   Three specific
terms that play a role here include the following: (a) inductive
effect; (b) mesomeric effect; and (c) steric effect.   These factors
are important because of their simultaneous influence on both the
amino acid and on the heterocyclic ring.   The inductive effect
(-I) will draw the electrons away from one of the carbons, making
it more electronegative.   By the same token, the nitrogen of the
amino acid will be less negative.   The inductive effect (+I) will
increase the base strength of the same product.   The mesomeric
effects (-M or +M) will play the opposite role on these carbons.

The combination of all of those factors, and the steric strain
(i.e., if the bulky group is in close proximity to an amino group
of the amino acid) must be taken into consideration in order to
justify the major product formation [10].

The cleavage of the C-X bond in a rigid system--heterocyclic
rings attached to cyclopentanes, cyclohexanes, decalins, and
steroids--is one of greater selectivity. The attack of the
nucleophilic reagents comes from the axial side; this produces a
trans-diaxial opening. This result is known as the Fürst-Plattner
principle [11]. One series of steroids gives an exception to this
rule [12], and the trans-diequatorial product is formed. Naturally,
if the open trans-diaxial product is not completely rigid, it can
be converted by interconversion (i.e., chair-chair) into the more
stable trans-diequatorial product.

Thus several functions can influence the direction of the
opening of small rings by nucleophilic agents. From these remarks
it is clear that the three most important functions are steric
hindrance resulting from the volume of R, the nature of R, and the
solvent used in the reaction. From the classical point of view,
the nucleophility (basicity) of the reagent can also be used to
interpret the reaction. Some criticism of this point is taken up
later.

The possibility of stabilizing the transition state discussed
here has recently been supported by useful discussions on the
reaction mechanism [13] and by work involving kinetic studies [14]
or deuterated solvents (Table 1) [15].

In some cases the $E^{\ominus}$ reagent can be the amino acid used in
the reaction; sometimes the zwitterionic structure can also participate as the electrophilic reagent.

It was shown that for some epoxides (or episulfides) there
are two stereomeric forms possible for the products. Thus erythro-
and/or threo-isomers can be formed with respect to the stereochemistry

of the initial small ring. The S_N2 mechanism demands the inversion
of the configuration in the attack site by the nucleophilic reagent.

TABLE 1

Opening of the Propylene Oxide with (a) Diethylamine [14, 15]
and (b) Ethyl Sarcosinate [15]

Solvent	K^a (liter mole^{-1} sec^{-1})	
	(a)	(b)
Methanol	9.8×10^{-4}	6.4×10^{-4}
Methanol-d$_1$	7.0×10^{-4}	5.3×10^{-4}

^{a}t = 60 ± 1°C

A similar change for the S_N1 mechanistic attack is not absolutely
necessary. In this case, the intermediate carbonium ion can be
attacked from either side so as to give a mixture of products (see
Reaction 3).

The rapport between the inversion and retention of the
configuration of the products varies with the specific starting
product; therefore, the two stereoisomeric product yields are not
the same. The approach of the reagent from one side or the other is
not equally easy, and the substituents R, R', R'', R''', and $X^\ominus$ will
play the principal roles. For a small ring affixed to a rigid
steroid skeleton, an almost exclusive inversion of the configuration is involved at the attack site. It is sometimes easier to
represent the products of the opening not as a result of the
inversion in the attack site, but rather as the result of the
retention of the heteroatom function, which apparently gives the

(3)

same over-all result. Finally, it was shown that the addition of
even small catalytic amounts of strong bases will increase this
type of opening ring reaction.

The reaction of three-membered rings with the primary amino
group in amino acids can give a double opening. Thus the reaction
proceeds by two steps: first, the initial opening with the first
molecule of the heterocycle, and later an opening with a second
molecule of the heterocyclic compound with a N-substituted amino
acid. This second opening is naturally more difficult to realize
because of stereochemical reasons. It is always important to note
the manner in which reagents are added in these reactions. Because
of the possible formation of both products, both steps must be considered when equimolar qualities of amino acid and heterocycle are
used in the reaction. No formation of quaternary amine salts are
observed, although these would form in the next stage.

C. Electrophilic Ring-Opening Reactions

Normally occurring electrophilic ring-opening reactions are generally catalyzed by acids. In the case of the opening reactions with an amino acid, this catalytic effect is decreased because of its zwitterionic character. As in the preceding case, the most important factors influencing the direction of the opening of the small cycles are the solvent and the electron distribution in the unsymmetrical heterocycle. The stereochemistry of this opening will be characterized by inversion at the point of attack [2], but some carbonium ion intermediates can give rise to racemic products [3]. A stable carbonium ion seems to favor an attack on the ab-normal carbon atom. In this mechanism, the transition state retains the partial positive charge on the carbon atom, where the charge can be stabilized by electron-releasing factors.

The concept of a reversible conjugated acid helps one to under-stand the reasons for the acceleration of the reaction in an acid medium or in the presence of the proton donors. Some of the classical examples reported for the opening of epoxides with amines involve a total absence of a reaction in nonpolar solvents. The application of an acid-catalyzed reaction with amino acids was limited, probably because of the easier protonation of the amino function rather than that of the heteroatom. The competition between the formation of a conjugated acid of a small ring and the protonation of the amino acid reagent must be due to the absence of reactivity between the majority of heterocycles in acidic pH and the zwitterionic forms.

D. Extrusion and Rearrangement Reactions

In the presence of strong bases, some epoxides and episulfides gave extrusion of the heteroatom. The deoxygenation (or desulfuri-zation) was generally observed as a side reaction in aromatic epoxides or episulfides [16]. The extrusion of SO_2 by some episulfones was

the other typical reaction. In this case, the reaction is not accel-
erated by the presence of the amino acids and proceeds well even in a
neutral solvent [17]. There is some evidence for a free radical
course in this reaction. By contrast, a wide series of seleno compounds show a typical extrusion reaction [18].

The thermal rearrangement of epoxides is well known and is
sometimes used for synthetic purposes in the steroid field. It was
shown that substituted aliphatic and aromatic three-member rings
with one heteroatom, as well as the majority of those with two
heteroatoms, isomerize to carbonyl compounds (if the heteroatom is
oxygen) in the absence of the normally applied Lewis acid-type
catalysts [19]. In this case, the amino acid reacts immediately
with the new isomerized function (Schiff base formation). Any
catalytic role assigned to the amino acid in terms of a rearrangement of a small heterocycle is doubtful, even if a base-catalyzed
isomerization is proposed [20]. More substituted small rings can
undergo rearrangement by themselves (enolization), but the presence
of a strong base, which, is unreactive in the opening sense is always
necessary. The temperature required for some reactions (60-100°C)
is a sufficient factor to produce the partial rearrangement. This
same temperature, unfortunately, decreases the quantity of the amino
acid because of the side reactions that sometimes give major products [21]. The reaction in the presence of solvent usually decreases
or eliminates the rearrangement products.

E. Amino Acids as Opening Reagents

There are some important analogies between the observed reactions
of amines and amino acids. The basicity of the amino acids plays a
similar role to the basicity of the amines, even if the amino acids
are weaker bases. However, no direct correlation is noted between
the opening product formation and the basic strength of the initial
amino acids. The steric factor is important in terms of yield for

some amino acids, especially for those having N-substituents and α-carboxyl substituents.

Because of their zwitterionic structure, amino acids react as amphoteric products. Both functional groups--amine and carboxylic-- are, by definition, ionic. The aqueous solution for most amino acids is slightly acidic (pH = 6 for alanine) and is 99% zwitterionic. In other solvents (water-ethanol, 86% or 65%), such a structure is still preponderant, but the percentage of this form is smaller than in water [22]. At the same time, the zwitterion formation in nonpolar solvents is much more difficult, but the solubility of free amino acids is extremely poor in solvents other than water. Water is a bad solvent for opening-type reactions in view of its well-known reactivity with small heterocycles (diol formation from epoxides). Thus the four structures possible for the amino acids represent a pH-regulated system:

$$NH_3^\oplus\text{-CHR-COOH} \rightleftharpoons NH_2\text{-CHR-COOH} \rightleftharpoons NH_3^\oplus\text{-CHR-COO}^\ominus \rightleftharpoons NH_2\text{-CHR-COO}^\ominus$$

conjugated acid   neutral   zwitterion   conjugated base

This scheme seems to show that amino acids with an unblocked carboxylic group are unable to open small rings because of the reduced nucleophilicity of the amino group. The protection of the carboxylic group of amino acids will increase the nucleophilic character, but the basicity of the amino group is still only comparable to that of the aziridines (see Table 2).

The pH of a solution of 0.1 M zwitterionic glycine (isoelectric point) is 5.97. Because of its diprotic base character, sodium glycinate has a pH of 0.1 M solution at 11.3 [23]. The degree of ionization of the amino group in a 0.1 M solution of sodium glycinate is only 2%. For an ethyl glycinate solution, the concentration of amine and its conjugate acid is approximately equal (0.1 M, pH of 6.2).

The protection of the carboxylic group in the form of an ester breaks the zwitterionic structure. Moreover, esterification of the amino acid creates a better leaving group on the carboxylic function.

TABLE 2

Equilibrium Constants of Amino
Acids, Derivatives, and Organic Bases[a]

Amino acids, derivatives, and organic bases	K_1	K_{12}	K_{13}
Alanine	9.69	2.35	
β-Alanine	10.19	3.60	
Aniline	4.58		
Aziridine	8.01		
Azetidine	11.29		
Aspartic acid	9.82	3.63	2.12
tert-Butyl glycinate	6.42		
Butyric acid, 4-amino	10.55	4.05	
Cystine	10.81	8.35	
Cysteine	10.28	8.36	1.96
Ethanolamine	9.51		
Ethyl alanate	7.80		
Ethyl β-alanate	9.13		
Ethyl aspartate	6.50		
Ethyl glycinate	7.73		
Ethyl glycylglycinate	7.79		
Ethyl sarcosinate	8.40		
Glutamic acid	9.10	4.21	2.39
Glycine	9.77	2.35	
Glycine, N-acetyl	3.67		
Glycinamide	8.06		
Glycine, N,N-bis (2-hydroxyethyl)	8.35		
Glycine, N,N-diethyl	10.47	2.04	
Glycine, α,α-dimethyl	10.30		
Glycine, N,N-dimethyl	9.92	2.15	
Glycine, N-ethyl	10.10	2.30	

Glycine, α-phenyl	4.39	1.80
Glycine, N-phenyl	9.25	3.17
Glycylglycine	8.12	3.33
Methionine	9.15	2.12
Methyl glycinate	7.66	
Morpholine	8.33	
Phenylalanine	9.17	1.89
β-Phenylalanine	9.18	2.21
Piperazine	9.83	
n-Propyl glycinate	7.68	
Pyridine	5.25	
Sarcosine	10.12	2.21
Triethylamine	10.76	

[a]Protonation of ligands $H^{\oplus} + L \rightleftharpoons HL$ $K = \dfrac{[HL]}{[H^{\oplus}][L]}$

$H^{\oplus} + H_{n-1}L \rightleftharpoons H_n L$ $n \geqslant 2$ $K_{1n} = \dfrac{[HnL]}{[H^{\oplus}][H_{n-1}L]}$.

Such esters are extremely susceptible to dimerization, and even after standing at 0-5°, 2,5-diketopiperazines (amino acids anhydrides) are formed. Methyl esters dimerize especially quickly, but the ethyl or higher homologs also give a good yield of those products. As expected, free amino acids dimerize relatively slowly at room temperature. Solvents (ethanol or ether-ethanol) only slightly decrease the rate of dimerization for those amino acid esters that dimerize instantly without solvents. The tert-butyl esters are less susceptible to dimerization, probably because of their diminished reactivity to basic reagents. Dimerization is extremely bad because the resulting 2,5-diketopiperazines are not very soluble, hard to separate, possess high melting points, and cause a diminution in the quantity of the ester reagent.

In summary, the formation of 2,5-diketopiperazines is known to be a function of the solvent, the type of ester, and the substituents of the nitrogen and α-carbon of amino acids (see Table 3).

TABLE 3

Spontaneous "Instant" Dimerization of the Amino Acid Esters [27][a].

Ester	Solvent	Dimer	Yield (%)
Methyl glycinate	None	2,5-Diketopiperazine	75
	Ethanol	2,5-Diketopiperazine	60
	Ethanol-ether 1:9	2,5-Diketopiperazine	55
Ethyl glycinate	None	2,5-Diketopiperazine	70
	Ethanol-ether 1:9	2,5-Diketopiperazine	52
n-Propyl glycinate	None	2,5-Diketopiperazine	41
tert-Butyl glycinate	None	2,5-Diketopiperazine	51
	Ethanol-ether 1:9	2,5-Diketopiperazine	43
Ethyl alanate	None	3,6-Dimethyl-2,5-diketopiperazine	65
Ethyl sarcosinate	None	N,N-Dimethyl-2,5-diketopiperazine	35
Ethyl α-phenylglycinate	None	3,6-Diphenyl-2,5-diketopiperazine	32
Ethyl N-phenylglycinate	None	N,N-Diphenyl-2,5-diketopiperazine	47
Ethyl phenylalanate	None	3,6-Dibenzyl-2,5-diketopiperazine	41

[a]After 24 hr at 0-5°.

It has been shown that 2,5-diketopiperazines are not reactive with small heterocyclic compounds. The 2,5-diketopiperazines have amide properties, but their basicity constant is too small to be measured [24]. The presence of tautomeric forms and the enolization of those products are possible [25], yet such intermediates cannot be found, even under catalytic conditions.

The opening of a small heterocyclic compound with the carboxyl group in an amino acid was also mentioned. There are two ways to achieve this: (a) by an opening as in the normal carboxylic acid case, or (b) by an opening with an anionic carboxylate (metal salt). An "acidic" amino acid can protonate a heteroatom and react as an autocatalyzer within restricted pH limits. Normally the moderate acidity of zwitterionic amino acids do not guarantee good opening yields. Successive substitution of the amino group increases the acidity of the carboxylate, but even in this example, just traces are found in an opening.

In some particular cases it is suitable to use (in addition to the opening agent) a blocking agent--i.e., an acetylation agent (acetic anhydride)--to protect the amine or hydroxyl formed from the aziridine or epoxide, respectively, and to stabilize the product of the opening. Acetic anhydride can be a competitive reagent, and, in a secondary reaction, will give an opening with CH_3COO^- and acetylation as CH_3CO^+. The presence of the "base" (amino acid) is used to catalyze this reaction, which normally works better in the presence of stronger organic bases, such as pyridine.

Trifunctional amino acids can open small rings with a second carboxyl (aspartic acid) or even a mercapto (cysteine) group [26]. The second carboxylic group of aspartic acid is completely unreactive to epoxides and episulfides. In contrast, the mercapto group of cysteine in the presence of unblocked zwitterion (or when the zwitterion is blocked) gives the preponderant opening, and an opening with the usual amino or carboxylic groups does not take place. The selective blocking of either the amino or the carboxyl again gives a major product related to mercapto opening. The competition between SH and NHR, if the carboxylic group is blocked, is completely shifted to the SH group opening. Similar cases are observed for some derivatives of the aminothiols or mercaptoacids [26-28]. In these, the addition of an organic base increases the yield of the open product. Thus the necessity for the formation of an ionic RS⁻ group is proved, and the reaction proceeds according to the general nucleophilic ring-opening mechanism [26].

F. Scope and Application of the Reactions

The products from the opening of three- or four-membered rings were studied for different amino, carboxylic, or mercapto functions of amino acids. The reactions were examined for both three-membered rings, such as epoxides, aziridines, episulfides, episelenirenes, as well as their derivatives (α-lactams), and four-membered rings, such as oxetanes, azetidines, thietanes, selenates, and their derivatives. Only the first three rings and some thietanes possess sufficient reactivity to merit attention. The yields in other reactions are too poor to be useful in organic synthesis. Among these, as far as opening goes, are the selenium derivatives, which, because of their instability, lack any application. The epoxides and aziridines may have some synthetic application, but even under catalytic conditions the yields are extremely poor with amino acids.

The addition of catalytic amounts of pyridine or triethylamine plays an important role in the case of the opening of small rings with cysteine derivatives. It is well known that this specific base catalysis involves a proton transfer, followed by a slow step without proton transfer. Depending on the nature of the compound employed as the catalyst, a two-step mechanism begins with the slow formation of an anion and a relatively fast reaction with the reagents. It is also possible that the catalytic reaction in this case depends more on the nucleophilicity of the catalyst than on their basicity. Generally, tertiary amines are twice as effective as are primary amines of the same K_1 value [29,30]. Unfortunately, the reactivity of a zwitterionic amino acid to a small heterocycle does not change under the base-catalyzed reaction condition.

The most common products obtained in these series of reactions are the N-substituted amino acids and some heterocycles. The opening of epoxides, episulfides, and aziridines with amino acids is a short and easy route for the synthesis of some more complex six- or seven-member heterocycles. Several cyclopropane derivatives were used as potential N-alkylation agents in a reaction with N-bromo amino acids. Opened products were obtained in small yields in photocatalyzed reactions under similar conditions.

The introduction of an α-carbonyl group in the heterocyclic ring increases the yield of the opening reaction. This reaction does not have a typical hydrolysis character and can give two kinds of derivatives (see Reaction 4). The first product has a peptide

$$(\underline{4})$$

bond that may be of potential application in peptide synthesis or in terms of a blocking group. The second product has the character of an alkylation agent.

The reaction of amino acids and simple peptides with small rings has been studied with emphasis on their application as the materials for the synthesis of biologically important derivatives and as potential blocking agents. As may be expected from their chemical structure and reactivity, the compounds resulting from the amino

acids and small heterocycle reactions are potential pharmacolo-
gicals. The important N-(2-halogeno alkyl) amines are known as
mustard compounds with a powerful biological action. Analogous
mustards can be easily obtained by the conversion of the N,N-bis-
(2-hydroxyalkyl) amino acids, which are formed by the opening of
the epoxides to the corresponding N,N-bis-(2-chloroalkyl) amino
acids by treatment with simple chlorination agents [31, 32] (see
Reaction 5).

$$(\underline{5})$$

Similarly, other halogeno derivatives may be employed, and a
variety of active alkylation agents can be synthesized using these
materials. It is hoped that these compounds may show more favorable
toxicities and more selective action on tumors than the many known
mustards. Several cytostatically active compounds were found, but
none warranted clinical experiments [33]. This investigation is
not considered complete or finished; it is regarded, rather as a
series of pilot experiments on which systematic exploration of many
amino acids can be based at a later date.

Concerning the application of small heterocycles as blocking
agents for amino acids, it was demonstrated that the amino group is
blocked selectively (by the carboxylic group). However, the
secondary reaction of inter- or intramolecular condensation (dimerization or internal cyclization of the primary opening product)
excludes the possibility of a return to the starting amino acid.
A good blocking agent ought to be easily converted to a starting
material in a deblocking reaction. This is not the case with

small heterocycles and amino acids. Therefore only some N-substituted, stable, primary open products can be converted into the initial amino acids by an elimination reaction in a basic medium. The blocking of the SH group in the presence of other groups (NH$_2$, COOH) is much more selective. Yet, for the cysteine function, many other selective SH-blocking groups are known at this time. By contrast, the selectivity of protection of the functional groups is observed where one uses the α-carboxylated ring.

II. REACTION OF THREE-MEMBERED RINGS WITH AMINO ACIDS

A. Reactions of Epoxides

1. Introduction

The amino acids as bases normally yield two different primary products, a and b (Scheme 1) corresponding to the normal or abnormal mono-openings of epoxides. Subsequently the products of mono-openings react with another molecule of epoxide giving three more products, c, d, and e, which correspond again to the normal, abnormal, and mixed openings. Naturally, the products a to e can exist in two isomeric series--erythro and threo--and these multiply by the factor of two the number of different products. The products c to e are found just as in the case of an opening with the primary amino group of an amino acid. The presence of additional hydroxyl, amino, and carboxylic groups on the same molecule is the reason for the instability of the products a to e. Sometimes various substituents can stabilize the product; the material can be isolated in such cases.

The amount of product b (abnormal opening) is less than the quantity of a. Consequently the yield of products d and e is also

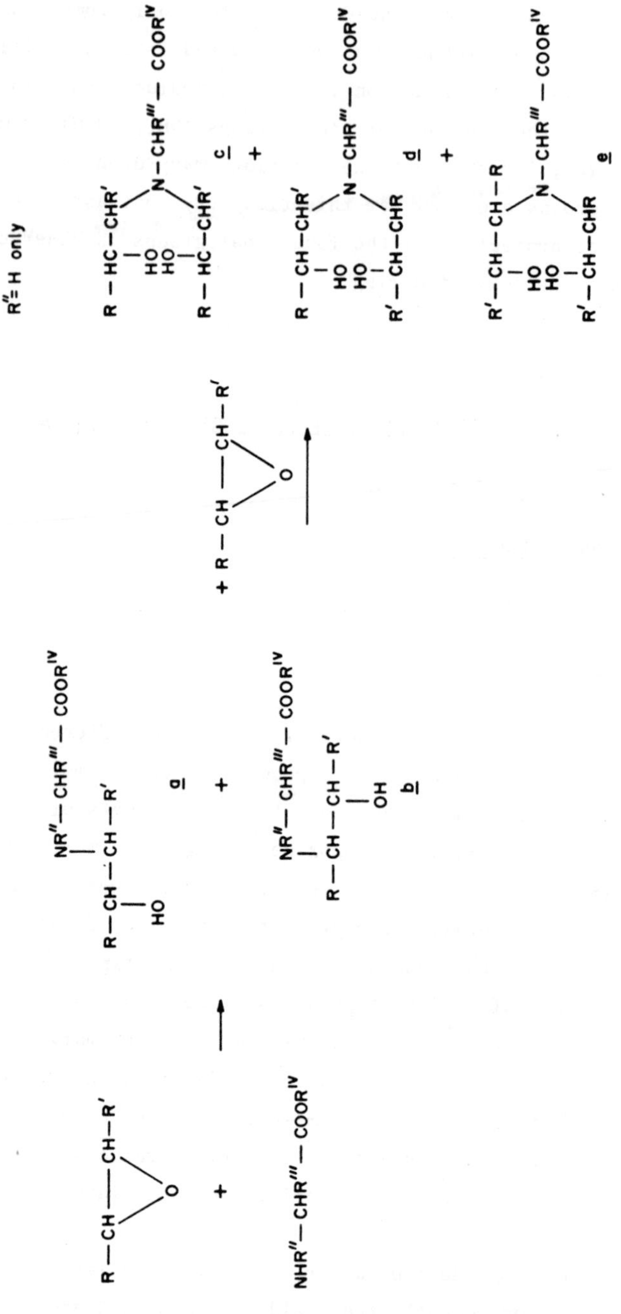

Scheme 1.

smaller than c. The important factor in regulating the quantity of
the a to e products is the order in which the reagents are combined
in the reaction. If the amino acid is added to the solution of
epoxide, then there is obtained N,N-disubstituted rather than mono-
substituted products. By contrast, the addition of epoxide to the
amino acid solution affords monosubstitution materials.

The reaction of halohydrins with the sodium salts of amino
acids is only a variation of the opening of the epoxides because
this reaction passes through the epoxide formation step (elimination
in basic medium) and then proceeds into the normal path. As a
result, the conversion of the epoxide is accompanied by the dimerization of the various intermediates. For example, amino acids
produce 2,5-diketopiperazines, epoxides form dioxane derivatives
or other polymers, while in some cases, with ethylene glycol as a
solvent, chain polymerization of the amino acids is preponderant.

2.   Inter- and Intramolecular Transposition of Products a - e

The products of mono- or di-opening can react in the inter- or
intramolecular sense giving the compounds f to p (Scheme 2). These
materials are sometimes the major products of the reaction. Compounds f and g are formed by the reaction of the molecule of a or
b with the starting amino acid. However, this type of 2,5-diketo-
piperazines is isolated just for one aliphatic epoxide [34]. This
is the only way to obtain an unsymmetrical N-monosubstituted 2,5-
diketopiperazine. A much more general reaction is the "dimerization"
of two molecules of a or the combination of one molecule, a, with
one b. This path gives symmetrically substituted N,N'-2,5-
diketopiperazines. The product resulting from the dimerization of
two molecules of b is not observed because of the small amount of
initial compound.

All products f to j, which result from intermolecular reaction,
have high melting points and are difficult to separate from the

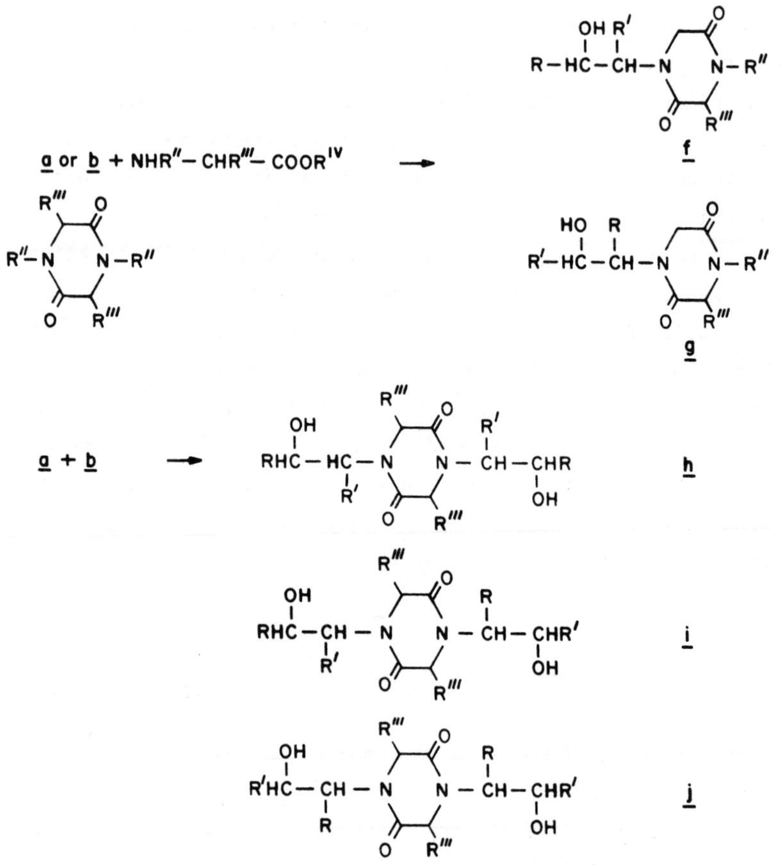

Scheme 2.

2,5-diketopiperazines. The intramolecular transformation on products $\underline{a}$ to $\underline{e}$ produces oily or low-melting 2-morpholones
(Scheme 3). The loss of $R^{iv}OH$ from the products is much easier
for the compounds of type $\underline{a}$ or $\underline{b}$ than for $\underline{c}$ to $\underline{e}$. This route
provides an easy method for the synthesis of a whole family of
those products. In addition, some compounds, $\underline{k}$ to $\underline{p}$, were isolated
here.

There are many possibilities for the formation of isomeric
products, so the choice of a structure is difficult at times. The
various assignments were proven by degradation, spectroscopic methods

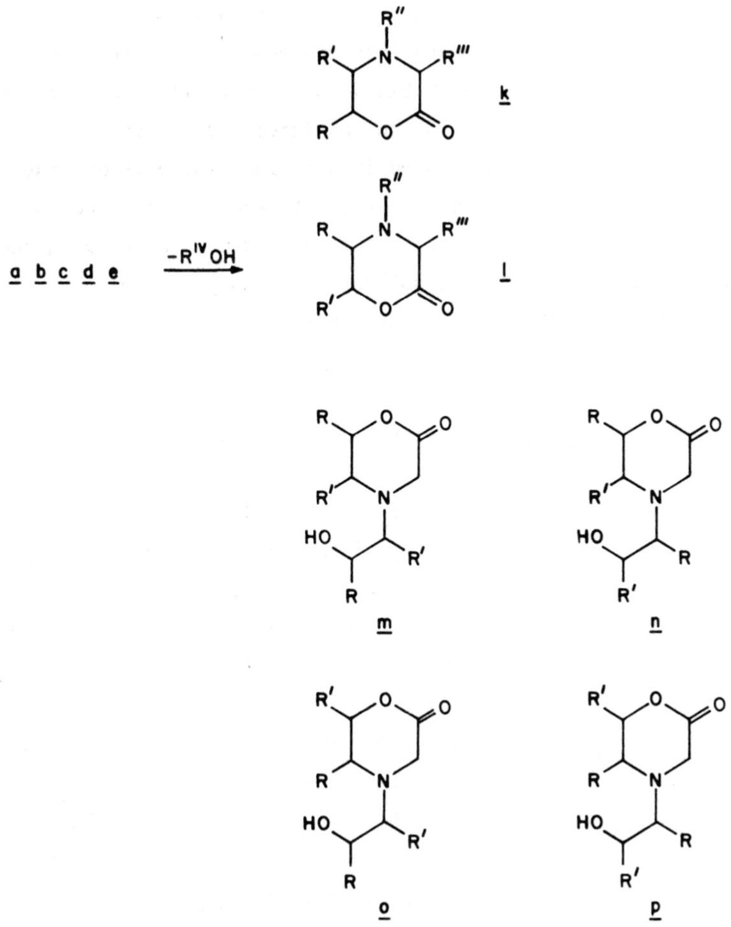

Scheme 3.

(NMR, mass spectrometry), and synthesis from the corresponding amino alcohols. Because of their small reactivity, the opening of epoxides with the 2,5-diketopiperazines was not realized. The relation between the inter- and intramolecular yields depends on the concentration of the products used in the reaction. In low concentration solutions, the intramolecular reaction is predominant.

Side reactions (polymerization of the amino acids and polymerization of the epoxides), especially with aromatic epoxides,

do not permit the generalization of certain considerations concerning
the relationship of intra- and intercondensations to the concentration.
The partial data available (Table 4) show that dilution increases the
yield of product corresponding to the intercondensation.

The same opening was tested in different solvents and the data
indicate that polarity is of great importance in the direction of
the ring opening (see Table 5). This dependency varies with the
dipole moment of the various solvents. In nonpolar solvents, the
normal product is the major constituent.

3.   Review of the Products of the Epoxide Openings

a.   Aliphatic Epoxide Openings. The aliphatic epoxides in the
reaction with the amino acids, generally give derivatives of type
c, d, k, l, m, n, and o [34, 27, 35, 36, 37]. In some cases, there
are isolated N-hydroxyalkyl amino acids [38, 27]. The majority of
the openings are normal, with the exception of 1,2-epoxybutene [34].
Primary products of the openings a or b were not isolated, but a
Schiff-base-type product was also formed. Amino acids containing
SH groups give a preponderant opening with this group only with
presence of a basic catalyst [26]. Opening by a carboxylate was
noted in one case [26, 27]. The epichlorhydrin reaction with an
excess of ethyl glycinate gives an interesting polymeric product
in the presence of inorganic base (Scheme 4). Generally, the
introduction of such bases increases this polymerization.

b.   Structures of Products. Two methods were applied to
determine product structures [41]. Acid hydrolysis of the 2,5-
diketopiperazine-type products gives N-hydroxyalkyl amino acid
hydrochlorides. Basic hydrolysis of the same product does not
produce a deblocking of the amino function, and these products are
quite stable in the basic medium (with the exception of the steroidal
series). The synthesis of the degradation products uses the amino
alcohols (obtained by the opening of the epoxides with amines) and

TABLE 4

Reaction of Ethyl Glycinate and Styrene Oxide [39]

Solvent	Concentration (%)	Total percentage of opening products	Inter:intra-relation
None	100	21	1000:1
Ethanol	20	17	58:1
	10	14	13:1
Chloroform	20	31	32:1
	10	15	28:1

TABLE 5

Solvent Effect on the Direction
of the Opening of Styrene Oxide with Ethyl Glycinate [39, 27]

Solvent	Total percentage of isolated opening product	Relation normal: abnormal
None	21.0	8:1
Methanol	25.5	13:2
Ethanol	22.0	9:1
Chloroform	23.5	11:1
Ether	20.0	12:1
Toluene	17.0	12:1
Benzene	20.0	14:1

the halogenated acids as starting materials. Some special synthetic methods give 2-morpholones (or N-hydroxyalkyl 2-morpholones), which normally are instantly formed after dehydrohalogenation of the amino

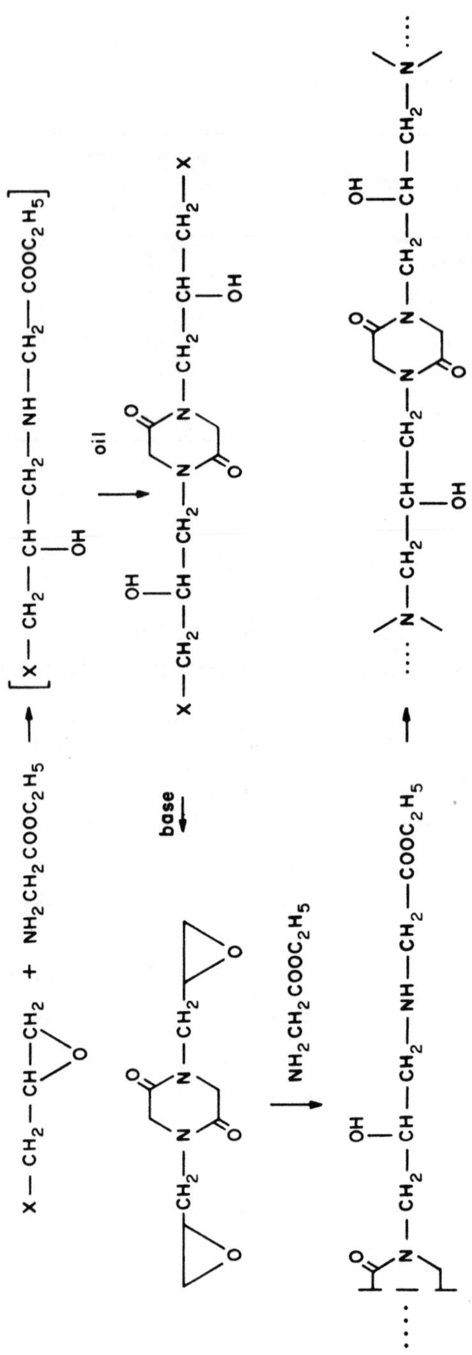

Scheme 4.

function. The 2-morpholones are easily hydrolyzed. The substitution of the N-function increases the stability of the 2-morpholones.

Some 1,2-halohydrins combine well with the metallic salts of the amino acids. This reaction is placed with the epoxides because of the reasons mentioned previously. The 2-morpholone derivatives yield decomposition products such as N,N-hydroxyalkyl amines, or N-alkylmorpholine, and N-morpholine carboxylic acids, especially in a hydrolytic medium [26, 41-43].

    c. Spectroscopic Methods. Nuclear magnetic resonance (NMR) and mass spectrometry are commonly used for the classification of the products of these openings [44, 45]. The insolubility of 2,5-diketopiperazines in the majority of common spectroscopic solvents is the main reason for the limited application of NMR to such structural problems. Recently, silylation has been used to increase volatility, but it is limited because of the deficiency of good model products [46]. The major fragmentation proceeds by ionization of the nitrogen of the heterocyclic ring [47]. Some other spectroscopic methods include optical rotatary dispersion for 2,5-diketopiperazines and, naturally, infrared and ultraviolet spectroscopy. The most important problem in the classification of the structures is the nature of the hyroxyl function of the products. The quickest method for the identification of an amino alcohol is infrared spectroscopy. Primary alcohol groups yield two absorptions at 3180-3190 and 1060-1075 cm^{-1}; secondary ones give two corresponding absorptions between 3300-3375 and 1110-1180 cm^{-1} [27]. The next quickest method consists of comparing the NMR chemical shifts of CH-OH proton (or CH_2-OH) with the corresponding acetylated or benzoylated derivatives (CH-OCOR and CH_2-OCOR) [27, 48]. However, this method is more complicated because of the use of insoluble derivatives [27].

    d. Epoxides Attached to a Ring. Epoxides on a five- or six-membered ring or in a steroid system can be opened by amino acids. Epoxides on cyclopentene, cyclohexene, and dihydropyran rings generally produce 2-morpholones attached to these rings or

2,5-diketopiperazines derivatives [48, 27]. Steroidal epoxides
(in 5,6 or 16,17 positions) yield an open product of type <u>a</u> or
<u>b</u>, and no inter- or intramolecular condensation was observed [49].
The inert steroidal system is probably the reason for good stability
in these products, but their low yields are explained by a competing
basic elimination reaction [49]. The special case of 16,17-epoxide
openings leads to the 17-hydroxy steroids, which, for the pregnan-
20-ones series, give D-homoannulation (Scheme 5) [49, 40]. This is
the first time that such a big substituent has been transported by
a D-homoannulation reaction.

The openings of 5α,6α- and 16α,17α-epoxides seem to follow
the trans diaxial opening principle with an inversion of configuration at the point of attack. In contrast, the 5β,6β-epoxides (cis
A-B junction) give a trans diequatorial product. Only a few epoxy-
steroids have been opened with amino acids. The structure proofs
were obtained by detailed studies of NMR spectra. The application
of Zürcher's corrections at C-18, C-19, and C-21 (pregnane derivatives), and studies on the long-range coupling of these methyls
(width of the angular methyl signal at its half-height), support
the suggested structures [40, 49, 50]. Electronic factors that
play a competetive role facing steric hindrance were found to be
of secondary importance in a study of ketal-epoxide steroidal reactions [40].

Amino acids do not attack the carbonyl function on the
steroidal skeleton, and no proof has been given for the formation
of Schiff bases. In spite of the stability of steroidal amino
esters and their well-established stereochemistry, the application
of steroids as protecting agents for the amino group is not promising
because of poor yields.

e. <u>Openings of Aromatic Epoxides</u>. Aromatic epoxides are much
more reactive than aliphatic epoxides. The primary opening products
are those of the type <u>a</u> or <u>b</u>, which are close analogs to adrenalin
and ephedrine. Unfortunately, the similarity of their structures
does not guarantee a physiological activity. Generally, products

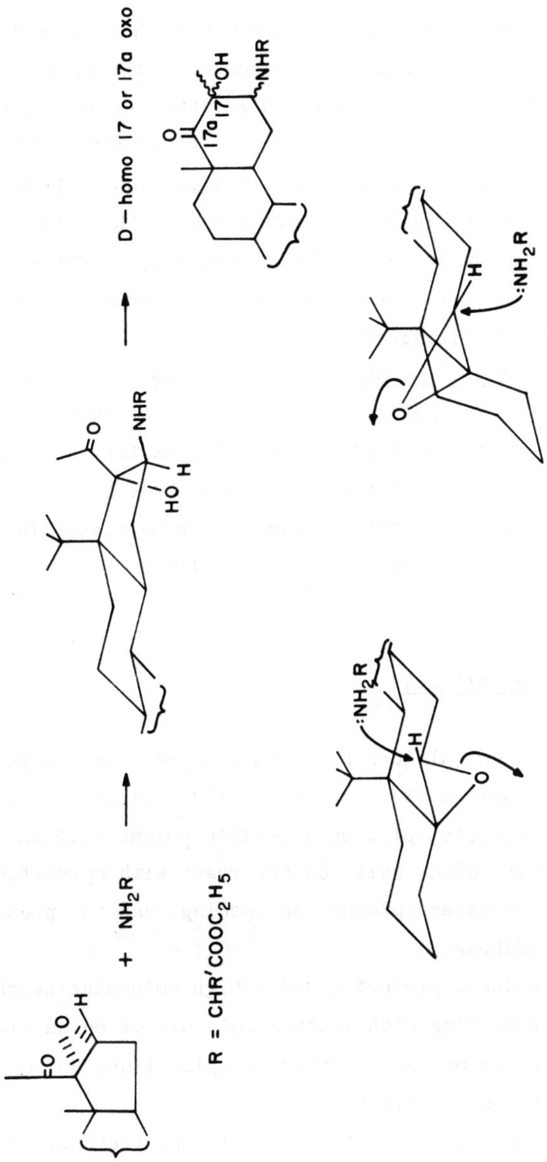

Scheme 5.

a or b are rather unreactive [27]. The principal products obtained
by this reaction are the 2,5-diketopiperazines of type h and i. The
majority of these compounds correspond to a normal opening. The
modifications of the preceding types give 2-morpholone derivatives
related to k-p. In some cases, substitution on the primary
opening products (a or b) increases their stability and it is
possible to isolate the materials (epoxystilbene) [34]. Of course,
one will obtain erythro and threo isomers from such compounds.
Some of the aromatic epoxides react quite well with lactams [51]
and cysteine-type amino acids or gluthatione [52]. However, all
aromatic epoxides polymerize quickly to give a considerable amount
of unidentified polymeric fractions.

     f. Preparation of 2-Morpholone Derivatives. Modification of
these procedures and an application of a different amount of epoxide
increases the yield of 2-morpholones or N-hydroxyalkyl 2-morpholones
[41]. Several derivatives of these compounds have been described
and a study has been made of their chemical properties. The products were also screened pharmacologically [41].

B. Reactions of Episulfides

     The opening of episulfides has been studied relatively less
than that of the epoxides [53-57]. Episulfides yield only a small
quantity of open products, plus considerable polymerization. Free
amino acids or their sodium salts do not react with episulfides. By
contrast, amino acid esters produce an opening, and two products are
generally formed (Scheme 9).

     Compound r produces product s and a high-molecular-weight
polymer by further opening with another molecule of episulfide.
The yield of s depends on the quantity of episulfide, amino acids,
and on the solvent used (Table 6).

     The decrease in the yield of r while in the presence of a large
quantity of episulfide is normal and predictable. The reactivity of

TABLE 6

Solvent Influence on the Yield of r and s [53]

Solvent	Episulfide: amino acid ester[a]	Proportion (r:s)
None	1:1	8:1
	2:1	3:4
Benzene	1:1	4:0.1
	2:1	3:2
Chloroform	1:1	8:0.1
	2:1	3:4

[a]Epithiocyclohexene: ethyl glycinate.

the SH bond in the type s product can be lowered if the reaction is attempted on the oxygenate (sulfonyl) derivative of an episulfide. Some sulfur extrusion reactions will follow, especially for products of the form

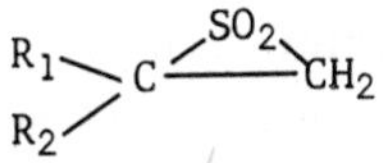

$$R_1 \diagdown C \diagup^{SO_2} \diagdown CH_2$$
$$R_2 \diagup$$

even in low-boiling-point solvents such as benzene [58, 59]. A second possibility is the S-acetylation of product r with acetic anhydride, but in such a case a strong competitive reaction is an opening with the anhydride itself, i.e.,

$$R-CH-CH_2-OAc$$
$$|$$
$$S-Ac$$

Aliphatic or aromatic episulfides give similar products (r or s). The steroidal episulfides have an extremely low reactivity to the amino acid esters, yet epithiocyclohexene gives a moderate amount of open trans ester r [53]. There was practically no difference in the reactivity of the amino acids or N-substituted amino acids. The generally low yields of the opening products, the high polymerization of the starting material, and the relatively good stability of products r or s in the basic mdeium do not permit the use of episulfides as protective agents for amino groups.

The sulfhydryl group of amino acids reacts exclusively with episulfides, what is similar to the epoxides [55, 56], but with even more polymerization. No traces of N,N-disubstitution (product t), 2,5-diketopiperazines, or p-2-thiazinones were detected here. Openings by the carboxylate ion of free amino acids were not found either.

C. Reactions of Aziridines

Aziridines have often been postulated as the starting materials in several syntheses. Yet, their rings are barely touched by nucleophilic reagents because of the basicity of the aziridines themselves. The openings of the aziridines can be realized with a C-C bond, C-N on the left, or on the right cleavages, while the cleavage is dependent on the ring's substitution pattern (Scheme 6).

It is apparent that aziridines differ markedly from epoxides and episulfides. The direct reaction of an amino acid ester with an aziridine gives a low yield of N-substituted product u, which can be converted into a derivative of an imidazole (v) by a simple condensation with an aldehyde [60]. No traces of 2,5-diketopiperazines or 2-piperazones were found, but large polymeric fractions always accompanied these reactions.

The potential opening with the NHR function of product u was not observed. Nevertheless, amino acids with other functional groups react at room temperature with the aziridines, and a strong selectivity was reported in the opening [61-67]. The application of aziridines as a blocking group was studied and a deblocking product was recorded for some cysteine derivatives [63, 67]. However, the reaction is not selective for other groups.

Some unsymmetrical aziridines gave an abnormal open major product, but generally a normal opening is the predominant one [61-74].

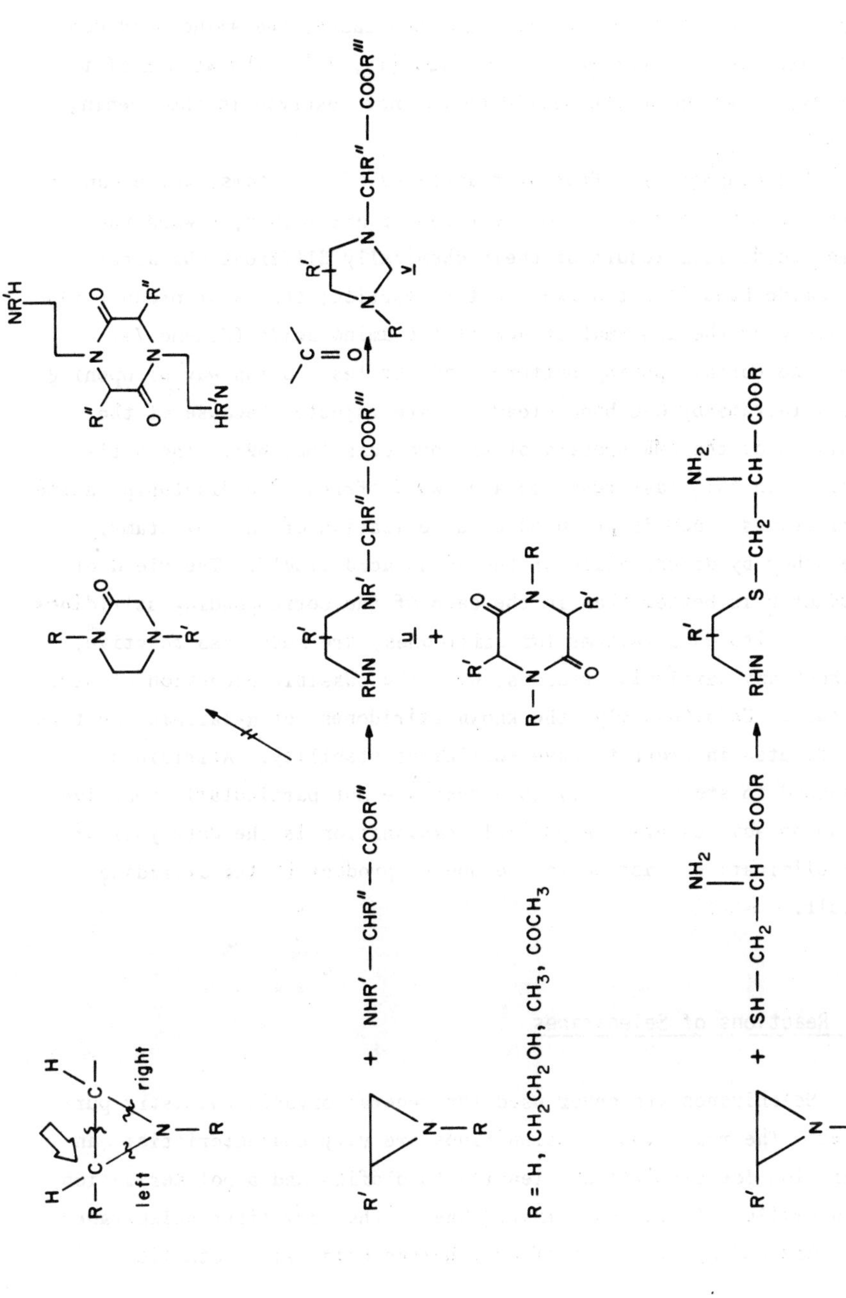

Scheme 6.

In a polar solvent medium, the dimerization of the aziridine to the piperazine derivatives occurs. In such cases, the amino acid does not catalyze this particular reaction [15, 75]. The attack of the carboxylic group on the aziridine was not observed in the opening sense.

The completely different reactivity of α-lactams, which can be considered to be carbonyl derivatives of aziridines, toward the amino acids is a result of their chemically different character. The amide bond is not broken in the reaction; the major product is obtained by the abnormal attack of the amino acids (Scheme 7).

The normal opening pattern, and the less common way of opening the α-lactams by C-C bond cleavage, are rejected because of the analysis of the NMR spectra of the products [62, 69]. The derivatives in this last reaction are two different 2,5-diketopiperazine derivatives: one is produced by dimerization of the α-lactams, the other by dimerization of the amino acid itself. The yield of product u is better than in the case of the corresponding aziridines. The substituted α-lactams (or aziridones) are much less reactive, without any particular reasons, with the possible exception of steric factors. Unfortunately, the known aziridones, or α-lactams, must be substituted in order to have sufficient stability. Aziridines attached to steroids or cyclohexanes are not particularly reactive to amino acid esters. A possible explanation is the catalysis of the elimination reaction to the opened product by the aziridine itself.

D. Reactions of Seleniranes

Seleniranes are never used for general organic chemistry purposes. The reactions of seleniranes are very characteristic: an extrusion (deselenization) leading to olefins and a polymerization to a cyclic 1,4-diseleno dimer. One of the very first seleniranes was obtained by treatment of cyclohexene oxide with potassium

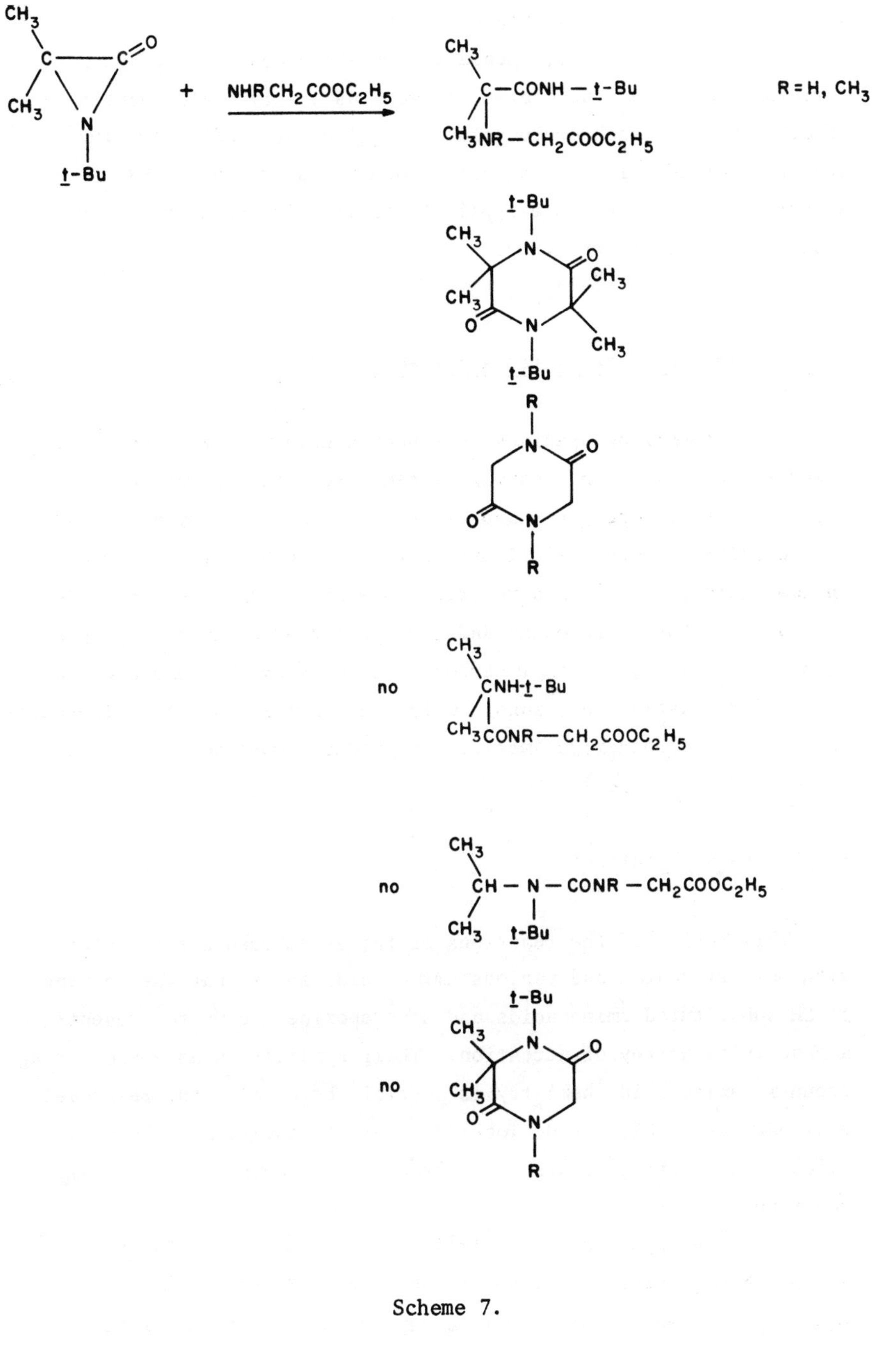
CH3
C
C=O
CH3
N
t-Bu
+
NHR CH2 COOC2H5
CH3
C—CONH — t-Bu
CH3 NR — CH2COOC2H5
R = H, CH3
no
CNH-t-Bu
CH3 CONR— CH2COOC2H5
CH — N — CONR — CH2COOC2H5
O
R

Scheme 7.

selenate [76]. This unstable selenirane, on treatment with an
amine or ethyl glycinate, opened to the above-mentioned polymeric
fraction and formed the olefin, as well as a very small quantity
of ethyl N-(2-selenohydrylocyclohexyl) glycinate [75, 77]. The
structure of this latter unstable product was proven by NMR
spectrometry. No suitable applications of this reaction can be
proposed at the present time.

E. Reactions of Other Three-Membered Rings

Many other ring openings have been studied by other workers.
However, only a few interesting systems are mentioned at this
point. Some cyclopropanes are cleaved by N-halogeno amino acids
in a catalytic photochemical procedure to give N-alkylated products
in low yield [15, 78]. No reaction appears to occur between 1,1-
dichloro-2-methycyclopropane and 1,1-dichloro-2-phenylcyclopropane
with ethyl glycinate, ethyl sarcosinate, or ethyl N-bromosarcosinate,
even under drastic conditions. Only some normal 2,5-diketopiperazine
derivatives and other polymerization products were observed here.

F. Future Applications

This review of the reactions of three-membered heterocycles
with one heteroatom and various amino acids shows that the opening
of SH-substituted amino acids and some epoxides seems to present
a reactivity worthy of attention. Their application as a protecting
group in amino acid chemistry is possible because of the relatively
easy path available for deblocking. Unfortunately, many less complicated protecting agents are currently available for these same
functions.

All other openings, especially because of side reactions,
which include polymerizations or condensations, are of little
practical interest. Some promise may be attached to the reaction

of α-lactams with amino acids, although a peptide bond formation is
not found in this case. It is necessary to test other less substituted α-lactams in order to explore the full potentiality in
this area. Other unknown three-membered rings with a carbonyl
function, such as α-lactones and α-thiolactones, might be good
materials for reactions with amino acids. Yet, despite the small
yields, all these reactions can be applied to the synthesis of
physiologically active compounds or their precursors.

III. REACTION OF FOUR-MEMBERED RINGS WITH AMINO ACIDS

A. Introduction

The four-membered heterocycles are, in general, more stable
and less reactive than their three-membered analogs. The substitution of four-membered rings by a carbonyl, especially alpha to a
heteroatom, gives a better opening reaction. Unsubstituted
heterocycles, and those substituted by alkyl, halogen, or aromatic
radicals, are practically inert. However, β-lactones and β-lactams
can easily react in other than an opening reaction meaning [79].
The most typical reaction of those compounds is an attack by a
nucleophilic reagent. This step is very slow, slower by a factor
of 10^3-10^5 times than for the three-membered analogs [80]. No
catalytic openings of the four-membered rings were observed in the
presence of an acid function. All these reactions are accompanied
by polymerization, which appears to be the major process.

B. Openings of Carbonyl-Containing Four-Membered Rings

A carbonyl function changes the reactivity of four-membered
rings considerably. An isolated carbonyl beta to a heteroatom

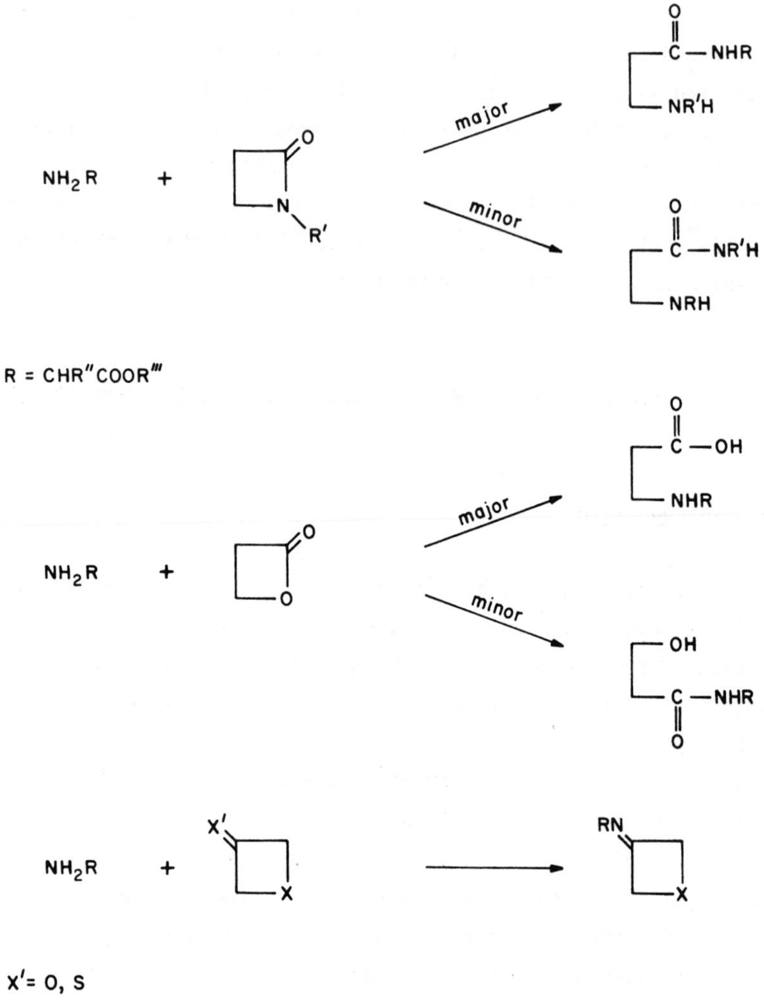

Scheme 8.

can react with amino acids, as in the formation of Schiff bases or
polymers, but no opening reactions were observed. The α-carbonyl
derivatives, β-lactams and β-lactones, are easily hydrolyzed in an
acid medium, yet in a basic medium they are stable, even under

drastic conditions [81]. If the hydrolysis does not take place,
then the amino acids can react with the α-carbonyl rings as acylation agents similar to the mixed anhydrides. Two types of products
can be formed: first, the opening of the β-lactams produces a
preponderant cleavage of the N-CO bond, and secondly, a cleavage
of the N-C bond.

The opening of β-lactones gives a major reaction with O-C bond
cleavage and a minor one with O-CO cleavage. It was mentioned   .
earlier that both reactivities are characteristic for the "hydrolysis"
of lactones or lactams (Scheme 8).

The reactions of β-lactones with amino acids (i.e., polymers, the
opening by the SH function of amino acids under catalytic conditions,
small yields) have been especially well studied [82, 83]. The
ketene dimer (diketene) is reactive in a basic medium with amino
acids, and the double bond is probably the basis of this reactivity.
One product of the reaction results from the addition to the double
bond after tautomeric stabilization of the system [84]. A second
product is a N-acetoacetyl derivative (the SH group must be blocked)
[84, 85]. β-Lactams usually cause the formation of peptide bonds

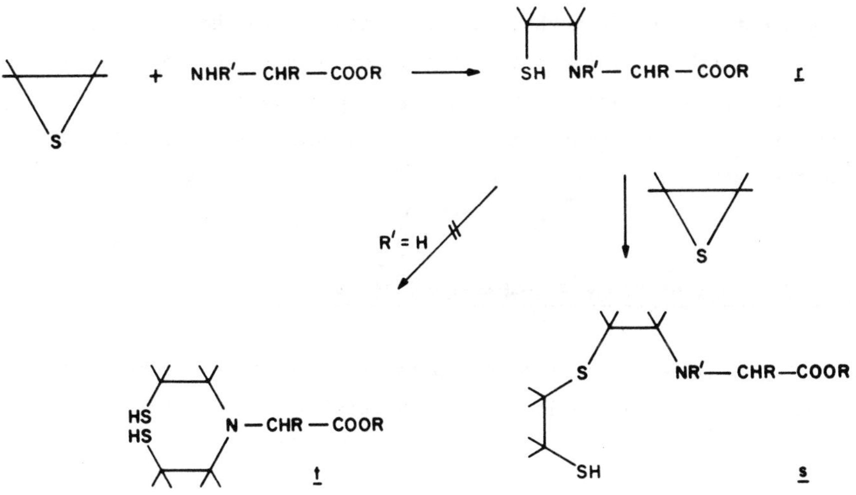

Scheme 9.

and lead to the derivatives of N-substituted amino acids [86]. It
is difficult to generalize this reactivity because of the limited
quantity of material studied in the past. In some cases, under
favorable conditions of solvent and temperature, the reaction
produces one of two "openings." Because of the small yield, no
applications exist in terms of blocking agent. However, β-lactones
or β-lactams protect the functional groups of amino acids.

By contrast, the opening with cysteine amino acids may be of
use in future applications [87]. For example, albumin reacts well
with β-propiolactone [88]. One typical side reaction is the
copolymerization of β-lactones by amino acids [89]. Generally,
the amines catalyze the polymerization of the lactones. The
selenous analogs of β-lactones are not reactive to the amino acid
derivatives.

The addition of amino acids to β-thiolactones gives an acylation
of the NH group, and is used for polypeptide synthesis [90]. Just
one year after the synthesis of β-thiolactones [91], such an
application was made by the Knunyant's research group (Scheme 10).
They found that the specific blocking of an NH-function can be
realized even in the presence of cysteine-type amino acids.

The remarkable acylating properties of propiothiolactone
represent a valuable method for the synthesis to the cysteine
polypeptides. Unfortunately, this reaction has been applied only
in penicillin chemistry. Finally, β-propiothiolactone polymerizes
more slowly than β-propiolactone, but is much more difficult to
synthesize.

C. Openings of Other Four-Membered Rings

Some different four-membered heterocycles were studied for
their reactions with amino acids. Generally, openings of a normal

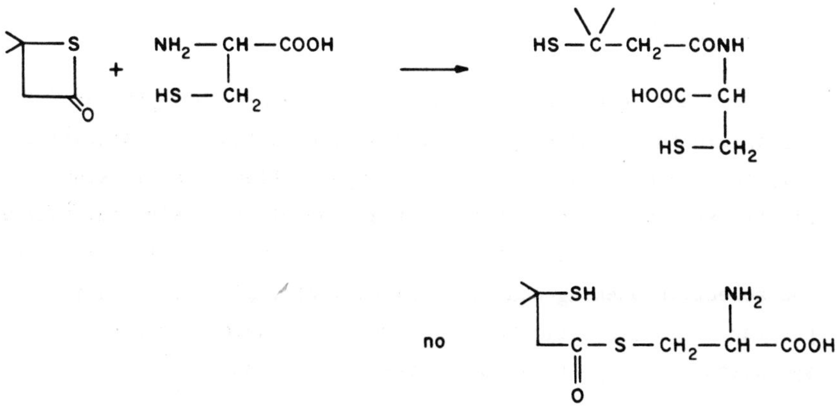

Scheme 10.

type (part I) were observed, but for nitrogen heterocycles, a small
quantity of abnormal product was reported [89]. The substitution
of a heterocyclic structure with aromatic or aliphatic substituents
does not change their reactivity and is of secondary importance.
Even in drastic, catalytic conditions, the opening yields are poor.
A ring containing more than one heteroatom does not react with
amino acids. The other selenous and sultone rings are completely
unreactive in the opening sense. Cyclobutane does not react with
N-bromo amino acids in a photochemical reaction [78, 92].

D. Polymerization

Specific polymerizations have been reported for β-propiolactone
in the presence of amines or amino acids [87-89]. These reactions
give many polymeric products according to the applied experimental
conditions. Cysteine does not catalyze the polymerization of
β-propiolactone. Some of these polymers have been applied as
corrosion inhibitors [93].

E. Conclusion

It seems necessary to study in more detail the opening of β-propiolactones and propiothiolactones by amino acid derivatives. Only this unique reaction offers an application as a blocking agent. Kinetic studies of the reaction of β-propiolactone with sulfhydrylo amines or cysteine derivatives show that in pH-controlled reactions, the selective opening can be realized (see Table 7). Even if polymerization is still the major reaction, some interesting syntheses may be possible using these reactions.

TABLE 7

Reactions of Cysteine-Like Products [a,b]   [15,26,28,94]

Starting material	Product with		
	β-Propiolactone	β-Propiolactam	β-Propiothiolactone
$SH-CH_2-\overset{\oplus}{\underset{\underset{R}{\vert}}{CH-NH_3}}$	S-opening	None	N-opening
$NH_2-\underset{\underset{R}{\vert}}{CH}-CH_2-SH$	N-opening	S-opening	N-opening
$\overset{\oplus}{NH_3}-\underset{\underset{R}{\vert}}{CH}-CH_2-\overset{\ominus}{S}$	S-opening	None	N-opening and S-opening
$NH_2-\underset{\underset{R}{\vert}}{CH}-CH_2-\overset{\ominus}{S}$	S-opening	None	N-opening

[a] R = H, COOH.

[b] Where R = H, K_1 = 10.81 and K_{12} = 8.35. Where R = COOH, k_1 = 10.28; K_{12} = 8.36 and K_{13} = 1.96.

IV. ILLUSTRATIVE PROCEDURES

Procedures are reported for reactions involving the opening of epoxides (solid, liquid, and gaseous), aziridine, episulfide, selenirane (episelenate), β-lactone, β-lactam, and cyclopropane by amino acids. Many specific modifications are given in the general procedure for the preparation of different cyclic products (2-morpholone and others) [26, 37, 38].

A. Reaction of 3β-Hydroxy-16α,17α-epoxy-5-pregnen-20-one with Ethyl Glycinate [49]

A solution of 3β-hydroxy-16α,17α-epoxy-5-pregnen-20-one (2 g, 7 mmoles) in anhydrous chloroform (25 ml) was heated on a steam bath for 12 hr with ethyl glycinate (0.5 g, 7 mmoles) followed by extraction with 2% hydrochloric acid (2 × 20 ml) and water. The solution was then dried over anhydrous potassium carbonate and evaporated to dryness. The residue was passed through a chromatographic column of silica gel. The following fractions were collected: (a) mp 260° (subl.), yield of 93 mg (21%); (b) mp 189-191°, yield 1.35 g (67, 5%); and (c) mp 220°, yield 26 mg (1%).

Product (a) was identified as 2,5-diketopiperazine from infrared spectra and mixed melting point determinations. Compound (b) could be identified as the starting material. Product (c) was crystallized from absolute ethanol and was identified as ethyl N-(oxo-20-dihydroxy-3β,17α-pregnen-5-yl)-16β-glycinate. Anal: calc. for $C_{25}H_{39}O_5N$ (433.6): C, 69.25%; H, 9.00%; N, 3.23%. Found: C, 69.16%; H, 8.80%; N, 3.01%.

B. Reaction of Styrene Oxide with Ethyl Glycinate [49]

Ethyl glycinate (20.6 g, 0.20 moles) and styrene oxide (24 g, 0.20 moles) were refluxed for 12 hr. The mixture, cooled to 0°, afforded crystals that were ground in absolute ethanol and filtered. The precipitate (9.3 g, 20.9%) was recrystallized from absolute ethanol. Four fractions separated:

Fraction	Melting point (°C)	Yield
a	260° (sublimation)	0.20 g (2.2%)
b	275°	6.98 g (74.0%)
c	206°	2.04 g (22.0%)
d	120-121°	0.037g (0.4%)

Product (a) was identified as 2,5-diketopiperazine. Product (b) was identified as 1,4-bis-(2-hydroxy-2-phenylethyl)-2,5-diketopiperzaine. Infrared spectrum: 1635 cm^{-1} and 3350 cm^{-1}. Anal: calc. for $C_{20}H_{22}O_4N_2$ (354.39); C, 67.77; H, 6.25; N, 7.89. Found: C, 68.03; H, 6.10; N, 8.23. Product (c) was identified as 1-(2-hydroxy-2-phenylethyl)-4-(hydroxy 1-phenylethyl)-2,5-diketopiperazine. Infrared spectrum: 1635 cm^{-1}, 3350 cm^{-1}, and 3180 cm^{-1}. Product (d) could be identified as the lactone of N-(2-hydroxy-2-phenylethyl) glycine. Infrared spectrum: 1720 cm^{-1} and 1100 cm^{-1}. Anal: calc. for $C_{10}H_{11}O_2N$ (177.19); N, 7.89. Found: N, 7.67.

C. Reactions of Epoxyethylene with Ethyl Esters of Amino Acids [26]

A solution of an ethyl ester of an amino acid (0.1 mole) in anhydrous chloroform at 0-5°C in a flask fitted with a dry ice condenser,

is saturated with epoxyethylene and kept at that temperature for 5 hr.
The mixture is then left at room temperature for 12 hr after which
the solvent and the excess of epoxide are distilled off under vacuum.
The residue is taken up in water containing about 0.1 mole of copper
sulfate and refluxed for 2 hr. The aqueous solution is extracted
three times with ether. The solvent and the products are distilled
off. The products are purified by fractional distillation (in
vacuo). The residue consisted of polymeric material and cupric salts.

D. Reaction of Epithiocyclohexane with Ethyl Sarcosinate [53]

An equimolar mixture of epithiocyclohexane and ethyl sarcosinate was refluxed during 12 hr. Separation of the three reaction
products was effected on silica column: (a) 1,4-dimethyl-2,5-
diketopiperazine, mp 144-146°, yield 20%, (b) ethyl N-(2-mercapto-
cyclohexyl) sarcosinate, bp 80-82°, yield 13%; (c) ethyl N-[2-
(2'-mercaptocyclohexylthio)-cyclohexyl] sarcosinate, bp 120°/0.1
mm, yield about 20%.

E. Reaction of Propylenimine with Sodium Glycinate [60]

An alcoholic solution of propylenimine (0.1 mole) and sodium
glycinate (0.1 mole) was heated with vigorous stirring for 12 hr.
The mixture was allowed to stand for 4 hr, and then evaporated to
dryness under vacuum. The residue was acidified to a pH of 6, the
solvent evaporated, and the mixture of products passed over a
column of alumina. The eluted fractions were concentrated and
esterified (with the ethanol-gaseous hydrogen chloride method),
then distilled several times under reduced pressure. Further
purification gave ethyl N-(2-amino-propyl) glycinate, bp 132-8°/0.05
mm; finally, thin-layer preparative chromatography gave 2.7% of
product.

F. Reaction of Episelenate with Ethyl Sarcosinate [75]

An ethanolic solution of ethyl sarcosinate (about 0.05 mole) and an equimolar quantity of episelenocyclohexene were refluxed on a steam bath for 2 hr under a nitrogen atmosphere. Anhydrous ether, twice the volume of the original solution, was added and the whole was heated another 2 hr. Evaporation of solvent under vacuum yielded an oily fraction that was taken up in ether. Five products separated: (a) 1,4-dimethyl-2,5-diketopiperazine (25%); (b) perhydroselanthrene (dodecahydro-1,4-diselenanthracene); (c) polymeric fraction (40%); (d) cyclohexene (22%); and (e) ethyl N-(2-selenylocyclohexyl) sarcosinate, purified by thin-layer chromatography, no boiling point. The nuclear magnetic resonance spectrum was: CH_2-CO 3.12, N-CH_3 2.45, SeH 2.67, CH_2-CH_3 4.27 and 1.19 (δ, ppm).

G. Reaction of β-Lactams with Ethyl Sarcosinate [86]

A solution of 2 g of N-phenyl-4,4-dimethyl-2-azetidone in benzene* (100 ml) was heated for 48 hr under reflux with 0.4 g of ethyl sarcosinate in the presence of 2 ml of pyridine. The mixture was extracted with 10 ml of water and dried over anhydrous magnesium sulfate. After removal of benzene and pyridine under reduced pressure, a colorless solid was obtained. Recrystallization gave 1,4-dimethyl-2,5-diketopiperazine, as well as an oily residue composed of 40 mg of ethyl N-(oxo-3-methyl-4-anilino-butyl) sarcosinate, identified by nuclear magnetic resonance spectrum: gem-CH_3 1.37, NH 5.17, CH_2-CO 2.35 (δ, ppm).

H. Reaction of β-Lactone with Ethyl Phenylglycinate [95]

To a stirred solution of β-propiolactone (0.1 mole) in 10 ml of ether, 0.1 mole of ethyl phenylglycinate in ether (20 ml) was

---

*An alcoholic solution can also be used.

added, the temperature being maintained at 0°. The solid fraction
that formed was filtered [3,6-dibenzyl-2,5-diketopiperazine and
traces of ethyl N-(2-carboxethyl)-phenylalanate] and the solution
was distilled to remove the solvent. The product distilling at
162-7°/0.2-0.4 mm (110 mg) was identified as ethyl N-(oxo-3-
hydroxypropyl)-phenylalanate.

I. Reaction of Methyl Cyclopropane with Ethyl N-Bromoglycinate [78]

Ethyl N-bromoglycinate (0.1 mole) in 150 ml of tetrahydrofuran
was stirred with methylcyclopropane (0.1 mole) under ultraviolet
irradiation (lamp, short wave) during 24 hr. The solvent was removed
under reduced pressure. The residue contained starting materials,
polymers (95-97%), a small quantity (2%) of ethyl N-(3-bromobutyl)
glycinate distilling at 170-180°/0.2-0.4 mm, and two other unidentified products.

V. TABULAR SURVEY OF THE REACTIONS

Selected reactions of three-member and four-member rings
with amino acids, are shown in the following two tables:

Table 8, see pages 192 through 197.

Table 9, see pages 198 and 199.

TABLE 8

Selected Reactions of Three-Member Rings with Amino Acids[a]

Ring	Amino acid or derivatives	Type of product	Reference
Oxides			
Ethylene	Glycine	Hydroxyamino acid	36, 38
	Glycine	2-Morpholone	38
	Glycine	N-(Hydroxyethyl)-2-morpholone	26, 37, 38, 41
	Glycine	2,5-Diketopiperazine	27,35
	Alanine	N-(Hydroxyethyl)-2-morpholone	26,96
	β-Alanine	N-(Hydroxyethyl)-2-morpholone	38
	Sarcosine	2-Morpholone	26, 38, 67, 96
	N-Phenylglycine	2-Morpholone	38
	2-Phenylglycine	N-(Hydroxyethyl)-2-morpholone	37, 38

	Aspartic acid	N-(Hydroxyethyl)-2-morpholone	38
	Phenylalanine	N-(Hydroxyethyl)-2-morpholone	26, 38
	Cysteine	S-(Hydroxyethyl)-amino acid	78
Propylenes	Glycine	2,5-Diketopiperazine	27
	Glycine	2-Morpholone	35
	Glycylglycine	Hydroxyamino acid	15
	Glycylglycine	2,5-Diketopiperazine	15
Butenes (1,2, or 2,3)	Glycine	2,5-Diketopiperazine	27, 34
	Glycine	2-Morpholone	96
	Sarcosine	2-Morpholone	38
	Sarcosine	N-(Hydroxyalkyl)-2-morpholone	38
	N-Phenylglycine	2-Morpholone	38, 96
	Glycine	N-(Hydroxyalkyl)-2-morpholone	38
	Glycine	Hydroxyamino acid	27, 36, 38
	Alanine	N-(Hydroxyalkyl)-2-morpholone	26, 27
	β-Alanine	N-(Hydroxyalkyl)-2-morpholone	26, 27

TABLE 8 (Continued)

Ring	Amino acid or derivatives	Type of product	Reference
Epihalohydrin	Glycine	2,5-Diketopiperazine, polymeric	27
Cyclopentenes	Glycine	2,5-Diketopiperazine	49
	Glycine	2-Morpholone	38
Cyclohexenes	Glycine	2,5-Diketopiperazine	34
	Glycine	2-Morpholone	28, 34
	Glycine	Hydroxyamino acid	26, 28, 35
	Alanine	2-Morpholone	28, 35, 97
	N-Phenylglycine	2-Morpholone	35
16,17-Epoxysteroids	Glycine	Hydroxyamino acid	49
	Alanine	Hydroxyamino acid	49, 78
	Sarcosine	Hydroxyamino acid	78
	Cysteine	S-(Hydroxyalkyl)-amino acid	78
5,6-Epoxysteroids	Glycine	Hydroxyamino acid	27, 49
2,3-Epoxysteroids	Glycine	Hydroxyamino acid	78
	Alanine	Hydroxyamino acid	78
Styrenes	Glycine	2,5-Diketopiperazine	39, 97
	Glycine	2-Morpholone	38, 39
	Glycine	Hydroxyamino acid	39

	Alanine	2,5-Diketopiperazine	27
	Sarcosine	2-Morpholone	38
	N-Phenylglycine	2-Morpholone	38
	2-Phenylglycine	2,5-Diketopiperazine	27
	Cysteine	**S-(Hydroxyalkyl)-** amino acid	58
Stilbenes	Glycine	Hydroxyamino acid	34
	Glycine	2-Morpholone	34
	Glycine	**N-(Hydroxyalkyl)-** 2-morpholone	25, 38
	Glycine	Decomposition products	34
Episulfides			
Ethylene	Glycine	Mercaptoamino acid	15
	Alanine	Mercaptoamino acid	15, 58, 59
	Sarcosine	Mercaptoamino acid	58
	Cysteine	S-opening	58
Propylenes	Glycine	Mercaptoamino acid	53-57
	Cysteine	S-opening	15
	Glycylglycine	Mercaptoamino acid	15
Cyclohexenes	Glycine	Mercaptothioalkyl amino acid	53-57
	Glycine	Mercaptoamino acid	53
	Sarcosine	Mercaptothioalkyl amino acid	53
	Cysteine	S-opening (amino acid)	15

TABLE 8 (Continued)

Ring	Amino acid or derivatives	Type of product	Reference
	Valine	N-opening (amino acid)	15
16,17-Epithiosteroids	Sarcosine	Mercaptoamino acid	15
Episulfonyl			
Dialkylethylene	Sarcosine	Unsaturated product	15
	Glycine	Unsaturated product	15
Aziridines			
Ethylenimine	Glycine	Aminoethylamino acid	60
	Sarcosine	Aminoethylamino acid	60, 61-74
	Cysteine	S-opening	
Propylenimine	Glycine	Aminoalkylamino acid	60
	Cysteine	S-opening	61
16,17-Steroids	Glycine	Aminoalkylamino acid	15
		2-3 products, N-opening	15
α-Lactams, aliphatic	Glycine	Aminoalkylamino acid	62, 69, 86
	Arginine	3-4 products, N-opening	15
	Glutamic acid	2 products, N-opening	15, 62

	Various amino acids	N-opening	
Seleniranes			
Cyclohexene	Various amino acids	N-opening	86
Cyclopropanes			
Alkylcyclopropanes	Glycine	Hydroxyamino acid	75
	Sarcosine	Hydroxyamino acid	75
	Glycine (halogeno)	N-Propylamino acid	15, 78
	Sarcosine (halogeno)	N-Propylamino acid	15, 78
Halogenocyclo-propanes	Glycine	---	15, 78
	Sarcosine	---	15
Arylcyclopropanes	Various amino acids	---	15

[a]Racemic.

TABLE 9

Selected Reactions of Four-Member Rings with Amino Acids[a]

Ring	Amino acid or derivatives	Type of product	Reference
Oxetanes			
Alkyl	Glycine	Hydroxyamino acid	15
	Sarcosine	Hydroxyamino acid	15
	Arginine	Hydroxyamino acid	15
	Glutamic acid	Hydroxyamino acid, N-opening	15
	Cysteine	S- or N-openings	15, 26, 28, 94
β-Lactones	Glycine	Hydroxyamino acid	81-3
	Sarcosine	Hydroxyamino acid	15, 81
	Cysteine	S-opening	15, 28, 82, 83, 94
	Different amino acids	S-, N-openings	81-92
β-Propiolactone	Different amino acids	Polymers	85, 89, 90
	Different S or N Peptides	Polymers	83-90
Thietanes			
Alkyl	Glycine	Mercaptoalkylamino acid	15
	Alanine	Mercaptoalkylamino acid	87, 90

	Sarcosine	---	15
Aryl	Cysteine	N-opening	15, 23, 26, 90, 94
	Cysteine	N- and S-opening	15, 91
β-Thiolactones	Glycine	N-opening	15, 90, 91
	Alanine	N-opening	90, 91
	Cysteines	N- or S- opening	15, 26, 28, 91, 94
	Peptides	S-opening	88
	Various amino acids	Polymers	78-92, 95
Azetidines			
Alkyl	Glycine	N-opening, N-CO cleavage	15
	Arginine	---	15
	Cysteine	S-opening	15, 26, 28, 94
Aryl	Sarcosine	N-opening product	86
β-Lactams	Glycine	N-opening, N-CO cleavage	86
	Sarcosine	N-opening	86
	Various amino acids	Polymers	87-89, 93
Selenates			
	Various amino acids	---	15
Cyclobutanes			
	Various amino acids	---	15, 78, 92

[a]Racemic.

ACKNOWLEDGMENTS

This work was supported by the National Research Council of
Canada. The author gratefully acknowledges the Université de
Moncton and the Université de Montréal for the use of their
facilities.

REFERENCES

1. V. Prelog, H. C. Brown, R. S. Fletcher, and R. B. Johannesen,
    J. Am. Chem. Soc., 73, 212 (1951).

2. R. E. Parker and N. S. Isaacs, Chem. Rev., 59, 737 (1959).

3. C. A. Stewart and C. A. VanderWerf, J. Am. Chem. Soc., 76,
    1259 (1954).

4. E. L. Eliel, in Steric Effects in Organic Chemistry (M. S.
    Newman, ed.), Wiley, New York, 1956, pp. 97-118.

5. S. Winstein and R. B. Henderson, in Heterocyclic Compounds
    (R. C. Elderfield, ed.), Vol. 1, Wiley, New York, 1950, Chap. 1.

6. R. Ghirardelli and H. J. Lucas, J. Am. Chem. Soc., 79, 734
    (1957); J. S. Fenton, in Heterocyclic Compounds (R. C.
    Elderfield, ed.), Vol. 1, Wiley, New York, Chap. 2; P. E.
    Fanta, in The Chemistry of Heterocyclic Compounds (A.
    Weissberger, ed.) Part I, Vol. 19, Wiley-Interscience, New
    York, 1964, Chap. 2.

7. C. Berse and K. Jankowski, Abstracts of papers, 52nd Canadian
    Chemical Conference and Exhibition, Chemical Institute of
    Canada, 1969, Org. 17, p. 53.

8. K. Krassusky, Comptes Rend., 146, 236 (1908).

9. N. S. Isaacs and K. Neelakahtan, Can. J. Chem., 45, 1597 (1967).

10. D. J. Cram and G. S. Hammond, Chimie Organique, Les Presses de l'Universite Laval, Quebec, 1963, pp 186-8, 195; E. S. Gould, Mechanism and Structure in Organic Chemistry, Holt, Reinhart and Winston, New York, 1963, p. 259.

11. A. Fürst and Pl. A. Plattner, abstracts of papers, 12th International Congress of Pure and Applied Chemistry, New York, 1951, p. 405; A. S. Hallsworth and H. B. Henebest, J. Chem. Soc., 1957, 404; A. Fürst and R. Scotoni, Helv. Chim. Acta, 36, 1332 (1953).

12. A. A. Akhrem, A. V. Kamernitskii, V. A. Dubrovskii, and A. M. Moiseenkov, Izv. Akad. Nauk SSSR, Ser. Khm. 9, 1726 (1964).

13. A. Feldstein and C. A. VanderWerf, J. Am. Chem. Soc., 76, 1626 (1954); C. A. VanderWerf and R. Fuchs, ibid., 76, 1631 (1954); See also Ref. 9.

14. L. A. Paquette, Principles of Modern Heterocyclic Chemistry, Benjamin, New York, 1968, Chap. 1, p. 27.

15. K. Jankowski, abstracts of papers, 8th International Symposium on the Chemistry of Natural Products, IUPAC, New Delhi, February 1972. See also Ref. 9.

16. M. Sander, Chem. Rev., 66, 297, 325 (1966); M. J. Boskin and D. B. Denney, Chem. Ind. (London), 1959, 330; C. C. J. Culvenor, W. Davies, and N. S. Heath, J. Chem. Soc., 1949, 282; R. E. Davis, J. Org. Chem., 23, 1768 (1960); S. Searles, H. R. Hays, and E. F. Lutz, J. Am. Chem. Soc., 80, 3168 (1958).

17. L. A. Paquette, J. Am. Chem. Soc., 86, 4085, 4089 (1964).

18. K. Jankowski, Tetrahedron Letters, submitted; K. W. Bagnall, The Chemistry of Selenium, Thellurium and Polonium, Elsevier Publish. Co., New York, 1966, p. 158-162; Handbook of Organo-metallic Compounds, Benjamin, New York, 1968, p. 748; N. M. Cullinane, A. G. Rees, and C. A. J. Plummer, J. Chem. Soc., 1939, 151.

19. H. B. Henebest and T. I. Wrigley, J. Chem. Soc., 1957, 4596.

20. H. W. Heine and F. Scholer, Tetrahedron Letters, 1964, 3667, K. Jankowski, Ph.D. Thesis, Université de Montréal, 1968.

21. H. Schott, J. Larkin, L. Rockland, and M. Dunn, J. Org. Chem.,
    12, 490 (1947); H. Sannier, Bull. Soc. Chim. France, 1942,
    487; H. Lenormant, ibid., 1948, 33.

22. A. Neuberger, Proc. Roy. Soc. (London), 115, 130 (1934); T. H.
    Jukes and C. L. A. Schmidt, J. Biol. Chem., 105, 359 (1934).

23. I. H. Segel, Biochemical Calculations, Wiley, New York, 1968,
    pp. 29, 123; See also Ref. 20; A. E. Martell, Stability Constants
    of Metal Ion Complexes, Sec. II: Organic Ligands, The Chemical
    Society, Burlington House, 1964; J. E. Purdie and N. L. Benoiton,
    Can. J. Chem., 49, 3468 (1971).

24. V. Orlander and V. Euler, Z. Physiol. Chem., 134, 381 (1928);
    Y. T. Pratt, in Heterocyclic Compounds (R. C. Elderfield, ed.)
    Vol. 6, Wiley, New York, 1957, p. 441; I. H. Sanborn, J. Phys.
    Chem., 36, 5799 (1932); H. Lenormant, Bull. Soc. Chim. France,
    1948, 487; H. Sannier and J. Poremski, Comptes Rend., 212, 786
    (1944); R. B. Corey, J. Am. Chem. Soc., 60, 1598 (1938).

25. E. Abdelhalden, Z. Physiol. Chem., 149, 100, 298 (1925); 155, 195
    (1926); 157, 140 (1926); 152, 88, 125 (1926); 153, 53 (1926).

26. K. Jankowski and C. Berse., Bull. Acad. Polon, Sci., 18, 183 (1970).
    K. Jankowski, C. Berse, and R. Coulombe, Bull. Acad. Polon. Sci.,
    19, 651 (1972).

27. K. Jankowski, Ph. D. Thesis, University of Montreal, 1968, p. 30.

28. R. Janowski, Tetrahedron Letters, 1974, in press.

29. R. P. Bell and G. L. Wilson, Trans. Faraday Soc., 46, 407 (1950);
    R. D. Gilliom, Introduction to Physical Organic Chemistry,
    Addison-Wesley, Reading, Mass., 1970, p. 172.

30. C. K. Ingold, Structure and Mechanism in Organic Chemistry,
    Cornell Univ. Press, Ithaca, N. Y., 1953, p. 345-352.

31. K. Jankowski, abstracts of papers, Chemical Institute of Canada,
    54th Canadian Chemical Conference, Org. VIII. 5, p. 36.

32. J. B. Jones and J. D. Leman, Can. J. Chem., 49, 1604 (1971).
    J. B. Jones and J. D. Leman, abstracts of papers, Chemical Institute
    of Canada, 53rd Canadian Chemical Conference, Medi. 8.

33. R. B. Livingston and S. K. Carter, Single Agents in Cancer
    Chemotherapy, IFI/Plenum, New York, 1970, p. 10-257.

34. K. Jankowski and C. Berse, Can. J. Chem., 45, 2865 (1967).

35. A. Kiprianov and G. Kiprianov, CA, 26, 108 (1932); 24, 1082
    (1930); 25, 5143 (1931); 22, 3134 (1928).

36. K. Maekawa and S. Tsumura, CA, 51, 2917 (1957).

37. M. L. Pascal, Bull. Soc. Chim. France, 1960, 435.

38. K. Jankowski and C. Berse, Can. J. Chem., 46, 1939 (1968);
    see also references cited in the above paper.

39. K. Jankowski and C. Berse, Can. J. Chem., 44, 1513 (1966).

40. P. Cohn and P. Friedlander, Chem. Ber., 37, 3034 (1904); H
    Gilman, C. S. Sherman, C. C. Price, R. C. Elderfield, J. T.
    Maynard, R. H. Reitsema, L. Tolman, S. P. Massie, F. J.
    Marshall, and L. Goldman, J. Am. Chem. Soc., 68, 1291 (1946);
    D. I. Weisblat, B. J. Magerlin, A. R. Hanze, D. R. Meyers,
    S. J. Rolfson, and E. I. Fairburn, ibid., 75, 3625 (1953);
    B. R. Baker, M. V. Querry, and A. F. Kadish, J. Org. Chem.,
    15, 400 (1950); T. Fukagawa, Chem. Ber., 68, 1344 (1935).

41. Typical Ref: 27, 34-39; K. Jankowski and C. Berse, Can. J.
    Chem., 47, 751 (1969); M. Pascal, Bull. Soc. Chim. France,
    1960, 465.

42. A. I. Kiprianov and G. I. Kiprianov, J. Gen. Chem. USSR, 2, 585 (1932
    (1932).

43. A. Rashkovan, Trav. Inst. Kim. Kharkov, 2, 41 (1936).

44. H. J. Svec and G. A. Junk, J. Am. Chem. Soc., 86, 2278 (1964).

45. K. Jankowski and L. Varfalvy, Can. J. Chem., 49, 1583 (1971).

46. K. Jankowski and L. Varfalvy, Bull. Soc. Chim. France,
    in press.

47. K. Jankowski and L. Varfalvy, Bull. Acad. Polon. Sci., 20, 423
    (1972).

48. L. M. Jackman, Application of NMR Spectroscopes in Organic
    Chemistry, Pergamon, London, 1959, p. 97.

49. K. Jankowski and C. Berse, Can. J. Chem., 46, 1835 (1968).

50. K. L. Williamson, T. Howell, and T. A. Spencer, J. Am. Chem.
    Soc., 88, 325 (1966).

51. H. Adkins and A. Rockbuck, J. Am. Chem. Soc., 70, 4041 (1948).

52. E. Boyland and K. Williams, Biochem. J., 94, 190 (1966).

53. K. Jankowski, R. Gauthier, and C. Berse, Can. J. Chem., 47, 3179 (1969).

54. L. G. Bulavian, Izv. Akad. Nauk S.S.S.R., Otd. Khim. Nauk, 2103 (1961).

55. H. R. Snyder, J. M. Stewart, and J. B. Ziegler, J. Am. Chem. Soc., 69, 2672, 2675 (1947), and later papers.

56. W. Reppe and F. Nicolai, U. S. Pat. 2, 105, 843 (1938).

57. M. Sander, Chem. Rev., 66, 297 (1966).

58. G. I. Braz, Zh. Obshch. Khim., 21, 688 (1951).

59. L. A. Paquette, J. Am. Chem. Soc., 86, 4383 (1964).

60. K. Jankowski and C. Berse, abstracts of papers, Chemical Institute of Canada, 52nd Canadian Chemical Conference, 1969, Org. 17, p. 52.

61. I. Lengyel and J. C. Sheehan, Angew. Chem. Intern. Ed., Engl., 7, 31 (1968).

62. J. C. Sheehan and I. Lengyel, J. Am. Chem. Soc., 86, 1356 (1964).

63. M. A. Raftery and R. D. Cole, J. Biol. Chem., 24, 3456 (1966).

64. L. B. Capp, J. Am. Chem. Soc., 70, 184 (1948); 73, 2121, 2584 (1951).

65. H. Gilman, J. Am. Chem. Soc., 67, 2106 (1945).

66. J. S. Fruton, W. H. Stein, and M. Bergmann, J. Org. Chem., 11, 559, 571 (1946).

67. L. A. Knorr, Chem. Ber., 37, 3507 (1904).

68. G. H. Coleman and J. E. Callen, J. Am. Chem. Soc., 68, 2006 (146).

69. L. Hellerman, Johns Hopkins Univ., personal communication.

70. A. Gilman and F. S. Philips, Science, 103, 490 (1946).

71. H. G. B. Marckwald and M. Frobensus, Chem. Ber., 34, 3544 (1901).

72. M. A. Raftery and R. David. J. Biol. Chem., 421, 3457 (1966).

73. Badische Anilin Soda Fabrik Ges., Ger. Pat. 857, 495 (1949).

74. T. L. Cairns, J. Am. Chem. Soc., 63, 871 (1943).

75. K. Jankowski, to be published.

76. E. E. van Tamelen, J. Am. Chem. Soc., 73, 3444 (1951).

77. K. Jankowski and R. Harvey, Synthesis, 1972, 627.

78. A. Rosowsky, in Heterocyclic Compounds with Three- and Four-Membered Rings (A. Weissberger, ed.), Part I, Interscience, New York, 1964, p. 2; P. E. Fanta, p. 524; D. D. Reynolds and D. L. Fields, p. 576; M. Sander, Chem. Rev., 66, 341 (1966).

79. Same openings of activated aziridines in O. C. Dermer and G. E. Ham, Ethylenimine and Other Aziridines, Academic, New York, 1969, p. 260.

80. J. Cark and D. D. Pervin, Quart. Rev., 18, 295 (1964); J. G. Pritchard and F. A. Long, J. Am. Chem. Soc., 73 4273 (1951).

81. Y. Etienne, R. Soulas, and H. Lumbroso, in Heterocyclic Compounds with Three- and Four-Membered Rings (A. Weissberger, ed.), Part II, Wiley (Interscience), New York, 1964, p. 647; Y. Etienne and N. Fischer, p. 729; J. A. Moore, p. 885; W. D. Emmons, p. 978; S. Searles, p. 983; S. Searles and V. P. Gregory, J. Am. Chem. Soc., 76, 2789 (1954).

82. S. Searles, J. Am. Chem. Soc., 73, 4515 (1951).

83. G. Machell, Ind. Chemist, 36, 13 (1960); R. J. W. Reynolds and E. J. Vickers, U. S. Pat. 2, 853, 476 (Sept. 23, 1958); See also Ref. 62.

84. F. D'Angeli, F. Filira, and E. Scoffone, Tetrahedron Letters, 1965, 605.

85. G. Buchi and G. Lukas, J. Am. Chem. Soc., 86, 5654 (1964).

86. I. Korosi, J. Prakt. Chem., 23, 212 (1964).

87. T. Kagiya, T. Sato, and K. Fukui, Kogyo Kagaku Zasshi, 67, 951 (1964).

88. S. K. Majumder, S. Godovaribai, M. Muthu, and K. Krishnamistly, J. Chromatog., 17, 373 (1965); M. K. Johnson, Biochem. J., 87, 8 (1963); C. E. Searle, Biochem. Biophys. Acta, 52, 579 (1961).

89. G. Talbot, R. Gandry, and L. Berlinguet, Can. J. Chem., 34, 1440 (1958); R. J. W. Reynolds and E. J. Bickers, U. S. Pat.

2,977,373 (March, 1961); F. Dickens and J. Cooke, Brit. J. Cancer, 19, 404 (1965); T. Kagiya, T. Sato, and K. Fukui, Kogyo Kagaku Zasshi, 68, 1141, 1144 (1965).

90. M. G. Lin'kova, O. V. Kil'disheva, and I. L. Knunyants, Bull. Acad. Sci. USSR, 401, 507 (1955); I. L. Knunyants, O. V. Kil'disheva, and E. Ya. Petrova, ibid., 401, 619 (1955); (1955); I. L. Knunyants and V. V. Shekina, Bull. Acad. Sci. USSR, 409 (1955).

91. D. Fles, A. Markovac-Prpic, and V. Tomasic, J. Am. Chem. Soc., 80, 4654 (1958); D. Fles and A. Markovac-Prpic, Croat Chem. Acta, 29, 183 (1957); J. C. Sheehan, Ann. N. Y. Acad. Sci., 88, 665 (1960); J. C. Sheehan and I. Lengyel, unpublished results.

92. M. K. Johnson, Biochem. Pharmacol., 14, 1383 (1965).

93. R. B. Thompson, U. S. Pat. 2, 968, 629 (Jan. 17, 1961); U. S. Pat. 3, 062, 631 (Nov. 6, 1962).

94. R. E. Davis, R. Nehring, S. P. Molnar, and L. A. Suba, Tetrahedron Letters, 1966, 885.

95. T. L. Gresham, J. E. Jansen, F. W. Shaver, F. T. Fiedorek, and R. A. Bankert, J. Am. Chem. Soc., 73, 3169 (1951).

96. E. Knorr, Ann. Chem., 307, 199 (1899); Chem. Ber., 32, 732 (1899).

97. C. Mannich and E. Thiele, Arch. Pharm., 253, 181 (1915).

CHAPTER 4
THE ISOXAZOLIUM SALT
METHOD OF PEPTIDE SYNTHESIS

Darrell J. Woodman

Department of Chemistry
University of Washington
Seattle, Washington

I. INTRODUCTION.   208
II. SYNTHESIS OF 3-UNSUBSTITUTED ISOXAZOLIUM SALTS. . .   211
III. REACTION WITH CARBOXYLIC ACID ANIONS.   217
    A. Acylketenimine Intermediates.   218
    B. Ring-Opening Process.   222
    C. Product Structure   224
    D. Postulated Mechanism of Enol Ester Formation. .   226
    E. Evidence Concerning the Addition Process. . . .   226
    F. Evidence for an Intermediate--Side Reactions in
        Enol Ester Formation.   228
IV. ENOL ESTER STABILITY AND THE IMIDE REARRANGEMENT. .   233
V. ENOL ESTER ACYLATING AGENTS   238
VI. ACYLATION BY-PRODUCTS   240
VII. RACEMIZATION.   242
    A. Control of the Azlactone Side Reaction During
        Activation.   244
    B. Decomposition of Enol Esters to Azlactone . . .   246
    C. Racemization-Resistant Esters from Isoxazolium
        Salts   248

VIII.   APPLICATIONS TO PEPTIDE SYNTHESIS. 250

A.   Use of N-Ethyl-5-phenylisoxazolium-3'-sulfonate. . . 250

B.   Guide to Abbreviations in Tables and Text. 253

C.   Early Use of Benzisoxazolium Salts 275

D.   Use of N-t-Butylisoxazolium Salts. 276

E.   Use of 4,5-Tetramethyleneisoxazolium Salts 278

F.   Use of the N-Ethyl-7-hydroxybenzisoxazolium Cation . 278

IX.   SPECIAL APPLICATIONS 280

A.   Solid-Phase Peptide Synthesis. 280

B.   Cyclic Peptides. 281

C.   Steroidal Peptides, Alkaloidal Peptides, Glycopeptides,
and Nucleopeptides 284

D.   Polypeptides 284

E.   Peptide and Protein Modification 285

F.   Preparation of Acylating Agents. 286

REFERENCES 287

I. INTRODUCTION

The isoxazolium salt method of peptide synthesis has been
developed during the last decade almost exclusively by R. B.
Woodward and his students. This approach to the formation of
peptide bonds is based on the reaction between carboxylic acid
anions and 3-unsubstituted isoxazolium cations ($\underline{1}$) to form enol
esters ($\underline{2}$). Since esters of the type ($\underline{2}$) are reactive acylating

agents, the reaction provides a way of converting free carboxylic acid groups of N-protected amino acids or peptide acids (3) to the corresponding derivatives (4) of the "active" ester type, which will generate a peptide linkage upon reaction with the amine function of another amino acid or peptide (5). As is discussed in later chapters, the isoxazolium salt technique is of special

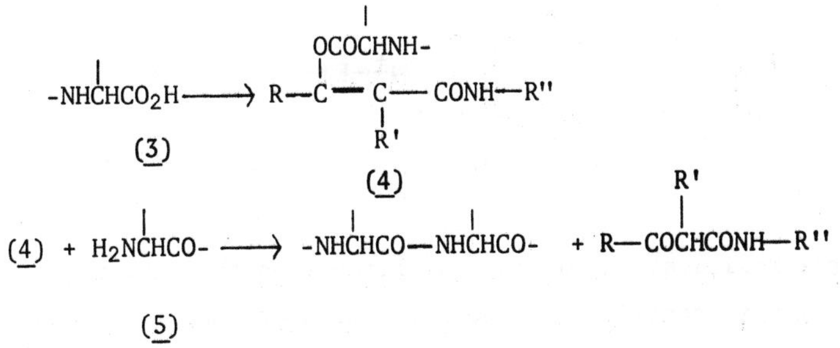

$$
-NHCHCO_2H \longrightarrow R-C \stackrel{\text{OCOCHNH-}}{=} C-CONH-R''
$$

(3)

(4)

$$
(4) + H_2NCHCO- \longrightarrow -NHCHCO-NHCHCO- \ + \ R-COCHCONH-R''
$$

(5)

interest for peptide synthesis primarily because it provides an especially efficient route to the intermediate acylating agents (4).

The reactions of 3-unsubstituted isoxazolium salts were studied near the turn of the century by Claisen, Mumm, and co-workers [1-14]. However, the potential value of the reaction with carboxylic acids as a method of making acylating agents was not exploited until nearly 60 years later, when Woodward and Olofson [15-17] reinvestigated the chemistry of these heterocyclic compounds and correctly elucidated the reaction course. A further extension of the reaction to N-protected amino acids and peptide acids by Woodward, Olofson, and Mayer [15, 18-19] culminated in the synthesis of a valuable new peptide reagent, N-ethyl-5-phenylisoxazolium-3'-sulfonate (6).

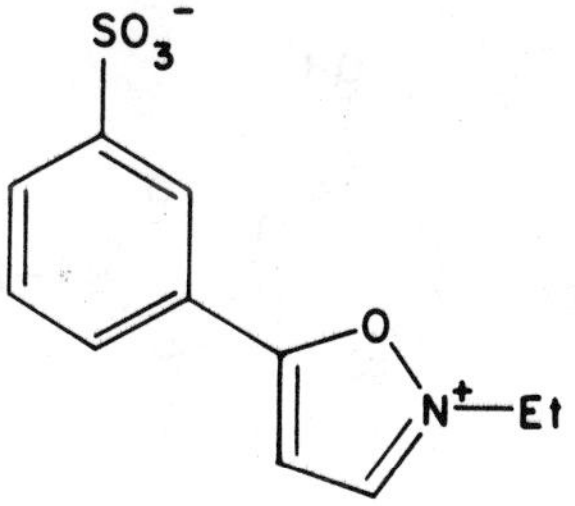

(6)

Since the development of the reagent (6), there has been continuing activity directed toward the design of new isoxazolium
salts that might give even better results in peptide synthesis.
The first attempt along these lines involved an extremely thorough
examination of the benzisoxazolium cation (7) by Woodward and
Kemp [20-22].

(7)

Next, several modifications of the substituent on the quaternary
nitrogen atom were considered. Woodward, Woodman, and Kobayashi
[23-25] investigated N-arylisoxazolium salts (8), while the
N-t-butyl compounds (9) were studied by Woodward, Woodman, and
Davidson [23, 24, 26-29]. Later approaches have been based on a
dichloro-substituted benzisoxazolium cation (10) by Rajappa and

(8)   (9)

Akerkar [30], the 7-hydroxy-2-ethylbenzisoxazolium cation (11) by
Kemp and Chien [31], and several 4,5-tetramethyleneisoxazolium
cations (12) by Olofson and Marino [32].

(10)   (11)   (12)

II. SYNTHESIS OF 3-UNSUBSTITUTED ISOXAZOLIUM SALTS

The renewed interest in isoxazolium salts for possible use in peptide chemistry has stimulated the elaboration of several preparative approaches to the heterocyclic system. Traditionally, the cations have been prepared by quaternization of the isoxazole nitrogen with primary alkylating agents [33]. In turn, the parent isoxazoles have been obtained by either of two generally useful routes to the five-membered ring: condensation to form either the 1-5 and 2-3 bonds (path A), or (path B) the 1-5 and 3-4 bonds [34, 35]. (An exception is the cyclization of β-ketovinylazides in which the 1-2 bond is formed last [36]).

For the preparation of 3-unsubstituted isoxazoles, the former strategy, which has been the more widely used, involves the condensation of hydroxylamine with hydroxymethylene carbonyl compounds or their stable derivatives. Here, one often obtains mixtures of isomeric isoxazoles that are hard to separate. However,

this difficulty has been overcome by Wilson and Burness [37], who found that improved selectivity is achieved in the condensation by using enamine derivatives of the carbonyl reactant.

The conditions for the cyclization step of the synthesis may also prove to be critical. For example, dehydration of the oxime

(13) with acetic anhydride leads to benzoylacetonitrile, while the
desired isoxazole (14) is obtained with acetyl chloride [15-17].

$$PhCOCH_2CN \xleftarrow{(MeCO)_2O} PhCOCH_2CH=NOH \xrightarrow{MeCOCl}$$

(13)   (14)

Moreover, nitrile formation is a general hazard as a result of the
base-sensitivity of 3-unsubstituted isoxazoles. In the case of
the specially reactive benzisoxazole (15), even simple heating
results in conversion to o-cyanophenol [38, 39]. The best procedure for the preparation of (15) is cyclization of the oxime
sulfonate (16) of salicylaldehyde [40] under mildly basic conditions involving sodium bicarbonate [20, 21].

(16)   (15)

A special problem associated with the synthesis of isoxazolium
salts for use as peptide reagents is the incorporation in the molecule of an anionic group, as in the zwitterionic sulfonate (6).
This is desirable so that the by-product of amide bond formation
will be soluble in water and thus can be easily separated from the
peptide. Unfortunately, chlorosulfonation of either 5-phenylisoxa-
zole (14) or the N-ethyl-5-phenylisoxazolium cation gives mixtures
of the m- and p-isomers, necessitating a separation prior to hydrolysis to the sulfonate in the synthesis [15-17] of (6). Although this
complication can be avoided with 5-p-tolylisoxazole (17), the
resultant zwitterion (18) is less effective than (6) as a peptide
reagent. Wilson and Burness [37] reported similar chlorosulfonation results with 5-phenylisoxazole (14) and also observed the

formation of the 3',4-bis(chlorosulfonyl) derivative (19) from
the p-tolyl compound (17).

Prior to the more recent studies, the standard methods for
the quaternization of 3-unsubstituted isoxazoles utilized primary
alkyl halides and dialkyl sulfates [8, 33]. Woodward, Olofson,
and Kemp [15-17, 20, 21] and Wilson and Burness [37] demonstrated

the value of primary trialkyloxonium ions [40] as extremely
efficient alkylating agents for this purpose. Triethyloxonium
fluoborate gives nearly quantitative yields of the N-ethylbenzisoxazolium cation (7), despite the low nucleophilicity of
benzisoxazole (15) and the great susceptibility of (7) to further
reaction. An interesting variation on the classic alkylation
techniques is the thermal conversion [15-17] of the methosulfate
(20) to the N-methyl zwitterion (21). Primary alkyl

dinitrobenzenesulfonates [37] and halides together with silver
fluoborate [41] have been used for alkylations, as well as 1,3-
propanesultone [42] in order to give the zwitterion (22).

In later work, it was demonstrated that the slight basicity
of the isoxazole ring permits a new type of alkylation, with
carbonium ions generated from branched alcohols in acidic media.
Eugster, Leichner, and Jenny [43] first postulated the intermediacy
of 3-unsubstituted isoxazolium salts to account for the N-substituted, ring-opened products they obtained from the treatment of
5-substituted isoxazoles with tertiary alcohols (and mesityl
oxide) in concentrated sulfuric acid. Woodward and Woodman
[23, 24] confirmed this possibility by the isolation of salts

of the cations and showed that N-t-butylisoxazolium perchlorates (23) could be prepared readily by mixing the
heterocycles with t-butyl alcohol

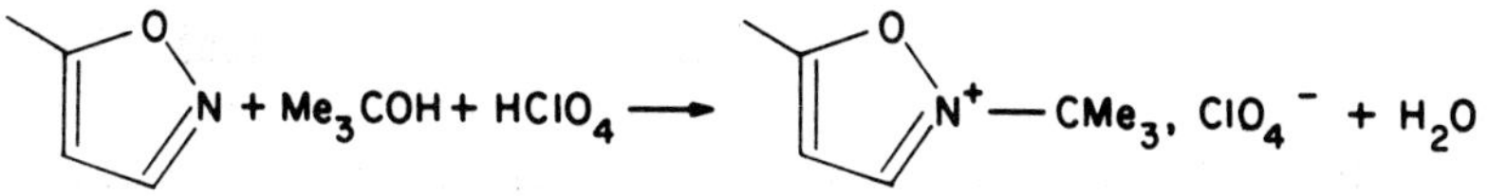

(23)

and perchloric acid. This S_N1 alkylation method has since been
demonstrated to be general for other alcohols that are efficient
sources of carbonium ions under acidic conditions [44]. Quaternization can be conducted in nonaqueous media [44] and, in some
cases, conversion of a 1,3-ketoaldehyde monoxime (24) to an
isoxazolium salt can be accomplished in one step, for example, by
addition of perchloric acid to a solution containing the alcohol
[45]. While considerable success has been achieved in the synthesis
of isoxazolium perchlorates, which are usually crystalline and
nonhygroscopic, precautions are required, since some of the
perchlorates are explosive [6, 37].

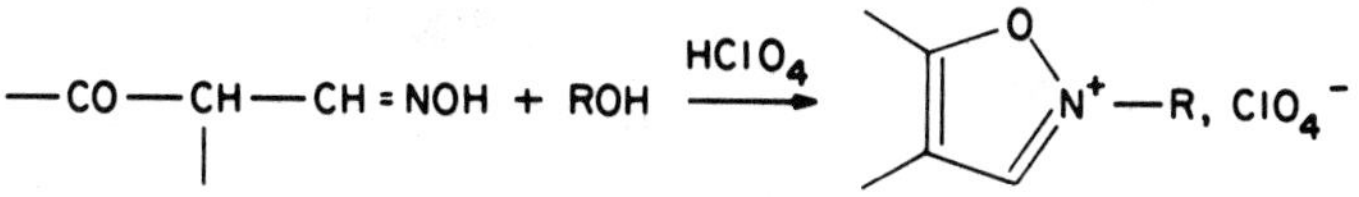

(24)

A variation on path A that eliminates the final quaternization
in the preparation of 3-unsubstituted isoxazolium salts is the
condensation of N-substituted hydroxylamines with hydroxymethylene
ketones. Although the condensation product 3-(N-hydroxyanilino)
acrylophenone (25) from hydroxymethyleneacetophenone and phenyl-
hydroxylamine had been reported to undergo hydrolysis in aqueous
acid [46], it was found that dehydrative cyclization of (25) to
the N-phenylisoxazolium cation (26) could be achieved in concentrated sulfuric acid [23, 24]. This approach should have general
value for the

PhCOCH = CHOH + PhNHOH ⟶ PhCOCH = CHNPh $\xrightarrow{H_2SO_4}$

$$(\underline{25})$$

$$(\underline{26})$$

preparation of N-arylisoxazolium salts because the N-substituted
hydroxylamines are easily accessible from the reduction of aromatic
nitro compounds. The method has been applied to the preparation
of the methyl-substituted derivatives (27) and (28) [47], as well
as the zwitterion (29) [23, 25]. Further refinements include an
extension to the N-alkylhydroxylamines and the demonstration that

$$, ClO_4^-$$

$$(\underline{27})$$

$$, ClO_4^-$$

$$(\underline{28})$$

$$(\underline{29})$$

both the condensation and cyclization reactions can be conducted in
a single step under mildly acidic conditions in nonaqueous media
[48].

—CO — CH — CHO + RNHOH + HClO$_4$ ⟶ N$^+$— R, ClO$_4^-$ + H$_2$O

III. REACTION WITH CARBOXYLIC ACID ANIONS

The investigation by Woodward and Olofson of the reaction between 3-unsubstituted isoxazolium salts (1) and carboxylic acid anions brought to light all the main features of the transformations that result in the formation of enol esters (2) [15-17]. The first stage of the reaction involves an abstraction of the proton from the 3-position of the heterocyclic cation, together with ring opening to produce an acylketenimine (30). Next, addition of the carboxylic acid to (30) gives an intermediate adduct, which is postulated to have the structure (31). Finally, O,O-acyl migration converts the initial adduct to the ester (2). Later studies have provided additional evidence in support of these conclusions and

have shed some further light on the mechanism. However, several details remain speculative, and the factors that are critical for the efficient production of enol esters (2) are not yet completely understood.

A. Acylketenimine Intermediates

The reactions of simple nucleophiles with 3-unsubstituted isoxazolium salts ($\underline{1}$) lead to ring-opening products similar to ($\underline{32}$), or a tautomer, in which the nucleophile has replaced the hydrogen originally on the carbon at the 3-position of the heterocyclic ring.

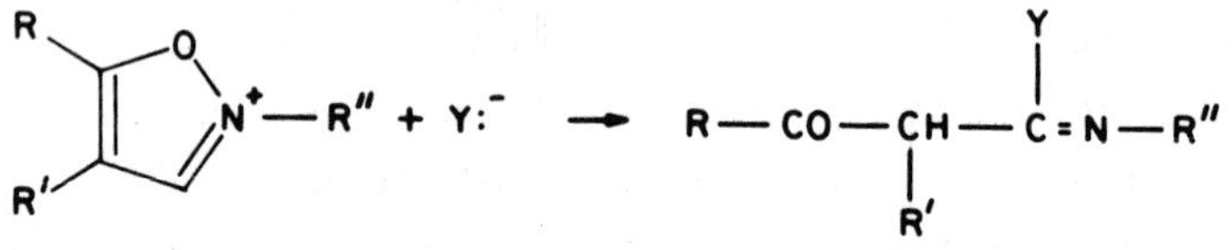

$$(\underline{1}) \qquad\qquad\qquad (\underline{32})$$

A mechanistic explanation (see Scheme 1) for such transformations was first offered by Kohler and co-workers [49, 50], who proposed that the reaction with hydroxide ion to give ketoamide ($\underline{33}$) involves nucleophilic addition at the 3-position of the isoxazolium cation ($\underline{1}$), followed by proton abstraction and ring opening. This postulate was refuted by demonstrating that the ring opening actually proceeded

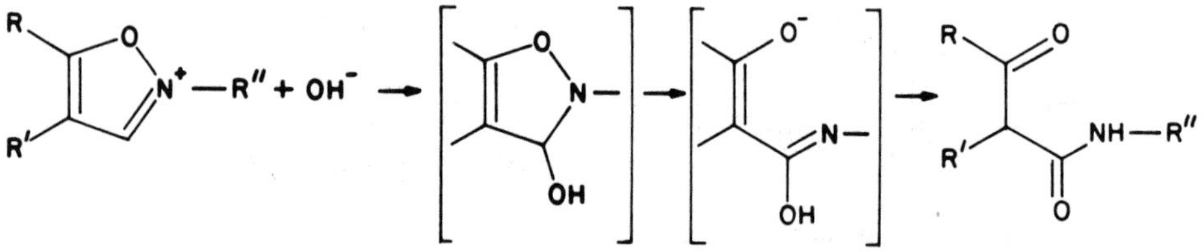

$$(\underline{1}) \qquad\qquad\qquad\qquad\qquad\qquad (\underline{33})$$

Scheme I

according to Scheme 2 with formation of acylketenimines ($\underline{30}$). The intermediates ($\underline{30}$) were detected in reactions with triethylamine by virtue of their characteristic absorption in the cumulene region of the infrared spectrum [15-17].

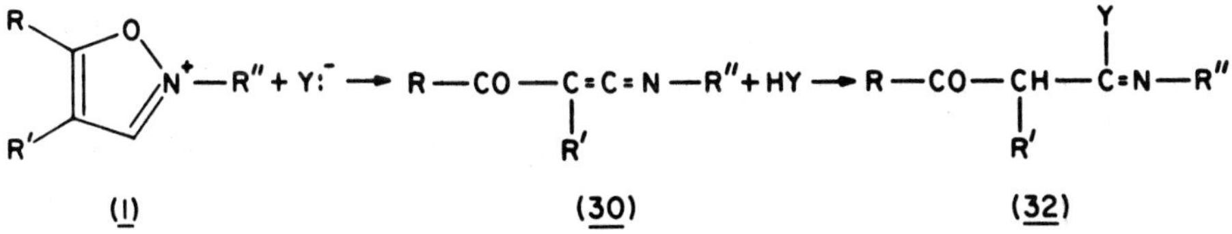

Scheme 2

Subsequently Eugster, Leichner, and Jenny [43] again invoked
a nucleophilic addition mechanism (see Scheme 3) to explain the
destruction of the isoxazolium cations generated in concentrated
sulfuric acid. However, Woodward and Woodman [23, 24] established
that the isoxazolium cations in question were, in fact, stable in

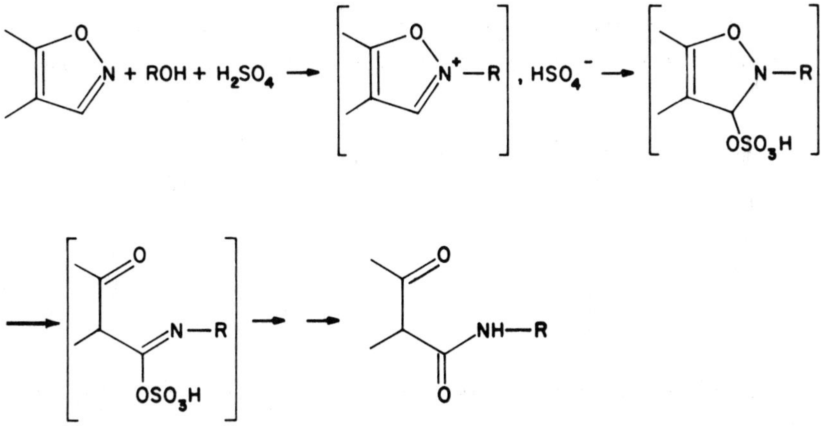

Scheme 3

sulfuric acid solution and that ring opening had occurred by the usual
mechanism upon neutralization in the work-up procedure.

Kemp [20, 22] obtained evidence for the same ketenimine mechanism
in the ring opening of the N-ethylbenzisoxazolium cation (7). Although
the corresponding cumulene (34) is especially unstable and cannot be
directly observed, kinetic results support its intermediacy as opposed
to the nucleophilic addition mechanism of Scheme 1. Specifically, the

(7)   (34)

ring opening of the cation (7) was found to be subject to general
base catalysis. However, with mixtures of bases, the product ratios
did not correspond to the contributions of the nucleophiles to the
observed rate, as would have been expected if rate-determining
nucleophilic addition had actually preceded ring opening. In addition, the isoxazoline (35), a model for the hypothetical intermediate
(36) in ring opening via prior nucleophilic addition, was found to be
relatively inert to attack by bases under the reaction conditions.

(35)   (36)

An investigation of the isoxazolium salts with branched quarter-
nizing groups laid to rest any uncertainty regarding the assignment
of the intermediate acylketenimine structures. Ring opening of the
N-t-butyl cations (37) with triethylamine gave N-t-butylacylketenimines (38) having sufficient stability to be both isolated and
characterized [23, 26]. Further, a variety of stable acylketenimines

(37)   (38)

with other branched groups on nitrogen have been obtained from the
ring openings of new isoxazolium salts [44]. Moreover, relatively
stable acylketenimines that possess a second substituent on the sp^2
carbon, but do not bear a branched alkyl group on the nitrogen, have
also been isolated [32, 45, 51], including the compounds (39), (40),
and the ring-fused cumulene (41).

Ph — CO — C = C = N — R
$\quad\quad\quad\quad$ |
$\quad\quad\quad\quad$ Me

(39): R = Ph

(40): R = CH$_2$Ph

$C{\displaystyle \atop \displaystyle ≈}$N — Et

(41)

The decomposition of acylketenimines in nonnucleophilic media
has not been closely scrutinized, but a polymerization is presumed
to be involved. In many cases, there is at least a qualitative
correlation between the ease of decomposition and the reactivity of
the acylketenimines in various addition reactions. It is tempting
to attribute the stability and low reactivity of acylketenimines
with N-t-butyl and other highly branched nitrogen substituents or
groups on the sp^2 carbon to a steric effect, such as one proposed to
account for the low reactivity of di-N-t-butylcarbodiimide [52].
However, the importance of inductive effects, as has been noted in
the case of simple N-alkylketenimines [53], must not be overlooked.
Qualitatively, the importance of this latter factor may be indicated
for acylketenimines by the instability of the N-benzhydryl compound
(42) in contrast to the stable N-α-methylbenzyl cumulene (43).

H-CO-C=C=N-R
$\quad\quad\quad$ |
$\quad\quad\quad$ Ph

(42): R = CHPh$_2$

(43): R = CHPh
$\quad\quad\quad\quad\quad$ |
$\quad\quad\quad\quad\quad$ Me

B. Ring-Opening Process

The ring opening of 3-unsubstituted isoxazolium salts may be
compared with the long-known reaction of the parent 3-unsubstituted
isoxazoles (44) by bases such as sodium ethoxide or sodium hydroxide [54]. The greater sensitivity of the isoxazolium salts,

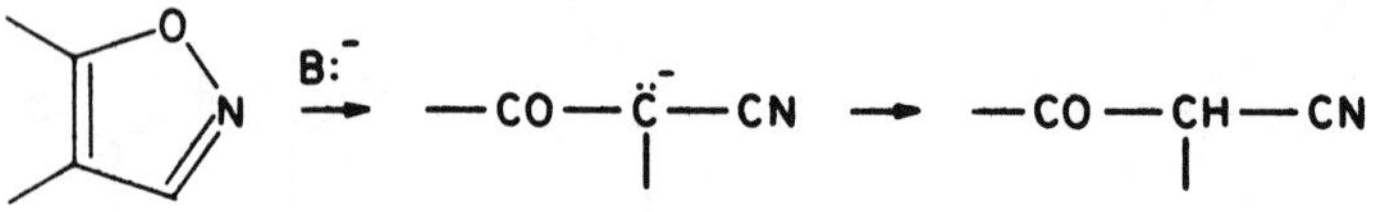

(44)

which react spontaneously with even weak bases such as acetate ion,
can be attributed to ylidic stabilization of the transition state
or intermediate (45) as a consequence of the positive charge on
nitrogen.

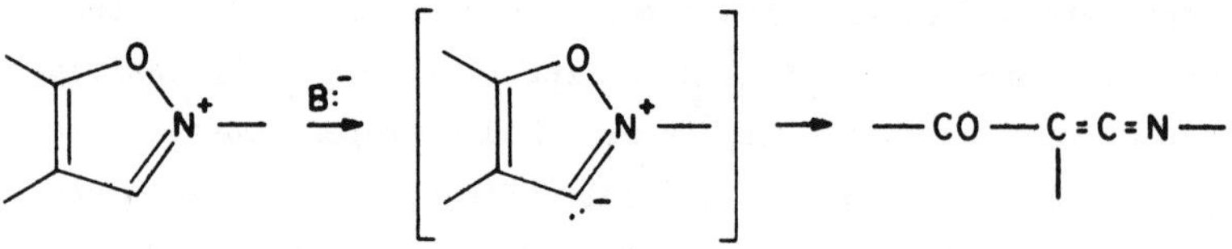

(45)

Ring openings conducted in the presence of deuterium oxide
[15-17, 23] failed to reveal deuterium incorporation upon nuclear
magnetic resonance (NMR) analysis of the recovered, unreacted isoxazolium salts. In a more sensitive NMR test for the reprotonation
of an ylide, there was no incorporation of hydrogen in the ring
opening in water of the deuterium-substituted benzisoxazolium
cation (46) [20, 21]. Thus if an ylide (45) is a full-fledged
intermediate, its ring opening is more rapid than the return to
starting material. The observation of a primary kinetic isotope

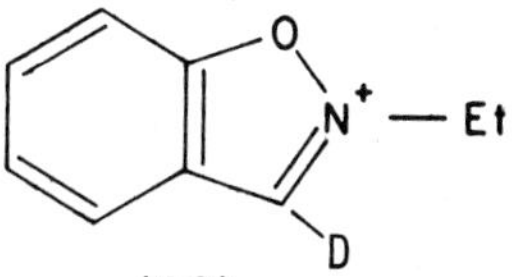

(**46**)

effect in the rate of reaction of (**46**) with bases supports the
abstraction of hydrogen from the 3-position in the slow step of
the ring opening. A very large rate enhancement for the reaction
of bases with the benzisoxazolium cations (**47**) and (**48**), which
have electron-withdrawing groups para to the oxygen atom, has been
interpreted as an indication of stabilization of developing negative
charge on the oxygen atom in the transition state. This result
suggests that, at least for these compounds, proton abstraction and
ring opening are concerted steps.

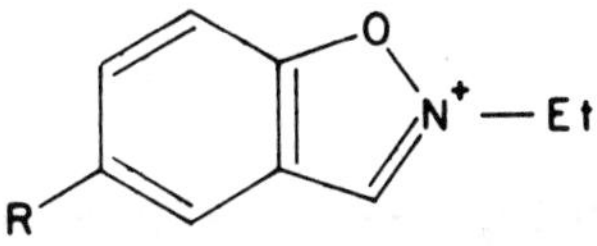

(**47**): R = NO$_2$

(**48**): R = SO$_2$Cl

The ring opening, regardless of whether it is concerted or
proceeds through an intermediate ylide, provides an interesting
case for speculation about the direction of electron flow. It was
argued that electron flow in the counterclockwise direction for
the ylide (**45**) would be preferred, as shown below. The reaction

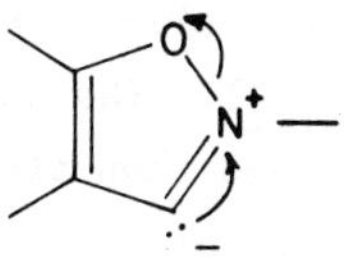

(**45**)

in this sense would be a simple trans elimination without disruption
of the π cloud of the ring [15-17].

C. Product Structure

On the basis of the reaction course with simple nucleophiles,
it is expected that the treatment of 3-unsubstituted isoxazolium
salts with carboxylate ions would result in addition of the carboxylate residue to the cumulene carbon of the intermediate acylketenimine. In the original work [1] with the N-methyl-5-phenylisoxazolium
cation (49), the structure of the product of the reaction with acetate
was assigned in this fashion as the iminoanhydride (50). However, the
initial product was found to rearrange to an isomer, formulated as the
imide (51).

$$Ph-CO-CH=C=N-Me + MeCO_2H$$

(49)

$$Ph-CO-CH_2-\overset{OCOMe}{\underset{|}{C}}=N-Me \longrightarrow Ph-CO-CH_2-CO-\overset{|}{\underset{Me}{N}}-COMe$$

(50)   (51)

Upon consideration of the various possibilities for such re-
arrangements, it is seen that there are three heteroatom sites to
which the acyl group might conceivably become attached to give rise
to any of the skeletons (52-54). The number of possible product
structures is further increased when the various tautomeric forms
and geometric isomers that can be envisioned for the different
candidate skeletons are taken into account in a systematic fashion.

$$\underset{(52)}{\overset{\displaystyle \overset{COMe}{\underset{|}{}}}{Ph-C-C-C-N-Me}}$$

$$\underset{(53)}{\overset{\displaystyle \overset{COMe}{\underset{|}{}}}{Ph-C-C-C-N-Me}}$$

$$\underset{(54)}{\overset{\displaystyle}{Ph-C-C-C-N-Me}}$$

In the modern reexamination of the product structures, spectroscopic evidence was employed, rather than chemical tests, in
order to avoid complications that might arise as a result of
possible facile interconversions between isomeric structures
[15-17]. The original assignment of the imide structure (51) to
the rearrangement product was confirmed by the absence of any
heteroatom-hydrogen or short-wavelength carbonyl absorption,
together with the presence of a simple acetophonone chromophore.
However, the initial product was found to be the enol ester (55),
instead of the proposed simple adduct (50). Here, infrared data
eliminated all candidates but (55) and (56), and the ultraviolet
spectra of model compounds indicated that structure (55) was
correct. Finally, the geometry about the double bond of (55)
was assigned on the assumption that the enol ester was formed by
an intramolecular 0,0-acyl migration from an initial adduct of the
type (52). In later work, no spectral evidence, including NMR data,
which is inconsistent with these structural assignments, has come
to light for the reactions of simple isoxazolium salts with carboxylic
acids. Only with the 7-hydroxybenzisoxazolium cation (11) is there
observed [31] a different outcome, in which, as a result of an
0,0-acyl migration, the usual type of enol ester (57) is converted
to the 3-acyloxy-2-hydroxy-N-ethylbenzamide product (58).

$$\underset{(55)}{\overset{\displaystyle MeCOO\diagdown \quad \diagup CONHMe}{\underset{Ph\diagup \quad \diagdown H}{C=C}}}$$

$$\underset{(56)}{\overset{\displaystyle PhCO\diagdown \quad \diagup OCOMe}{\underset{H\diagup \quad \diagdown NHMe}{C=C}}}$$

D. Postulated Mechanism of Enol Ester Formation

A very detailed picture has been proposed for the addition of carboxylic acids to acylketenimines [15-17]. Assuming addition of the acid to the cumulene carbon, as with simple nucleophiles, intra-molecular O,O-acyl migration would offer the most straightforward route to the product. However, a deuterium-incorporation experiment established that the CH_2 iminoanhydride tautomer (50) is not an intermediate in the reaction. The nonintervention of (50) was taken to indicate that the addition took place so as to provide directly an adduct with the requisite geometry for intramolecular rearrangement. Other modes of addition that would instead generate adducts such as (59) with the wrong geometry were considered less likely, since the most satisfactory route to enol ester (55) from this type of intermediate would be via (50). A coplanar, cyclic

addition of carboxylic acid to the S-cis conformation of the acyl-ketenimine was postulated as the most favorable process to give adducts (60) and (61), having the desired geometry for rearrangement to (55).

E. Evidence Concerning the Addition Process

Some rate data has been obtained on the actual addition step. Qualitative rate comparisons have shown acylketenimines react more rapidly with carboxylic acids alone in organic solvents than with the acid plus triethylamine, in accord with the proposed attack of

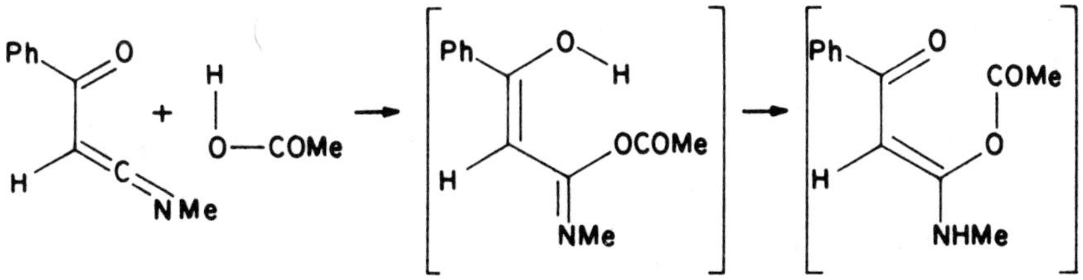

(60)   (61)

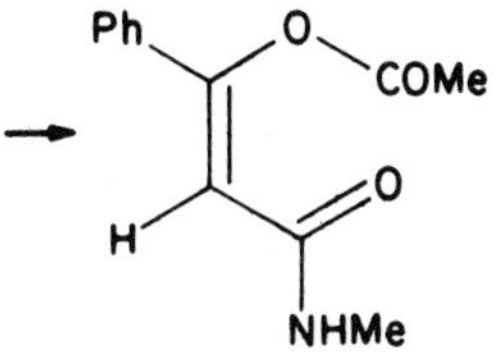

(55)

free carboxylic acid on the acylketenimine [15-17, 32]. The results
of later infrared studies with stable N-t-butylacylketenimines were
consistent with a reaction first order in each component, but diminished rates of reaction were observed in some polar solvents [23].
It should be noted that the related addition of carboxylic acids to
carbodiimides shows a similar solvent dependence, and acid dimers
have been implicated in that reaction [55]. Although the addition
step cannot be examined directly for the benzisoxazolium cation (7),
it was concluded that the addition of acetate ion to the derived
cumulene (34), or its conjugate acid, provided the most reasonable
explanation for the results of a product study in aqueous buffers
[20, 22]. Finally, the cis-fused ketenimine (41) was found to
react much more rapidly than the parent compound (62) [32]. Assuming
the S-cis conformation is involved in the addition reaction, the
rate difference could be an indication that the compound (62) does
not exist solely as the S-cis form in solution.

(41)   (62)

F. Evidence for an Intermediate--Side Reactions in Enol
    Ester Formation

    Side reactions during enol ester formation, while undesirable
for synthetic applications, have provided confirmation of the postulated intermediacy of an initial adduct that rearranges to the
observed product. The intermediates in the proposed mechanism
would be iminoanhydrides (31) or vinylogous anhydrides (63) and,
therefore, highly reactive acylating agents. As high-energy
acylating agents, such intermediates might be expected to decompose
by acyl migration to nitrogen to give imides (64) or by acylation
of carboxylic acids present in the reaction mixture to give

anhydrides. Such products are observed in the related additions
of carboxylic acids to simple ketenimines [56] or carbodiimides
[57, 58]. The failure to note the products of alternate modes of
decomposition of the postulated intermediates suggested that the
intramolecular rearrangement to give enol esters (2) is an extremely
efficient process [15-17].
The fact that the reaction pathway provides such an efficient, albeit
multistep, route to useful acylating agents, has been one of the
major practical advantages of the isoxazolium salt method of
peptide synthesis.

The first observation of a side reaction in enol ester
formation involved the detection of azlactones (65) in the reaction
of the N-ethylbenzisoxazolium cation (7) with N-acylamino acids
under relatively acidic conditions [20]. Later, it was found that
this result was not a peculiar characteristic of (7), but that
the side reaction was a general one for isoxazolium salts, although
enol ester formation was favored in a basic environment [23, 58].
Since the product esters (66) did not themselves undergo conversion
to azlactones under the reaction conditions, these results support
the intervention of a more highly reactive intermediate, as
proposed in the postulated mechanism. The simplest rationalization

(66)   (67)

(65)

of the acidity dependence of the course of reaction would be that
decomposition via the anion from the intermediate (67) preferen-
tially leads to the desired ester (66).

It has been argued that the ring-fused isoxazolium salt (12)
should be less subject to this side reaction on steric grounds.
Here, the ring fusion would favor enol ester formation by diminishing
the number of possible intermediate structures that do not have the
proper geometry for the desired intramolecular rearrangement.   Thus
it was felt that the rate of acyl migration via the iminoanhydride
(68) would be enhanced relative to the competing decomposition to
azlactone [32].   In accord with this view, spectral comparisons of

(12)   (68)

the extent of azlactone formation from hippuric acid and various
isoxazolium salts [23, 32] revealed much less azlactone with the
benz-fused cation (7), (12), or (69) than with the cations (70)
or, still worse, (71) and (72).   However, it may be that these

(7)   (69)

(70)   (71)   (72)

results do not actually indicate the influence of structural modifications on the competition between the alternative pathways for decomposition of the intermediates. Instead, the observed differences in the proportions of azlactone and enol ester might be explained on the basis of a medium effect [23]. That is, the slower the rate of addition of carboxylic acid to acylketenimine (step b), in comparison with the rates of ring opening (step a) and decomposition of the intermediate (steps c and d), the greater will be the acidity of the medium in which the intermediate is decomposing.

$$\begin{array}{c}\text{isoxazolium}\\\text{cation}\end{array} + RCO_2^- \xrightarrow{\text{step a}} \text{ketenimine} + RCO_2H \xrightarrow{\text{step b}} \text{intermediate}$$

step c / \ step d

enol   azlactone
ester

In view of the established sensitivity of the relative rates of steps c and d to acidity [20, 23, 59], it would be expected that a relatively fast addition step would favor enol ester formation, while a rate-determining addition would lead to a larger proportion of azlactone. Those ring-fused cations (7), (66), and (69) giving little azlactone are, in fact, also cases where the addition step is relatively rapid. Moreover, the N-t-butyl cations, (71) and (72), forming the greatest amount of azlactone are the ones for which the derived acylketenimine undergoes slow addition. Therefore, as a result of differences in the rates of addition, the acidity of the medium may influence, or even exert a predominant influence on, the outcome of such spectral tests. This same effect has been shown to be a crucial factor in the avoidance of racemization during peptide synthesis with the zwitterionic reagent (6) (see Sec. VII, Racemization).

In the case of N-t-butylacetylketenimine (73), for which the addition reaction is slow, the formation of enol esters does not proceed with such high efficiency. Recent work has implicated other modes of decomposition of such a reactive intermediate to

afford hydrolysis products or N-t-butylamides (74) of the starting
acids [28, 29]. The acidity of the medium does not seem to be a
controlling factor here, in contrast to the azlactone side reaction,
since the yield is not improved in basic media. A second type of
high-energy intermediate, which fails to convert to enol ester,
perhaps because it has the wrong geometry as in (75), could be
responsible for these side reactions. Consistent with this view,
when water is present in the reaction medium, hydrolysis products
are favored, as would be expected for interception of such an
intermediate. By conducting the reactions in wet acetonitrile and
using excess acylketenimine to combine with the acid presumably
liberated by the hydrolysis of the second intermediate, the yields
of enol ester may be improved. According to this interpretation,
in the absence of hydrolysis, the second intermediate decomposes
to (74). The simplest precursor of (74) would be the imide (76),
which could be formed from an adduct with the wrong geometry by
O,N-acyl migration. The assumption of spontaneous decomposition of
the imide (76) presents a problem with this argument because in
one case (see Sec. IV), the isolation of the derived imide has been
claimed [60]. Furthermore, the reason why the N-t-butyl compounds

are subject to this problem has not yet been clarified. A survey of
N-t-butyl acylketenimines has shown that the side reaction is a
fairly general one [29]. Even the ring-fused cumulene (77), for
which one would anticipate an efficient conversion to enol ester on
the basis of an earlier argument [32], does not escape this

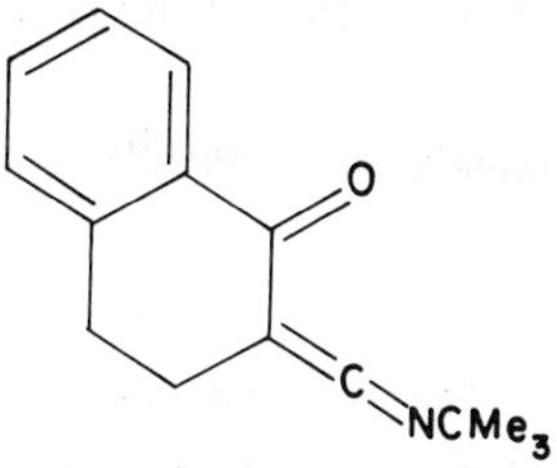

(77)

difficulty. Of all the acylketenimines examined to date, the
compound (78) has the least efficiency in the formation of enol
esters. With (78), even intermolecular interception of the

$$Ph-CO-\underset{\underset{Me}{|}}{C}=C=N-CMe_3$$

(78)

intermediate by the carboxylic acid reactant, so as to generate an
acid anhydride, competes with the desired pathway [29].

IV.   ENOL ESTER STABILITY AND THE IMIDE REARRANGEMENT

A major consideration in attempts to design improved isoxazolium salt reagents has been the elimination of the decomposition of
the enol esters (2) to imides (79). Since the imides from

$$\underset{\underset{R'}{|}}{R-C}=\overset{\overset{OCOR'''}{|}}{C}-CONH-R''$$

$$R-CO-\underset{\underset{R'}{|}}{CH}-CO-\underset{\underset{R''}{|}}{N}-COR'''$$

    (2)   (79)

simple ketenimines were known to react with nucleophiles [56], it
was expected that the rearrangement products (79) might themselves
be acylating agents, although of lower reactivity than the enol
esters (2). However, the imides (79) would be less useful for
peptide synthesis because of the possibility of a second mode of
cleavage that would not lead to the desired products.

$$R-CO-\underset{\underset{R}{|}}{CH}-CONH- \quad + R'''CONHR''\xleftarrow{H_2N-} (79)\xrightarrow{H_2N-} R-CO-\underset{\underset{R'}{|}}{CH}-CONHR'' + R'''CONH-$$

Some results suggested that the rearrangement (2)⟶(79) might
be a yield-limiting factor in peptide bond formations with the
reagent N-ethyl-5-phenylisoxazolium-3'-sulfonate (6) [15-19]. It
was initially observed that the N-methyl enol esters rearranged to the
corresponding imides under mild conditions, and N-methylisoxazolium salt
were found to give lower yields of peptides than the N-ethyl compounds. The improved yields with N-ethylisoxazolium salts could
be an indication that the rearrangement was slower with the N-ethyl
group. However, it was also found that if addition of the amine
component to the solutions of enol ester derived from (6) was
delayed, the yields of peptides were reduced. Therefore decomposition of even the N-ethyl enol esters did seem to take place under
the conditions of peptide synthesis. A reagent totally resistant
to this side reaction might, then, be expected to give still higher
yields of peptides than are obtained with (6). Furthermore, completely stable enol esters can provide the peptide chemist with
the option of isolation, purification, and storage of the intermediate acylating agents for later synthetic use.

One strategy considered to retard the rate of enol ester decomposition might be to modify further the nature of the group on
the isoxazolium salt nitrogen atom. The study of the N-phenylisoxazolium salts was then undertaken by Woodward and Woodman [23, 25] in
the hope that the steric and electronic influences of the aromatic
substituent would result in a reduction in the nucleophilicity of

the nitrogen atom in the corresponding enol esters (80). Unfortunately, rearrangement to the corresponding imide (81) actually is extremely facile in this case, even under the mildly basic conditions used for the preparation of the enol esters. The extreme base-sensitivity of the N-phenyl esters (80), compared to the N-alkyl compounds, has been interpreted as an indication that the aromatic group lowers the barrier to rearrangement by stabilization of the intermediate anion (82).

$$
\underset{(80)}{\overset{\text{OCOR}}{\underset{|}{-C}=CH-CONHPh}} \longrightarrow \left[\underset{(82)}{\overset{\text{OCOR}}{\underset{|}{-C}=CH-CO-\bar{N}-Ph}}\right] \longrightarrow \underset{(81)}{-CO-CH_2-CO-\underset{\underset{Ph}{|}}{N}-COR}
$$

Greater success in blocking the rearrangement of enol esters was achieved with the isoxazolium salts bearing the bulky N-t-butyl group [23, 27-29]. The N-t-butyl-β-acyloxycrotonamides (83) from the isoxazolium cation (72) were found to be sufficiently stable for isolation and storage, and could even be subjected to treatment with tertiary amines without decomposition. Shepard, Halczenko, and Cragoe [60] reported that the ester (84) from 3,5-diamino-6-chloropyra-

$$
\underset{(72)}{\overset{\text{Me}}{\underset{}{\boxed{}}}N^+{-}CMe_3} \xrightarrow{RCO_2^-} \longrightarrow \underset{(83)}{\overset{\text{OCOR}}{\underset{|}{Me}-C}=CH-CONHCMe_3} \xrightarrow{\;\;\times\;\;} \underset{\underset{CMe_3}{|}}{MeCOCH_2CONCOR}
$$

zinecarboxylic acid isomerized to the imide (85) on exposure to sodium methoxide in dimethylformamide, and it was concluded that the resistance to rearrangement of the N-t-butyl enol esters is the result of a large kinetic barrier rather than an unfavorable equilibrium.

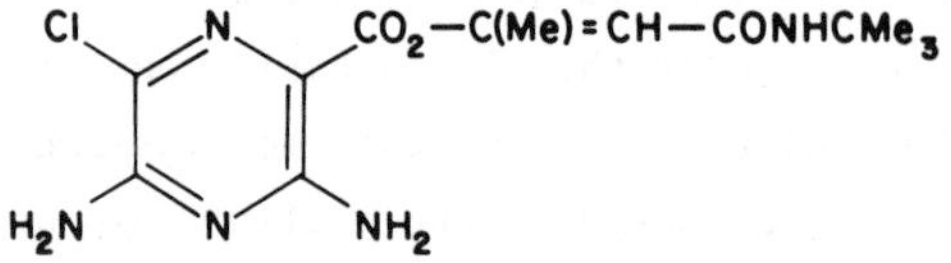

Cl
N
CO2—C(Me)=CH—CONHCMe3
H2N
N
NH2

(84)

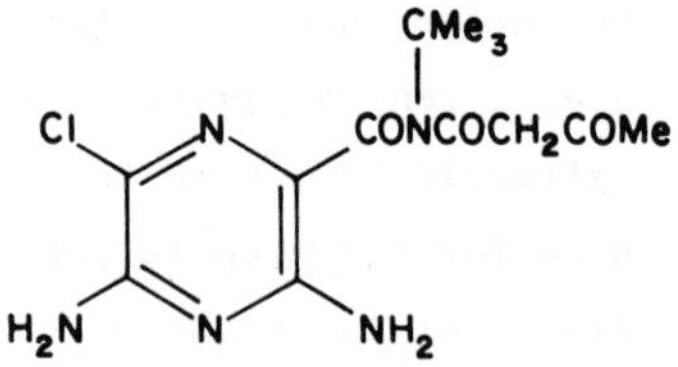

CMe3
Cl
N
CONCOCH2COMe
H2N
N
NH2

(85)

A second approach that has led to stable enol esters has been based on benzisoxazolium salts. Kemp and Woodward [20-22] found that the acylsalicylamides (86) from the N-ethylbenzisoxazolium cation (7) also were sufficiently stable for isolation and storage. In this instance the stability of the esters may be attributed to an unfavorable steric interaction in the most stable, planar, hydrogen-bonded conformation of the rearrangement product (87).

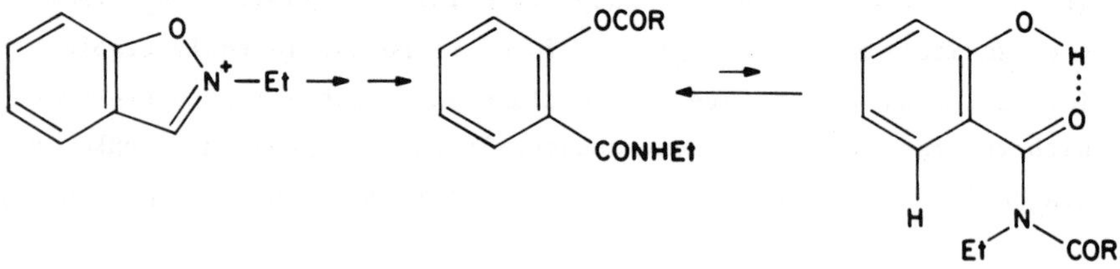

O
N+—Et
OCOR
CONHEt
O
H
O
H
N
Et
COR

(7) (86) (87)

The finding that with N-unsubstituted acylsalicylamides the equilibrium favors the corresponding imides, which are not subject to this destabilizing interaction, adds evidence in support of this view. A similar stability is observed [30] for enol esters derived from the dichloro-substituted N-ethylbenzisoxazolium cation (10). Decomposition under basic conditions have been reported for the esters of N-benzyloxycarbonylamino acids from the unsubstituted cation, and the reaction may proceed via the unstable imide, as shown below [61].

In their development of the tetramethyleneisoxazolium cations (12) as peptide reagents, Olofson and Marino [32] considered that destabilization of imides relative to enol esters, similar to the situation with the benzisoxazolium salts, might be achieved by other steric interactions. In particular, it was felt that a sub-stituent in the 4-position of the heterocyclic ring might lead to a conformational restriction appropriate for this purpose. The ring-fused isoxazolium cations, intended to provide efficient enol ester formation, incorporate a substituent in the desired position and, as was hoped, the derived esters, (88) and (89), were found to be resistant to O,N-acyl migration under a variety of conditions.

(88)   (89)

V. ENOL ESTER ACYLATING AGENTS

The enol esters derived from 3-unsubstituted isoxazolium salts
were expected to react as acylating agents toward nucleophiles,
since the leaving group would be the anion of a β-ketoamide [15-17].
Some additional stabilization of the leaving group might be derived
from intramolecular hydrogen-bonding to the forming O-anion during
attack by the nucleophile (:Y⁻).

The level of reactivity of the intermediate enol esters from
N-ethyl-5-phenylisoxazolium-3'-sulfonate (6) was found to be sufficient to allow acylation reaction to be run overnight in organic
solvents with exact equivalents of all reactants (0.1-0.2 M concentrations). It was estimated from some acylation rates that a reaction time of 2 to 3 days would be necessary for the stable esters
from the N-ethylbenzisoxazolium cation (7) [20]. The rather low
reactivity of the N-ethylsalicylamide O-esters from (7) is further
illustrated by the coupling of the ester of N-benzyloxycarbonyl-
glycyl-L-phenylalanine with ethyl glycinate. Here, the rate
constant in dimethylformamide was slightly less than 1/20 of the
value for the same reaction with the p-nitrophenyl ester [62].
Enhanced reactivity, as a result of the electron-withdrawing substituents, was expected for the enol esters made from the dichlo-
robenzisoxazolium cation (10) of Rajappa and Akerkar [30], but

no rate studies have been reported to date. The nitro-substituted
cation was also examined, but the derived enol esters were so
labile that their isolation was skipped [63]. The β-acyloxycroton-
amides from the 5-methyl-N-t-butylisoxazolium cation (72) may be
roughly equivalent to the p-nitrophenyl esters, since the coupling
of the enol ester of benzyloxycarbonylglycine with benzylamine in
acetonitrile was about ten times faster than the comparable reaction
using the ester from the benz-fused cation (7) [20, 23]. The
anionic group of the phenolic esters (90) from the 2-ethyl-7-
hydroxybenzisoxazolium cation (11) may serve as an internal basic
catalyst to accelerate aminolytic reactions [31, 64]. Couplings
with these reagents, carried out with the salts of amino acids
and peptides in the presence of the base tetramethyl-guanidine
(TMG), have times for 50% reaction (0.2 M reactants in dimethylformamide) ranging from 1/2 min (Z-Gly-OH ester + Gly$^-$, TMG$^+$) to
50 min (Z-L-Val-OH ester + L-Val$^-$, Me$_4$N$^+$).

(90)

The study of the N-t-butyl-β-acyloxycrotonamide (84) from
3,5-diamino-6-chloropyrazine carboxylic acid may shed some light
on the limits of applicability of enol esters from isoxazolium
salts as acylating agents [60]. The reagent (84) under all conditions examined failed to acylate very weak nucleophiles such
as 3-amino-1,2,4-triazole and 2-aminobenzimidazole. At the other
extreme, diminished yields were observed in acylations of very
strong bases (guanidine, aminoguanidine, sodium alkoxide, or
sodium urea) in polar solvents, perhaps because these conditions
favor decomposition of the enol esters to imides.

VI.   ACYLATION BY-PRODUCTS

An important practical consideration in peptide synthesis is
the ease of isolation of the desired product in a high state of
purity from the reaction mixture. The original isoxazolium salt
reagent (6) was designed to facilitate product isolation by providing an ionic ketoamide by-product (91). Moreover, any consumed

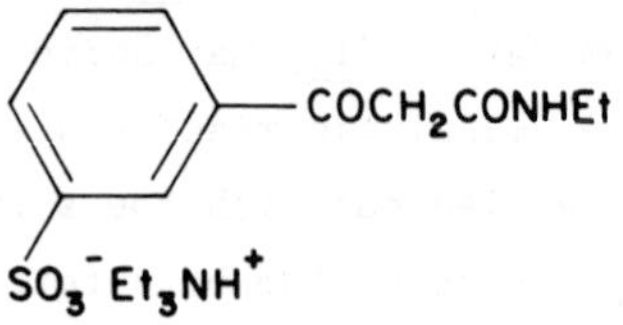

$$COCH_2CONHEt$$

$$SO_3^- Et_3NH^+$$

(91)

enol ester or side-reaction products connected to the sulfonated
framework would be removed from the peptide along with (91). With
the benzisoxazolium salts, the by-products are phenols that, in
general, may be separated by appropriate acid-base extraction
schemes. The ketoamide (92) from acylations with the β-acyloxy-
N-t-butylcrotonamides is freely soluble in both carbon tetrachloride
or water, so trituration of the product peptides with either
solvent can be used to eliminate the by-product.

$$MeCOCH_2CONHCMe_3$$

(92)

The by-product (92) presents a special danger for peptide
synthesis with the N-t-butyl reagent (72). With enol esters of
hindered acids, the rate of coupling may become slow enough so

that condensation of the amine component with (92) becomes a significant competing reaction [29]. In the limiting case of acylation of the primary amine benzylamine by the enol ester (93) of pivalic acid, this side reaction leading to (94) consumed approximately half of the amine. Although the desired acylation

$$\underset{(93)}{Me-\overset{OCOCMe_3}{\overset{|}{C}}=CHCONHCMe_3} \quad + \quad PhCH_2NH_2 \longrightarrow Me_3CCONHCMe_3 \quad + \quad \underset{(92)}{MeCOCH_2CONHCMe_3}$$

$$\underset{(92)}{MeCOCH_2CONHCMe_3} \quad + \quad PhCH_2NH_2 \longrightarrow \underset{(94)}{Me-\overset{NHCH_2Ph}{\overset{|}{C}}=CHCONHCMe_3}$$

reaction is several thousand times faster than the condensation with the ester of an unhindered acid, the side reaction may be a source of impurities, as well as a yield-limiting factor. It should be stressed that this problem is an unusual one for the isoxazolium cation (72) and applies neither to the benzisoxazolium cation reagents nor to the original 5-aryl zwitterionic reagent (6). As a model to examine the sensitivity of the benzoyl group of the by-product of peptide synthesis from (6) to such condensations, the enol ester (95) of pivalic acid was prepared. Only a few per-

$$\underset{(95)}{Ph-\overset{OCOCMe_3}{\overset{|}{C}}=CHCONHCMe_3}$$

cent of condensation product could be detected in the coupling of (95) with benzylamine, indicating that this side reaction would not be important in normal circumstances with (6) or other 5-arylisoxazolium salts.

VII. RACEMIZATION

In the original investigation of the synthetic utility of the reagent N-ethyl-5-phenylisoxazolium-3'-sulfonate (6), no racemization was detected in the Anderson test [65] in the coupling of benzyloxycarbonylglycyl-L-phenylalanine (96) with glycine ethyl ester under optimum conditions. The fragmentation of acylating agents to give racemization-prone azlactones (65) has been shown to be responsible for the loss of chirality with N-acylamino acids and peptide acids such as (96) [66-70]. Therefore the favorable test result implied that neither the enol ester (66) nor the intermediate adduct (67) underwent significant decomposition to azlactone (65) under the usual reaction conditions. Since the enol esters (66) are acylating agents of only moderate reactivity, it was not expected that they would be highly vulnerable to the azlactone side reaction. However the more highly energetic intermediates (67) must be more susceptible to fragmentation. Therefore the absence of racemization was taken as an indication that the intramolecular rearrangement of (67) to enol ester was much more rapid than azlactone formation. Thus the isoxazolium salt method seemed to offer an attractive solution to the problem of racemization in peptide synthesis by providing an efficient, safe pathway to chiral enol esters.

Further tests by other workers established that the zwitterionic reagent (6) actually did not completely eliminate racemization in many cases. Williams and Young [71] found the coupling of benzoyl-L-leucine with glycine ethyl ester in acetonitrile gave no racemization. However, some racemization was detected when the hydrochloride of the amine was used, while the amount increased with nitromethane as the solvent. Still worse results were obtained by Weygand [72, 73] in the preparation of methyl trifluoroacetyl-L-valyl-L-valinate (> 60% racemization), although only a low level of racemization (3.2%) was detected in the coupling of benzyloxycarbonyl-L-leucyl-L-phenylalanine with t-butyl L-valinate. Bodanszky and Conklin [74]

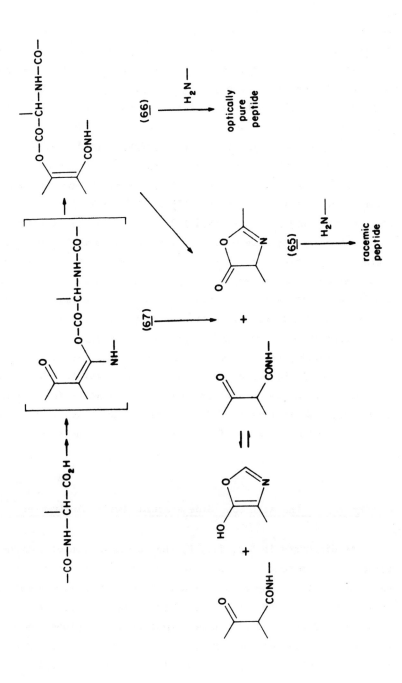

observed variable racemization (4-20%) in the formation of acetyl-
L-isoleucylglycine ethyl ester, depending on the tertiary base used
to liberate glycine ethyl ester from the hydrochloride salt. In
their NMR test for racemization in couplings between alanine and
phenylalanine, Halpern, Chew, and Weinstein [75] found that (6) and
carbonyldiimidazole both were superior to carbodiimide reagents.
The extent of racemization was below the level of detection (3%)
for most acylations with (6) and the acylamino acids, but some
racemization (5-15%) was seen with carboxyl terminal alanine or
phenylalanine peptide acids [76]. Only slight racemization was
reported by Izumiya and Muraoka [77] in the preparation of
benzyloxycarbonylglycyl-L-alanyl-L-leucine benzyl ester with (6), but
even better results were obtained with the reagents N-ethoxycarbonyl-
2-ethoxy-1,2-dihydroquinoline, isobutylchloroformate, or N-hydroxysuccinimide plus dicyclohexylcarbodiimide. Finally, Kemp [78] has
applied a sensitive isotopic dilution assay for racemization to the
Anderson and Young tests with (6) and found some racemate (2.1% and
8.5%, respectively), even under optimum conditions.

The results of these tests show that the isoxazolium salt (6)
does not afford the guarantee against racemization that had been anticipated on the basis of earlier work. Later studies (discussed below)
have clarified the extent of the hazard of racemization via azlactone
during both the formation of the enol esters and their reaction with
amines.

A. Control of the Azlactone Side Reaction During Activation

As discussed in Sec. III, F, the decomposition of intermediate
adducts to azlactone in competition with rearrangement to enol ester
was found to be a general side reaction in studies with a variety of
isoxazolium salts and N-acylamino acids. However, spectral tests
with hippuric acid did not reveal significant azlactone formation
with the original peptide reagent (6). The explanation for the
anomalous result with (6) was found to involve the acidity

dependence of the competition between azlactone and enol ester
formation [23, 58]. That is, the low rate of solution of the
relatively insoluble zwitterion (6) fortuitously maintains a
relatively basic reaction medium. This factor then favors the
conversion of the intermediate to enol ester, rather than to
azlactone. Most other isoxazolium salts dissolve more rapidly,

$$
(\underline{6}) \ + \ \text{acid anion} \xrightarrow{\text{slow}} [\text{ketenimine} + \text{free acid}]
$$

$$
[\text{intermediate}]
$$

$$
\text{azlactone} \qquad \text{enol ester}
$$

the concentrations of free acid and ketenimine increase, and the
intermediate in the more acidic medium gives a greater proportion
of azlactone. This interpretation was confirmed by an experiment
in which the ketenimine from the reaction of (6) with triethylamine
was combined with free hippuric acid. As expected, considerable
azlactone was produced in the acidic medium.

The hippuric acid spectral tests established that use of (6)
for activation of acids subject to the azlactone side reaction
provides a route to enol esters that keeps the attendant danger of
racemization to a minimum. Nevertheless, the recommended [15, 18,
19] reaction conditions (enol ester formation in acetonitrile at
0°C) must be followed, since variations on the technique, which
might alter the rate of solution of the reagent, can lead to
increased amounts of azlactone [23, 59].

With other isoxazolium salts, basic reaction media can be
selected that exert a comparable control over the azlactone side
reaction. The use of benz-fused isoxazolium salts in a two-phase
system buffered with pyridine gives high yields of the O-acylsali-
cyclamide esters to the exclusion of azlactone [20, 22, 31].
Somewhat more basic conditions--for example, 2-picoline as a
solvent--were found to be necessary during the formation of the
enol ester of hippuric acid from the N-t-butyl-5-methylisoxazolium
cation (72). The preliminary results with the 4,5-tetramethylene-

isoxazolium cation (12) indicate the cyclohexyl ring fusion may
obviate the requirement for such basic buffers [32], regardless of
whether the efficiency of enol ester formation with this system
is the result of an enhanced rate of rearrangement or of a rapid
addition step (see Sec. III, F).

The spectral assay with hippuric acid has been considered a
particularly sensitive test that exaggerates the problem of
azlactone formation [32]. Conditions resulting in the elimination
of azlactone in this test should, then, be safely applicable to the
formation of chiral enol esters in common racemization-prone cases.
Of course, this assumption may not be valid for other especially
critical situations. For example, with the Young test, a small
amount of azlactone was detected in the preparation of the
β-acyloxycrotonamide enol ester from the N-t-butyl reagent (72)
with the basic solvent 2-picoline [29].

B. Decomposition of Enol Esters to Azlactone

The careful investigations by Kemp and co-workers have implicated the fragmentation of enol esters of acylamino acids to
azlactones during the coupling reaction as a problem, which, in fact,
may also be a main source of racemization encountered with isoxazolium
salts. The N-ethylsalicylamide enol esters (97) formed from the
Anderson test acid (96) were found to undergo racemization in the
presence of triethylamine, with a linear rate dependence on the
concentration ratio $[Et_3N]/[Et_3NH^+]$. The most straightforward
interpretation of this result would involve a rate-determining
fragmentation to the azlactone (98) of the amide anion (99), formed
in a rapid prior equilibrium [62]. An extension of the isotopic
dilution technique has permitted measurement of the isotope effect
on the rates of racemization with α-deuterio-labeled benzoyl-L-leucine
in several coupling reactions [79]. The coupling of the N-ethylsalicylamide enol ester of this acid with ethyl glycinate was
carried out in the presence of a high concentration of

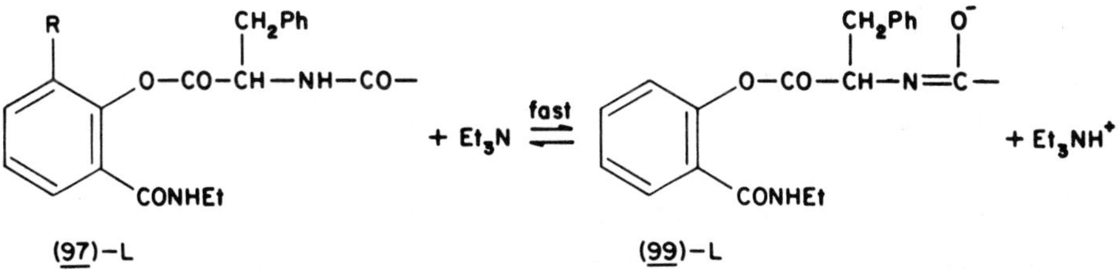

R

CH$_2$Ph

O—CO—CH—NH—CO—

CONHEt

(97)—L

+ Et$_3$N $\overset{fast}{\rightleftharpoons}$

CH$_2$Ph   O$^-$

O—CO—CH—N=C—

CONHEt

(99)—L

+ Et$_3$NH$^+$

R = H, OMe, Me

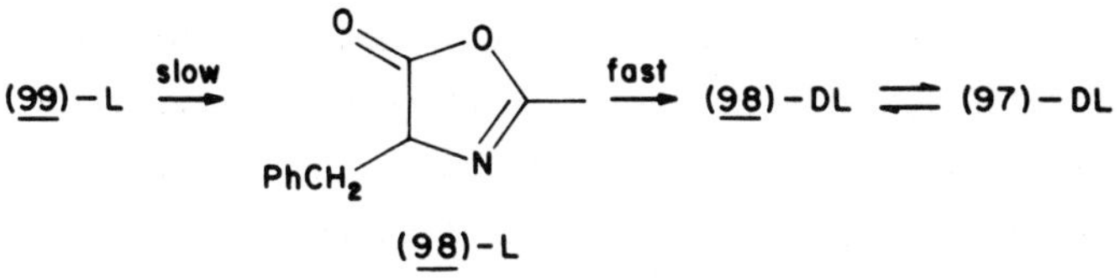

(99)—L $\overset{slow}{\longrightarrow}$

PhCH$_2$

(98)—L

$\overset{fast}{\longrightarrow}$ (98)—DL $\rightleftharpoons$ (97)—DL

triethylammonium ion to suppress the above type of specific-base catalyzed racemization. Still, some racemization, attributed to general base catalysis, resulted. An isotope effect of unity was also observed, in support of azlactone formation as opposed to direct enolization at the α-carbon. Further confirmation of the postulated decomposition of the enol esters to azlactones was provided by the direct spectral detection of the azlactone from hippuric acid when the N-t-butylacyloxycrotonamide enol ester was treated with triethylamine [59]. As proven by control experiments, azlactone is favored at equilibrium over the more reactive esters from the N-t-butylisoxazolium cation (72).

Although the above evidence establishes the hazard of racemization via azlactones for chiral enol esters in basic media, the crucial issue for peptide synthesis is the ratio of rates for the coupling and racemization reactions. Racemization is clearly competitive with couplings of purified O-acyl-N-ethylsalicylamide esters from (7) in the Anderson and Young tests, which gave variable amounts of racemate (1.0-1.4% and 7.0-13.8%, respectively)

under various conditions [78]. By way of comparison, the use of
optically pure p-nitrophenyl ester in the Anderson test gave only
about half as much racemate with dimethylformamide as the solvent.
The greater vulnerability of the phenolic esters from (7) to
racemization versus the p-nitrophenyl esters, was also
determined by comparisons of the rate constants for specific
base-catalyzed racemization and coupling with the two types of
acylating agents [62].

The possibility that racemization observed with the original
zwitterionic isoxazolium salt (6) may involve decomposition of the
intermediate enol ester, as opposed to azlactone formation during
the activation step, was supported in a modified preparation of the
Anderson test peptide [78]. The enol ester solution was concentrated
and the residue washed with ether to eliminate any azlactone that
might have been produced in competition with enol ester formation.
The amount of racemate (4.0%) from subsequent coupling reaction was
comparable with that obtained without the azlactone separation step.
Assuming the observed racemization does all result from ester
fragmentation, then the enol esters from (6) would seem to be still
more vulnerable to the side reaction in competition with coupling
than are the salicylamide esters from (7). This conclusion may be
open to question, however, since it seems to be based on the
assumption that the azlactone is favored at equilibrium over the
enol ester from (6). If this were not the case, azlactone formed
during the activation might give rise to racemic enol ester, which
could then account for the result.

C. Racemization-Resistant Esters from Isoxazolium Salts

The phenolic esters (100) of Kemp and Chien [31] derived from
the 2-ethyl-7-hydroxybenzisoxazolium cation (11) combine the desirable features of the isoxazolium salt approach to peptide synthesis
with a valuable modification in the structure of the ultimate

acylating agent. The intramolecular rearrangement pathway of the
isoxazolium salts permits the avoidance of azlactone and the
efficient, direct formation of chiral esters from acylamino acids
or peptide acids under properly controlled reaction conditions. The
esters (100) differ from acylating agents made with other isoxazolium
salts in that they favor aminolysis relative to racemization. (For
reviews of other racemization-resistant esters, see Refs. 80-82.) As
mentioned in Sec. V, it is proposed that the phenolate anion of (100)
accelerates the rate of aminolysis, while no such rate enhancement is
operative for intramolecular attack by the oxygen nucleophile in the
acyl group that leads to azlactone. Furthermore, specific base
catalysis of azlactone formation should be retarded for (100), which
already bears a negative charge. Thus the new esters provide a
rational strategy to eliminate racemization during both the
activation and coupling stages of peptide synthesis.

azlactone   (100)   amide

The resistance of the esters of the type (100) to fragmentation
to azlactone is demonstrated by the finding of only a low amount of
racemization (3.5%) upon exposure of these compounds to triethylamine (0.4%) in dimethylformamide for 12 hr [31, 78]. The practical
result is that couplings may be done with just traces of racemization.
In the Anderson and Young tests, the new phenolic esters at 0°C gave
results equivalent or, under strongly basic conditions, superior to

the azide method. Owing to the stability of the esters (100) in
basic media, the basic salts of amino acids and peptides may be
employed as the amine component in acylations, while the extent of
racemization remains low (< 1%). Racemization is observed to
increase at higher temperatures; this is not surprising in view of
the gross differences in transition state structure for coupling
versus racemization. Under optimum conditions the isoxazolium salt
reagent (11) appears to offer one of the best approaches yet
discovered for racemization-free peptide synthesis.

VIII. APPLICATIONS TO PEPTIDE SYNTHESIS

A. Use of N-Ethyl-5-phenylisoxazolium-3'-sulfonate (6)

In the decade since the publication of the original communication [18] illustrating the synthetic utility of the isoxazolium salt
method of peptide synthesis, the zwitterionic reagent (6) has seen
considerable use by peptide chemists. While the related p-tolyl
(18) and N-methyl (21) compounds also give good yields of peptides
in test couplings, they are not as attractive as the more effective,
commercially available reagent (6). The zwitterion (6) was first
manufactured by Pilot Chemicals (Watertown, Mass.), and is currently
also sold by Aldrich Chemical Co., Pierce Chemical Co., Regis Chemical Co., Schwarz-Mann, Sigma Chemical Co., and Wateree Chemical Co.
[83].

The detailed procedure for the use of (6), summarized below, is
given in Ref. 18. In brief, the enol ester formation (activation
step) is conducted by stirring the reagent in a solution of the acid
component and an equivalent of triethylamine in acetonitrile or
nitromethane as a solvent (usually) at 0°C (or room temperature)
until solution results. At this point, it is assumed that the
reactions leading to enol ester are complete, and the amine component
(or amine hydrochloride plus triethylamine) is added (sometimes in

dimethylformamide) for the acylation (coupling) step. After an
overnight (or longer) reaction period, the solvent is evaporated
and the product is isolated by the usual procedures. Recommended
isolation schemes include: (a) simple trituration with water to
remove the ketoamide and other impurities bearing the ionic sulfonate
function, or (b) separation of acids and bases along with ionic con-
taminants by (ethyl acetate) partition with water, bicarbonate, and
dilute hydrochloric acid (or citric acid) washings. The ionic nature
of the by-product has also facilitated product isolation by countercurrent distribution [84, 85] and chromatographic techniques [86].
Otherwise, few new modifications of the original recipe have been
introduced by users of the reagent (6).

The most frequent application of (6) has been for acylations
with N-benzyloxycarbonylamino acids or amino acids with other
nitrogen-masking groups, which provide comparable protection
against racemization via the azlactone mechanism [87]. Halpern,
Chew, and Westley [88] have shown that the t-butyloxycarbonyl group
also meets this qualification in their test for racemization
involving the gas-liquid chromatography (glc) separation of
diastereomeric 4-methyl-2-pentyl amides of protected amino
acids prepared from many amide-bond-forming reagents, including (6).

Couplings of this type, reported through the middle of 1971,
are summarized in Table 1. The table is organized, first, according
to the amino acid activated (regardless of the nature of the group
on nitrogen) and second, on the basis of the N-terminal amino acid
of the amine component (regardless of length of the peptide chain).
The same organization is used in subsequent tables so that potential
users of (6) will easily be able to determine if the reagent has
been applied previously in peptide bond formations between the amino
acids of interest. To provide information on the efficiency of the
recommended isolation techniques, yield and melting point data
(where available) for both directly isolated and recrystallized
products are listed. Comparative yields with other coupling
methods are given only when performed by the same workers.

The general picture that emerges from the data of Table 1 is
that yields are generally good (> 70%) and frequently, but not
invariably, better than with other coupling methods. One particular
advantage foreseen for the reagent (6), as a consequence of the
efficient intramolecular rearrangement pathway to enol esters, was
the avoidance of side reactions of unprotected functional groups of
amino acid side chains. For example, unprotected hydroxyl groups
of hydroxyproline (as the carboxyl component) and tyrosine (as the
amine component) were found not to interfere with the desired
acylation in the original examination of (6) [15, 18, 19]. In
addition, good yields were obtained in the presence of the ω-amide
groups of asparagine and glutamine in the carboxyl component. Subsequently Ruttenberg [89] has confirmed by infrared spectroscopy
the absence of dehydration of asparagine to β-cyanoalanine during
activation with (6). In the activation of serine with a free
hydroxyl group, a side reaction was detected by Costopanagiotis,
Handford, and Weinstein [90], when the reaction was conducted at
room temperature, rather than at 0°C. Theodoropoulos and Fruton
[91] noted that (6), as well as the mixed anhydride method,
eliminated difficulties encountered with carbodiimide couplings
with tryptophan. The selectivity of the derived enol ester
acylating agents has permitted the use of hydrazide as a
carboxyl-protecting group [92].

Activations using (6) have also been performed with other acyl
derivatives of amino acids, including simple acyl derivatives (see
Table 2), N-(2-quinoxaloyl) amino acids (see Table 3), and--in the
synthesis of modified actinomycin sequences by Brockmann, Lackner,
and co-workers [86, 131-138]--a variety of modified N-benzoyl
groupings (see Table 4). In these instances there have been no
reports of any appreciable racemization, although somewhat lower
yields are reported than for the protected amino acids of Table 1.

Finally, the isoxazolium salt (6) has found frequent application in couplings utilizing peptide acids, with considerable variation in yields. Most commonly, synthetic strategies that eliminate
the hazard of racemization via azlactones are employed by allowing

the C-terminal acid to be glycine (see Table 5) or proline (see Table 6). Other special acylations with the ω-acid functions of aspartic acid or glutamic acid, or with racemic peptide acids have been reported from time to time (see Table 7). Beyond this, a considerable number of acylations with peptide acids susceptible to the azlactone side reaction have been conducted (see Table 8), again with no reports of appreciable racemization. However, in many of these cases the possibility of some racemization was not excluded.

B. Guide to Abbreviations in Tables and Text

The following is a guide to the abbreviations in Tables 1 to 10 and in the text. An asterisk (*) signifies that the yield was for recrystallized product. This symbol —ξ— denotes the amide bond formed in Tables 5-8. All amino acid configurations are L unless otherwise specified.

Amino acids

Abu	α-aminobutyric acid	Leu	leucine
Ala	alanine	Lys	lysine
Arg	arginine	MePhe	N-methylphenylalanine
Asn	asparagine	Met	methionine
Asp	aspartic acid	MeVal	N-methylvaline
δAva	δ-aminovaleric acid	Phe	phenylalanine
2PhGly	C-phenylglycine	Pro	proline
Cys	cysteine	Sar	sarcosine
Gln	glutamine	Ser	serine
Glu	glutamic acid	Thr	threonine
Gly	glycine	Trp	tryptophan
Hyp	hydroxyproline	Tyr	tyrosine
Ile	isoleucine	Val	valine

Coupling agents			N-, O-, and S-Protecting groups	
CDI	carbonyldiimidazole		Ac	acetyl
DCC	dicyclohexylcarbodiimide		Boc	tert-butyloxycarbonyl
EDC	ethyl, dimethylamino-propylcarbodiimide		Btm	benzylthiomethyl
			Bu^t	tert-butyl
NHS	N-hydroxysuccinimide		Bz	benzoyl
			Bzh	benzhydryl
See also carboxyl derivative abbreviations.			Bzl	benzyl
			Cla	chloroacetyl
			For	formyl
Carboxyl derivatives			Pht	phthalyl
			Pz	p-phenylazobenzyloxy-carbonyl
OBu^t	tert-butyl ester		Qcb	quinoxaloyl
OCB	o-cyanobenzyl ester		Tfa	trifluoroacetyl
OCH_2CN	cyanomethyl ester		Tos	tosyl
ONBzl	p-nitrobenzyl ester		Trt	trityl
ONP	p-nitrophenyl ester		Z	benzyloxycarbonyl
OPip	1-hydroxypiperidine ester		Z(OMe)	p-methoxybenzyloxy-carbonyl
OTCP	2,4,5-trichlorophenyl ester			

TABLE 1

Acylations with (6) and N-Protected Amino Acids

Compound	Yield (%)	Reference
N-protected alanine		
Z-Ala-Gln-ONBzl	66	93
Z-Ala-Gly-Gly-NHNH$_2$	63*	92
Z-Ala-Met-OPip	80	94
Z-Ala-Phe-OPip	77	94
(vs 44% by EtOCOCl mixed anhydride)		
Z-Ala-Val-OPip	72	94

TABLE 1 (continued)

Compound	Yield (%)	Reference
N^{α}, N^{G}-protected arginine		
Tos Tos | | Z-Arg-Arg-NH$_2$	72	85
Tos Tos | | Z-Arg-Arg-δAva-OMe	78	95
Tos Tos | | Z-Arg-Arg-Pro-OBu^t	71	84
Tos Tos | | Z-Arg-Arg-Prolinol	81	96
Tos | Z-Arg-δAva-OMe	88	95
NO$_2$ | Z-Arg-Phe-Pro-OH	61	97
Tos Tos | | Z-Arg-Gly-Phe-Phe-Tyr-Thr-Pro-Lys-Ala-OMe	87	98
Tos Tos | | Z-Arg-Gly-Phe-Phe-Tyr-Thr-Pro-Lys-Thr-OMe		99
Tos | Z-Arg-Pro-OBu^t	76*	84
Tos | Z-Arg-Pro-OBzl	85	100
Tos | Z-Arg-Prolinol	85	96
N-protected asparagine		
NO$_2$ Me Bzl | | | Z-Asn-Arg-Val-Tyr-Val-His-Pro-Phe-OMe		101
Z-Asn-Asn-ONBzl	53	93, 102
Bzl | Z-Asn-Cys-OBzl	82*	103
Btm | Z-Asn-Cys-ONP	72	104

TABLE 1 (continued)

Compound	Yield (%)	Reference
Btm | Boc-Asn-Cys-ONP	60	104
Bzl | Z-Asn-Cys-Pro-Leu-Gly-NH$_2$ (vs 61% by Me$_3$CCOCl mixed anhydride, several other methods also poor)	poor	105
Z-Asn-Gln-ONBzl	53	93, 102
Boc-Asn-Gly-OMe		106
Z-Asn-Gly-OEt	80 mp 185.5 -187 (rcrd mp 186-187) 78	19 89
Z-Asn-Leu-OMe	76 mp 176.5-178 (rcrd mp 177.5-178.5)	19
Z | Z-Asn-Lys-OMe (vs 70% by ONP)	80	107
Bu^t | Z-Asn-Thr-OBu^t (better over-all yield than by ONP)	75	108
N-protected aspartic acid β-esters		
OBu^t | Z-Asp-Gly-OBu^t	88	109
N-protected aspartic acid α-esters		
Z-Asp-OBzl | Z-Lys-OBzl	65*	110

TABLE 1 (continued)

Compound	Yield (%)	Reference

N,S-protected cysteine

```
  Trt      Z        Bzl
   |       |         |
Z-Cys-Gly-Lys-Gly-Gly-Cys-OMe
```
                                      89   111

N,N'-protected cysteine

```
Tos-Cys-Gly-OEt
     |
Tos-Cys-Gly-OEt
```
                                            112

```
Z-Cys-Gly-OEt
   |
Z-Cys-Gly-OEt
```
                                            112

N-protected glutamine

Compound	Yield (%)	Reference
Z-Gln-Asn-ONBzl	65	93, 102

```
      Bzl
       |
Z-Gln-Cys-OEt
```

Z-Gln-Cys-OEt	56	93
Z-Gln-Gly-OMe	56*	113
(vs 57% by ONP and 51% by OTCP)		
Boc-Gln-Gly-OMe		106
Z-Gln-Gly-ONBzl	68	93
Z-D-Gln-Gly-OEt	79*	114
Z-Gln-Phe-OPip	78	94
Z-Gln-Pro-OBzl		115
Z-Gln-Tyr-OMe	75	19
	mp 198-199 (rcrd	
	mp 197.5-198.5)	
Z-Gln-Val-OMe	77	19
	mp 172.5-173 (rcrd	
	mp 172.5-173)	

N-protected glutamic acid γ-esters

```
   OMe OEt OMe OEt OMe OEt OMe OEt
    |   |   |   |   |   |   |   |
Z-Glu-Glu-Glu-Glu-Glu-Glu-Glu-Glu-Val-ONBzl
```

Z-Glu-Glu-Glu-Glu-Glu-Glu-Glu-Glu-Val-ONBzl	54*	116

TABLE 1 (continued)

Compound	Yield (%)	Reference
OMe OEt OMe OEt OMe OEt Z-Glu-Glu-Glu-Glu-Glu-Glu-Val-ONBzl	55*	116
OEt OMe OEt OMe OEt Z-Glu-Glu-Glu-Glu-Glu-Val-ONBzl	60*	116
OMe OEt OMe OEt Z-Glu-Glu-Glu-Glu-ONBzl	40*	116
OEt OMe OEt Z-Glu-Glu-Glu-ONBzl (ONP, DCC also poor, side reactions of dipeptide suggested)	60	116
OEt OMe OEt OMe OEt Z-Glu-Glu-Glu-Val-Glu-Glu-ONBzl	60*	116
OEt OMe OEt Z-Glu-Val-Glu-Glu-ONBzl	63*	116

N-protected glutamic acid α-esters

Btm ⌐Cys-Gly-OBu^t ⌐ Z-Glu-OBu^t	24*	117

N-protected glycine

Compound	Yield (%)	Reference
Z-Gly-NHCH$_2$Ph	96 mp 119-119.5 (rcrd mp 119-120)	19
Z-Gly-Gln-ONBzl	68	93
Z-Gly-Gly-NHNH$_2$	62*	92
Boc-Gly-Gly-OBu^t	59	110
Pht-Gly-Gly-OEt	88 (mp unchanged after rcrn)	19
Z-Gly-Gly-OBu^t		103

TABLE 1 (continued)

Compound	Yield (%)	Reference
Z-Gly-Gly-Gly-OEt	92	19
	mp 167–168	
	(best mp 167.5–168)	
Z-Gly-Pro-Gly-Gly-OBu^t		103
Z-Gly-Pro-Leu-Val-OMe		118
N-protected hydroxyproline		
Z-Hyp-Gly-Gly-OEt	80	19
	(mp unchanged after rcrn)	
Z-Hyp-D-Phe-OMe		119
Z-Hyp-D-Val-OMe	74	119
N-protected leucine		
Z-Leu-Phe-OPip	86	94, 120
(vs 59% by EtOCOCl mixed anhydride)		
N^α, N^ε-protected lysine		
$\overset{\displaystyle Z}{\underset{\displaystyle |}{Z\text{-Lys}}}$-Gly-OEt	98	19
$\overset{\displaystyle Z}{\underset{\displaystyle |}{Z\text{-Lys}}}$-D-Met-OMe		119
N-protected methionine		
$\overset{\displaystyle Bu^t}{\underset{\displaystyle |}{Z\text{-Met-Asn-Thr}}}$-OtBu	52.4	108
(vs 85.4% with OTCP)		
Z-Met-Gly-Gly-OEt	90	19
	mp 128–131	
	(86, mp 131.5–133 after rcrn)	

TABLE 1 (continued)

Compound	Yield (%)	Reference
Z-Met-OPip	71	94
(vs 56% by DCC and 45% by EtOCOCl mixed anhydride, but 86% by the ynamine method)		
Z-Met-Val-OPip	61	94

N-protected phenylalanine

Compound	Yield (%)	Reference
Tos | Z-Phe-Arg-Trp-Gly-OBu^t	84*	121, 122
Bzl Tos | | Z-Phe-Gln-Asn-Cys-Pro-Lys-Gly-NH$_2$	87	123
Z-Phe-Gly-OEt	98 (93 after rcrn)	19
Z-Phe-Leu-OMe	94 (90 after rcrn)	19
Boc | Z-Phe-Lys-Trp-Gly-OBu^t	74 mp 160-175 (70, mp 183-184 after rcrn)	124
Tos | Z-Phe-Lys-Trp-Gly-OBu^t	92 (79 after rcrn)	124
Z-Phe-Met-OPip	77	94
(vs 53% by EtOCOCl mixed anhydride)		
Z-Phe-Phe-OPip	75	94
(vs 58% by DCC and 47% by EtOCOCl mixed anhydride)		
Z-Phe-Val-OPip	77	94

N-protected proline

Compound	Yield (%)	Reference
Z-Pro-Leu-Val-OMe		118
Z-Pro-Val-OPip	78	94

TABLE 1 (continued)

Compound	Yield (%)	Reference
<u>N-protected serine</u>		
Z-Ser-Lys-OMe ‹Boc›	49*	90
(side reaction noted with activation at room temp. rather than 0°)		
Z-Ser-Phe-OPip	78	94
Z-Ser-Val-OPip	50	94
<u>N-protected threonine</u>		
Z-Thr-Ala-OMe	poor	126
(EDC also poor, but high yield with DCC)		
<u>N-protected tryptophan</u>		
Z-Trp-Gly-OBzl	89 mp 115-117 (rcrd mp 117-119)	91
Z(OMe)-Trp-Ala-OPip	70	94
Z(OMe)-Trp-Phe-OPip	84	94
<u>N,O-protected tyrosine</u>		
Z-Tyr-Leu-OMe ‹Z›	55*	127
<u>N-protected valine</u>		
Z-Val-Glu-Glu-Glu-ONBzl ‹OEt OMe OEt›	19*	116
Z-Val-Met-OPip	82	94

TABLE 2

Acylations with (6) and Simple N-Acylamino Acids

Compound	Yield (%)	Reference
<u>N-formylamino acids</u>		
For-Cys-Gly-OEt		112
| For-Cys-Gly-OEt		
Bzl | For-Gln-Cys-OBzl	65 mp 169-173 (rcrd mp 176-180)	93, 102
<u>N-acetylamino acids</u>		

SH
  |
  CMe$_2$
  |
Ac-NH-CH-CO-Gly-OEt
		112

Ac-Cys-NMe$_2$ | Ac-Cys-NMe$_2$	30*	128
<u>N-benzoylamino acids</u>		
Bz-Cys-Gly-OEt | Bz-Cys-Gly-OEt		129
<u>N-chloroacetylamino acids</u>		
Cla-Cys-Gly-OEt | Cla-Cys-Gly-OEt		129
<u>N-trifluoroacetylamino acids</u>		
Bzl | Tfa-Ser-Gly-OBu^t		118

TABLE 3

Acylations with (6) and N-(2-Quinoxaloyl)amino acids

Compound	Yield (%)	Reference
Qcb-Ala-Ala-OMe	57.5*	130
Qcb-Ala-Met-OMe	62.8*	130
Qcb-Ala-Val-OMe	37.5*	130
Qcb-D-Ser-Ala-OMe	45.6*	130
(vs 33.3% by DCC)		
Qcb-D-Ser-Ala-OBu^t	55.1*	130
Qcb-Val-Met-OMe	14.3*	130

TABLE 4

Acylations with (6) and Modified N-Benzoylamino Acids

Compound	Yield (%)	Reference
2-nitro-3-benzyloxy-4-methylbenzoylamino acids		
(2-NO$_2$-3-OBzl-4-Me-Bz)-Ser-D-Val-Pro-Sar-MeVal-OBzl	66	86, 131
(2-NO$_2$-3-OBzl-4-Me-Bz)-D-Ser-D-Val-Pro-Sar-MeVal-OBzl	70	86
(2-NO$_2$-3-OBzl-4-Me-Bz)-Thr-D-Ala-Pro-Sar-MeVal-OBzl	68	86
(2-NO$_2$-3-OBzl-4-Me-Bz)-Thr-Gly-Ome	58	86
(2-NO$_2$-3-OBzl-4-Me-Bz)-Thr-D-allo-Ile-Pro-Sar-OBzl	65	132
(by-product separation more difficult with DCC)		
(2-NO2-3-OBzl-4-Me-Bz)-Thr-D-allo-Ile-Pro-Sar-MeVal-OBzl	53	86
(2-NO2-3-OBzl-4-Me-Bz)-Thr-D-allo-Ile-Pro-Sar-DL-MeVal-OBzl	50	86
(2-NO2-3-OBzl-4-Me-Bz)-Thr-D-allo-Ile-Pro-Sar-MeVal-OMe	66	133, 134
(2-NO2-3-OBzl-4-Me-Bz)-Thr-D-Leu-Pro-Sar-MeVal-OBzl	70	86
(2-NO2-3-OBzl-4-Me-Bz)-Thr-Val-Pro-Sar-MeVal-OBzl	55	86
(2-NO$_2$-3-OBzl-4-Me-Bz)-Thr-D-Val-Pro-Sar-MeVal-OBzl	68	86, 135
(2-NO$_2$-3-OBzl-4-Me-Bz)-Thr-D-Val-Pro-Sar-D-MeVal-OBzl	78	86
(2-NO$_2$-3-OBzl-4-Me-Bz)-D-Thr-Val-D-Pro-Sar-D-MeVal-OBzl		136
(2-NO$_2$-3-OBzl-4-Me-Bz)-Thr-D-Val-Pro-Sar-MeVal-OMe	65	134
(2-NO$_2$-3-OBzl-4-Me-Bz)-Thr-D-Val-Pro-Sar-Pro-OBzl	55	86
(2-NO$_2$-3-OBzl-4-Me-Bz)-Thr-D-Val-Pro-Sar-Sar-OBzl	60	86
2-nitro-3-benzyloxybenzoylamino acids		
(2-NO$_2$-3-OBzl-Bz)-Thr-D-Val-Pro-Sar-MeVal-OBzl		137
2-nitro-3-benzyloxy-4-methoxybenzoylamino acids		
(2-NO$_2$-3-OBzl-4-OMe-Bz)-Thr-D-Val-Pro-Sar-MeVal-OBzl		138

TABLE 5

Acylations with (6) and Glycine Terminal Peptide Acids

Compound	Yield (%)	Reference
Couplings to cysteine		
Acetone-oxytocinoyl $\rightarrow$ oxytocin		139
Deamino-oxytocinoyl $\rightarrow$ oxytocin		139
Trt Bzh Trt		
Z-Cys-Cys-Gly-Phe-Gly $\rightarrow$ Cys-Phe-Gly-OBu^t	51.8	140
(vs 91% by DCC)	(44.7 after rcrn)	
Couplings to glutamine		
Z-Gly-Gly $\rightarrow$ Gln-Gly-OEt	50*	114
	(yield includes prior deprotection step)	
Couplings to glutamic acid		
Bzl OBzl Tos		
Z-Ala-Leu-Tyr-Leu-Val-Cys-Gly $\rightarrow$ Glu- Arg-Gly-Phe-		
-Phe-Tyr-Thr-Pro-Lys-Ala-OMe	38	141
Tos		
(vs 56% by CDI)		

<u>Couplings to glycine</u>

Z-Ala-Gly⨍Gly-OCB	86	142
Z-Ala-Gly⨍Gly-Gly-OCB	73	142
(vs 81% by ONP)		
Boc-Gly-glycolyl-Gly⨍Gly-glycolyl-Gly-ONBzl	66	143
Z-Gly-Pro-Gly⨍Gly-Pro-Gly-OEt	62.8*	144

OBzl Bzl
 | |
For-Gly-Asp-Ser-Gly⨍Gly-Pro-Leu-Val-OMe   118
                                  Z-Ala-D-Glu-NH$_2$
                                           |
Boc-Lys-D-Ala-Gly-Gly-Gly⨍Gly-Gly-Lys-D-   68.5   145
-Ala-OBzl   mp 192-194
                                              (rcrd mp 195-196)

Z-Gly-Pro-Gly-Gly-Pro-Gly⨍Gly-Pro-Gly-Gly-   42.4*   144
-Pro-Gly-OEt

<u>Couplings to histidine</u>

OBzl
 |
Z-His-Ser-Asp-Gly⨍Thr-Phe-Thr-Ser-   unsatisfactory   146
-Glu-Leu-Ser-Arg-Leu-Arg-Asp-Ser-Ala-Arg-Leu-Gln-Arg-

TABLE 5 (continued)

Compound	Yield (%)	Reference
-Leu-Leu-Gln-Gly-Leu-Val-NH$_2$ (product mixtures interpreted to suggest decomp. of Asp side chain protection, similar results by CDI or mixed anhydride)		
Bzl OBzl Boc-His-Ser-Asp-Gly ⊣ Thr-Phe-Thr -Ser-Glu-Leu-Ser-Arg-Leu-Arg-Asp-Ser-Ala-Arg-Leu- -Gln-Arg-Leu-Leu-Gln-Gly-Leu-Val-NH$_2$ (same as above)	unsatisfactory	146
Couplings to leucine		
Z-Phe-Ile-Gly ⊣ Leu-Met-NH$_2$	54*	147, 148
Z-Phe-Trp-Gly ⊣ Leu-Met-NH$_2$	54*	147, 148
Couplings to lysine		
Boc Boc Boc Tos Z-Lys-Pro-Val-Gly ⊣ Lys-Lys-Arg-NH$_2$	71*	85
Tos Tos Tos Tos Boc-Lys-Pro-Val-Gly ⊣ Lys-Lys-Arg-OBzl (vs 63% by DCC)	78	149

$$\overset{NO_2}{\underset{|}{}}\quad \text{For}$$
Z-His-Phe-Arg-Trp-Gly $\dashv$ Lys-Pro-Val-NH$_2$ 16   150

(yield includes subsequent hydrogenolysis and purification, by-product detected, vs 49% by DCC and 41% by DCC + NHS)

Boc   Boc Boc Tos Tos
Z-Lys-Pro-Val-Gly $\dashv$ Lys-Lys-Arg-Arg-NH$_2$ 65   85

Boc   Boc Tos Tos
Z-Lys-Pro-Val-Gly $\dashv$ Lys-Arg-Arg-Prolinol   86*   96

Boc   Boc Boc Tos Tos
Z-Lys-Pro-Val-Gly $\dashv$ Lys-Arg-Arg-δAva-OMe   83*   95

Boc   Boc Boc Tos Tos
Z-Lys-Pro-Val-Gly $\dashv$ Lys-Lys-Arg-Arg-Pro-NH$_2$ 72   85

Tos   Tos Tos Tos Tos
Z-Lys-Pro-Val-Gly $\dashv$ Lys-Arg-Arg-Pro-OBu^t 72   151

OBzl   Tos   Boc Boc
Z-Ser-Tyr-Ser-Met-Glu-His-Phe-Arg-Trp-Gly $\dashv$ Lys-Lys-   84

Tos Tos
-Arg-Arg-Pro-OBu^t

TABLE 5 (continued)

Compound	Yield (%)	Reference

OBzl   Tos   Tos
    | | |
Z-Ser-Tyr-Ser-Met-Glu-His-Phe-Arg-Trp-Gly⊢Lys-Pro-
 	
    Tos Tos Tos
    | | |
-Val-Gly-Lys-Lys-Arg-OBzl
 	
(vs 58% by DCC)
	48.6	149

Couplings to tyrosine Z-Gly-Gly⊢Tyr-OMe	88 mp 157-160 (84, mp 159.5- 161.5 after rcrn)	19
Couplings to valine (2-NO₂-3-OBzl-4-Me-Bz)-Thr-Gly⊬D-Val-Pro-Sar- -MeVal-OBzl	85	86

TABLE 6

Acylations with (6) and Proline Terminal Peptide Acids

Compound	Yield (%)	Reference
$\overset{\overset{\text{Tos}}{\mid}}{\text{Z-Arg}}$-$\overset{\overset{\text{Tos}}{\mid}}{\text{Arg}}$-Pro $\big\{$ NH$_2$	96	85
Z-Gly-Pro $\big\{$ Ala-OMe	64	144
	mp 155-158	
	(rdrd mp 159-160)	
Z-Val-$\overset{\overset{\text{Tos}}{\mid}}{\text{Lys}}$-Val-Tyr-Pro $\big\{$ $\overset{\overset{\text{OBu}^t}{\mid}}{\text{Asp}}$-Gly-OBu^t	51	109
Z-Gly-Pro $\big\{$ Gly-OEt	99.1	144
Z-$\overset{\overset{\text{Tos}}{\mid}}{\text{Lys}}$-Pro-Val-$\overset{\overset{\text{Tos}}{\mid}}{\text{Lys}}$-$\overset{\overset{\text{Tos}}{\mid}}{\text{Lys}}$-$\overset{\overset{\text{Tos}}{\mid}}{\text{Arg}}$-Arg-Pro $\big\{$ Val-$\overset{\overset{\text{Tos}}{\mid}}{\text{Lys}}$-	50	109
-Val-Tyr-Pro-$\overset{\overset{\text{OBu}^t}{\mid}}{\text{Asp}}$-Gly-OBu^t		

TABLE 7

Special Acylations with (6) and Peptide Acids

Compound	Yield (%)	Reference
Boc-Gly-$\underset{\sim\!\sim\!\sim}{\text{Asp}}$-Gly-OBu^t Boc-Gly-Lys-Gly-OBu^t	51	110
$\lceil\big\{$-$\overset{\overset{\text{Z}}{\mid}}{\text{Lys}}$-D-Ala-OBzl Boc-Ala-D-Glu-NH$_2$	62*	152
$\lceil\big\{$-$\overset{\overset{\text{Z}}{\mid}}{\text{Lys}}$-D-Ala-D-Ala-ONBzl Z-Ala-D-Glu-OBzl	51*	153
Z-Gly-DL-Phe $\big\{$ Gly-OEt	92	19
	(89 after rcrn)	
Z-$\overset{\overset{\text{OBu}^t}{\mid}}{\text{Ser}}$-OCH$_2CH_2$CO $\big\{$ $\overset{\overset{\text{OBu}^t}{\mid}}{\text{Ser}}$-OCH$_2CH_2CO_2$Bzl	68	125
(vs 97% by DCC)		

TABLE 8

Acylations with (6) and Other Peptide Acids

Compound	Yield (%)	Reference
Alanine terminal peptide acids		
Z-Gly-Pro-Ala ┼ Gly-Pro-Ala-OMe	58, mp 202-2-4 (mp 208-210 after rcrn)	144
Z-Gly-Pro-Ala-Gly-Pro-Ala ┼ Gly-Pro-Ala-Gly- -Pro-Ala-OMe	35, mp 260-265 (mp 263-264 after rcrn)	144
Asparagine terminal peptide acids		
Bzl Z-Cys-Tyr-Ile-Gln-Asn ┼ Cys-Pro-Leu-Gly-NH (vs 80% by DCC, 70.5% by CDI;) also prepared with dimethoxydiphenylmethyl cysteine protection)	85	154 155
Bzl Z-Cys-Tyr-Ile-Gly-Asn ┼ Cys-Pro-Leu-Gly-NH (vs 39% by DCC)	89	154

Aspartic acid terminal peptide acids

Peptide	Yield	Ref.
Boc-Gly-Asp(ONBzl)—Gly-OBu^t	55	110
For-Gly-Asp(OBzl)—Ser-Gly-OBu^t		118
Z-Gly-Asp(OBzl)—D-Ser-Gly-OEt	50*	156

Glutamine terminal peptide acids

Peptide	Yield	Ref.
Z-Asn-Gln(Btm)—Cys-ONP	unsuccessful	104

(problem with insolubility of Et$_3$N salt of acid)

Glutamic acid terminal peptide acids

Peptide	Yield	Ref.
Z-Glu-Glu(OEt,OMe)—Glu(OEt)-Glu(OMe)-Glu(OEt)-Glu(OMe)-Val-ONBzl	65*	116
Z-Glu-Glu(OEt,OMe)—Glu(OEt)-Glu(OMe)-Glu(OEt)-Val-ONBzl	47*	116
Z-Glu-Glu(OEt,OMe)—Glu(OEt)-Glu(OMe)-Glu(OEt)-ONBzl	71*	116

TABLE 8 (Continued)

Compound	Yield (%)	Reference
| ```
    OMe OEt   OMe OEt
     |   |      |   |
Z-Glu-Glu ╅ Glu-Glu-Val-Glu-Glu-ONBzl
``` | 45* | 116 |
| ```
    OMe OEt   OMe OEt
     |   |      |   |
Z-Glu-Glu ╅ Glu-Glu-Val-ONBzl
``` | 70* | 116 |
| ```
    OMe OMe   OMe OMe
     |   |      |   |
Z-Glu-Glu ╅ Glu-Glu-Val-ONBzl
``` | 44* | 116 |
| ```
    OMe   OEt OMe OEt
     |     |   |   |
Z-Val-Glu ╅ Glu-Glu-Glu-ONBzl
                  |
                 OBzl
``` | 55* | 116 |
| ```
Z-Ala-Glu ╅ Lys-D-Ala-D-Ala-OMe
(vs 55% by ONP, poorer yields by DCC
or -OCH₂CN)
``` | 71* | 153 |
| ```
      OBzl  Z
       |    |
Z-Ala-D-Glu ╅ Lys-D-Ala-ONBzl
``` | 62 | 153 |
| ```
       OBzl Bzl
        |    |
For-Gly-Gl ╅ Ser-Gly-OBuᵗ
``` | | 118 |
| ```
      OBzl
       |
Z-Gly-Glu ╅ D-Ser-Gly-OEt
``` | 62* | 156 |

$$
\begin{array}{c}
\text{OME OEt} \quad \text{OMe OEt} \\
\mid \qquad\quad \mid \\
\text{Z-Glu-Glu} \;\text{—}\!\!\!\backslash\!\!\backslash\text{—}\; \text{Val-Glu-Glu-ONBzl}
\end{array}
$$

51   116

Leucine terminal peptide acids

$$
\begin{array}{c}
\text{NO}_2 \\
\mid \\
\text{Z-Leu-Leu} \;\text{—}\!\!\!\backslash\!\!\backslash\text{—}\; \text{Arg-OMe}
\end{array}
$$

157

Lysine terminal peptide acids

$$
\begin{array}{c}
\text{Boc Boc} \quad \text{NO}_2 \; \text{NO}_2 \\
\mid \quad\; \mid \qquad \mid \qquad \mid \\
\text{Trt-Lys-Lys} \;\text{—}\!\!\!\backslash\!\!\backslash\text{—}\; \text{Arg-Arg-Pro-OMe}
\end{array}
$$

(vs 64% by DCI)

81   158

$$
\begin{array}{c}
\text{Pz} \\
\mid \\
\text{Boc-Gly-Lys} \;\text{—}\!\!\!\backslash\!\!\backslash\text{—}\; \text{Gly-OBu}^{t}
\end{array}
$$

(vs 76% and higher rotation by DCC)

71   110

$$
\begin{array}{c}
\text{Z} \\
\mid \\
\text{Z-Asn-Lys} \;\text{—}\!\!\!\backslash\!\!\backslash\text{—}\; \text{His-OMe}
\end{array}
$$

(vs 65% by azide)

71   107

$$
\begin{array}{c}
\text{Z} \qquad\quad \text{Bzl Bzl NO}_2 \\
\mid \qquad\quad\; \mid \quad \mid \quad \mid \\
\text{Z-Asn-Lys} \;\text{—}\!\!\!\backslash\!\!\backslash\text{—}\; \text{His-His-Arg-ONBzl}
\end{array}
$$

50   107

Table 8 (continued)

Compound	Yield (%)	Reference
Phenylalanine terminal peptide acids		
Z-Gly-Phe $\dfrac{}{}$ Gly-OEt	98 (92 after rcrn)	19
Serine terminal peptide acids		
Z-Phe-Ser $\dfrac{}{}$ Gly-Phe-Arg-OMe (NO$_2$)	50*	159
Tryptophan terminal peptide acids		
Trt-Gly-Trp $\dfrac{}{}$ Gly-Trp-NH$_2$		91
Z-Ser-Tyr-Ser-Met-Glu-His-Phe-Arg-Trp $\dfrac{}{}$ Gly-Lys- (OBzl, Tos, Tos)	80	109
-Pro-Val-Gly-Lys-Arg-Pro-Val-Lys-Val-Tyr- (Tos, Tos, Tos, OBu^t)		
-Pro-Asp-Gly-OBu^t		
(vs 70% by DCC, acylurea contamination problem)		

C. Early Use of Benzisoxazolium Salts

The application of the N-ethylbenzisoxazolium cation (7) to
peptide synthesis has not been stressed by Kemp and co-workers, who
have found that the 7-hydroxy derivative (11), discussed below, has
superior properties. Nevertheless, the reagent (7) has several
desirable features in common with (11). As mentioned in previous
sections, the selectivity of the unstable cumulenes from the ring
opening of benzisoxazolium salts results in a very efficient
addition of carboxylate ions to give phenolic esters, rather than
the water addition product, in aqueous media. The same conditions,
achieved with a two-phase, pyridine-buffered reaction medium,
eliminates azlactone formation as a serious competing reaction.
Furthermore, the phenolic esters from (7) are sufficiently stable
for isolation and storage and give high yields in acylation reactions.
The by-product N-ethylsalicylamide is conveniently removed, either by
extraction from aqueous solution with aqueous sodium hydroxide or by
virtue of its high solubility in carbon tetrachloride. The chief
deficiencies of (7), relative to (11), are the lower reactivity
and resistance to racemization of the esters prepared from
(7). Despite these relative shortcomings, excellent results were
obtained in the synthesis of the hexapeptide (101) in 40% over-all
yield by a pyramidal strategy for fragment combination, using the
esters derived from (7) for all couplings [20].

$$Z-(Gly-Leu-Gly)_2OH$$

$$(101)$$

Attempts to increase the reactivity of the O-acyl salicylamides
from benzisoxazolium salts with electron-withdrawing substituents in
the benzene ring have not resulted in outstanding synthetic results
to date. A number of esters were successfully isolated from the
dichloro derivative (10), including those of several benzyloxycarbonylamino acids, such as serine [30] and glutamine or asparagine [63].

Coupling of the serine ester with glycine p-nitrobenzyl ester was
achieved in 78.5% yield, but only a 17% yield of Z-Gly-Asn-Pro-OMe
resulted from use of the glutamine ester. Esters from a 5-nitro-
substituted benzisoxazolium cation were used without isolation, but
poor yields were obtained in couplings with proline, asparagine,
glutamine, and glutamic acid [63].

D. Use of N-t-Butylisoxazolium Salts

The reagent N-t-butyl-5-methylisoxazolium (72) perchlorate,
commercially available from Aldrich Chemical Co. [83], offers a
second type of isolable enol ester, although some special sources
of difficulty must be kept in mind.

The formation of the enol esters is complicated by the side
reactions discussed previously, but these may be kept to a minimum
by a technique for activation that utilizes excess (72) in wet
acetonitrile [28]. With this special method, stable enol esters
have been prepared in yields of 73-94% from N-protected amino acids,
including serine and glutamine. In contrast to other types of
isoxazolium salts, the efficiency of the rearrangement process is
not adequate with (72) to avoid dehydration of the amide group of
asparagine. The esters are more reactive than the phenolic esters
from (7) and the by-product ketoamide is readily removed with water
or carbon tetrachloride.

Some couplings have been carried out (see Table 9) with the
purified β-acyloxy-N-t-butylcrotonamide esters of N-protected amino
acids, which gave yields of crude products somewhat higher than the
over-all yields (both activation and couplings) in comparable
couplings with the zwitterionic reagent (6). However, the crude
products were usually less pure, as judged by slight melting point
depressions, than in the reactions using (6). One possible factor
that might result in trace impurities could be the failure of the
isolation method in removing small amounts of unconsumed enol ester

from the product. The ionic enol esters from (6), however
would be separated from the peptides by either common isolation strategy.

In instances where the isolation, purification, and storage
of the intermediate acylating agents from N-protected amino acids is

TABLE 9

Acylations with β-Acyloxy-N-t-
butylcrotonamides from N-Protected Amino acids

Compound	Yield (%)	Reference
Z-Gln-Tyr-OMe	81	29
Z-Gly-NHCH$_2$Ph	98	27
Z-Gly-Gly-Gly-OEt	90	29
Pht-Gly-Gly-OEt	92	29
Z-Gly-Ser-OMe	80*	29

desired, the reagent (72) [or the benzisoxazolium cation (7)] appears
to have merit. Otherwise major practical advantages in peptide
synthesis relative to the more convenient reagent (6) do not seem to
be provided by esters from (72) [and (7)], despite the elimination
of losses in the activation step by using purified esters and the
prevention of enol ester rearrangement during the coupling step.

While basic conditions were found (2-picoline as the solvent),
which would avoid azlactone formation in the activation of hippuric
acid [58], the side reaction was not completely avoided in the case
of the acid of the Young test. Furthermore, good yields of enol ester
are not, in general, obtained with 2-picoline, so (72) does not
appear to be useful for acylations with racemization-prone peptide
acids.

E. Use of 4,5-Tetramethyleneisoxazolium Salts

While few results have been reported to date for this new type
of cation, the preliminary studies of stability of the esters and
the efficiency of enol ester formation appear quite promising [32].
The latter factor is indicated by the avoidance of azlactone in the
spectral tests with hippuric acid, as discussed in Sec. VII. Moreover, no racemate was detected in the Anderson test, suggesting
that enol ester fragmentation to azlactone also may not be a serious
problem.

F. Use of the N-Ethyl-7-hydroxybenzisoxazolium Cation

The results described to date for the modified benzisoxazolium
reagent (11), commercially available from Midway Bio-Organics (P. O.
Box 1804), Kansas City, Mo., 64140), are extremely good [31, 64].
The buffered activation medium mentioned previously (Sec. B) has
provided active esters (102) from more than 30 N-protected peptide
and amino acids, generally in 85-90% yields after purification.
Only a small reduction in yield (75-80%) is observed in the case
of asparagine and glutamine activations. Moreover, the reaction
conditions hold to a minimum the competing formation of azlactones
(less than 0.05% in the case of the Anderson test).

$$RCO_2^-$$

$$H_2O/EtOAc$$
+ pyridine, pH 4.5
15°, 20 min

(11)   (102)

Since these esters are stable to basic media with respect to
the imide and azlactone side reactions, they may be used directly

for coupling with peptide or amino acid salts of the base tetra-
methylguanidine (TMG). The major limitation for the esters is their

facile hydrolysis, the avoidance of which dictates the use of dipolar
aprotic solvents for the coupling reaction. It is reported that all
20 common amino acids have been found to be acylated in satisfactory
yields under these conditions. Glutamine and asparagine esters
present no problems, and even the slowest observed coupling (valine
to valine) gives 70-80% yields of product after a 12-hr reaction
period. The only difficulties mentioned so far have been with
activated serine, histidine, and arginine. Some specific yields are
given in Table 10.

TABLE 10

Acylations with the Phenolic Esters (102)

Compound	Yield (%)	Reference
Bzl Bzl | | Z-Cys-Cys-OH	80	64
Bzl | Z-Cys-Gly-OH	80	64
Z-Gly-Phe-OH	91*	31
	90*	64
Z-Gly-Phe-Ala—{— Gly-OH	89*	64

It has been noted in Sec. VII, C that the esters from (11) are one of the most racemization-resistant types of acylating agents yet devised for peptide synthesis. Even in salt couplings, racemization is held below tolerable limits, so long as low temperatures are employed for the reaction. Finally, the efficiency of the salt coupling approach, which obviates the repetition of deprotection steps after each coupling, is dramatically demonstrated by the synthesis of each of the peptides (103) and (104) in two days' time with over-all yields of 50-75%.

$$Z-(Ala)_5-OH \qquad Z-Gly-Leu-Gly-Gly-OH$$

$$(\underline{103}) \qquad\qquad . \qquad (\underline{104})$$

IX. SPECIAL APPLICATIONS

In addition to the role of isoxazolium salts in peptide synthesis by classical linear and pyramidal combination schemes covered in the previous section, the derived enol ester acylating agents from peptide and amino acids have found several other applications in protein chemistry, as described below.

A. Solid-Phase Peptide Synthesis

The isoxazolium salt approach may play a significant role in the refinement of the solid-phase method by extension of the technique to acylations of polymer-bound residues by activated derivatives of peptide acids. Omenn and Anfinsen [160] have used N-ethyl-5-phenylisoxazolium-3'-sulfonate (6), as well as the azide method and N-hydroxysuccinimide plus dicyclohexylcarbodiimide, as a racemiza-

tion-resistant coupling strategy to join peptide acids to the amine
terminus of peptide chains attached to the Merrifield polymer. Large
excesses of the enol ester acylating agents from (6) were found to
be necessary for coupling to the bound substrates in good yield for
the preparation of the peptides (105) and (106). Even though

$$\text{H-Glu-Lys-Lys-Ser}\!\!\left\{\!\!-\text{Leu-Pro-OH}\right.$$

(105)

$$\text{H-Leu-Ala-Tyr}\!\!\left\{\!\!-(\overset{\displaystyle\overset{\text{Tfa}}{|}}{\text{Lys}})_5\text{-OH}\right.$$

(106)

careful attention to reaction conditions may be necessary for complete solid-phase couplings, such acylations with peptide acids
would offer several desirable features, including more facile
separation of incomplete sequences than is the case for stepwise
solid-phase synthesis.

B. Cyclic Peptides

On a number of occasions isoxazolium salts have been applied
to peptide cyclizations. Bláha and Rudinger [161] employed a two-
stage cyclization technique with N-ethyl-5-phenylisoxazolium-3'-
sulfonate (6). The activation step was conducted at high concentration, keeping the amine group protected by protonation. After
dilution to favor the intramolecular cyclization, the free amine
group of the intermediate enol ester was liberated with base.
Starting with H-Gly-Phe-Leu-Gly-Phe-Leu-OH and H-Gly-D-Phe-Leu-
Gly-D-Phe-Leu-OH, cyclic hexapeptides were obtained in yields of 30
and 45%, respectively, similar to the results in cyclizations via
the azide route. Attempts to carry out a dimerizing cyclization,
in which the dilution method is less useful, gave poorer results.
Only 15% of the cyclic hexapeptide was obtained from H-Gly-D-Phe-
Leu-OH. Rudinger and Jošt [162, 163] conducted a high-dilution

cyclization of an octapeptide with (6) to prepare (107), an oxytocin
analog lacking a disulfide bridge. Although by-product removal from

$$CH_2CH_2CO \dashv Tyr-Ile-Gln-Asn-Cys-Pro-Leu-Gly-NH_2$$

(107)

cyclic peptides by ion-exchange desalting should be particularly
well-suited for the sulfonated reagent (6), product yields and
biological activities were low, compared with other techniques
(p-nitrophenyl ester, mixed carbonic anhydride, and azide methods),
which were pursued in later studies [164].

Low yields, in comparison with the use of carbodiimides, were
also obtained with (6) as the coupling reagent for the cyclization
step of the preparation of the peptide lactone (108), an analog
of vernamycin β_α, by Ondetti and Thomas [165]. Of the synthetic
methods examined for this cyclization, only the reagents (6), N,N'-
carbonylidiimidazole, and 1-cyclohexyl-3-(2-morpholinoethyl)-
carbodiimide gave sufficiently polar by-products for ready isolation
of (108). The isolation of (109) from the cyclization of o-(2-
aminoethyldithio)benzoic acid failed in attempts with (6) and

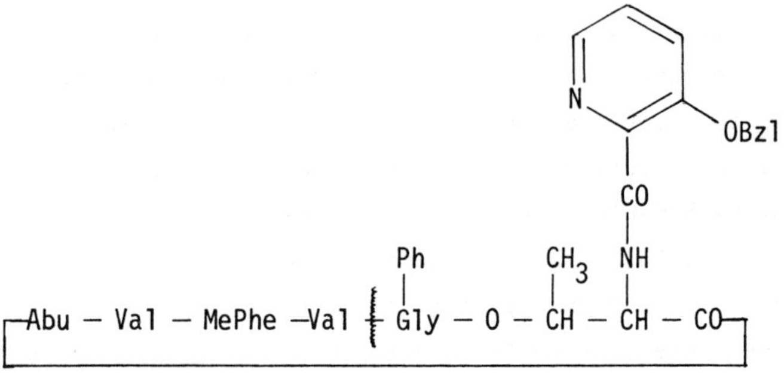

(108)

several other reagents, presumably because of product decomposition [166]. Finally, (6) was found to be useful in the preparation of the cyclic hexapeptide (110), because it was the only reagent tried that would bring the starting material, pentaglycyl-L-tyrosine, into solution in dimethylformamide [167].

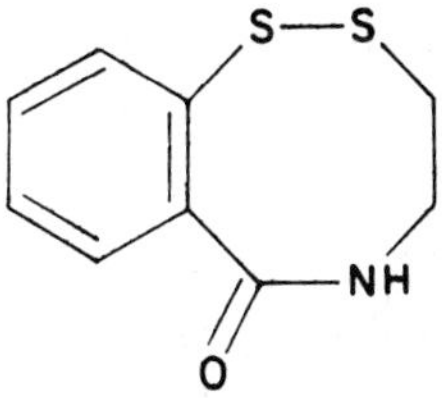

(109) 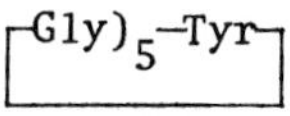   (110)

Superior results in dimerizing cyclizations have been obtained by Rajappa and Akerkar [168], using N-benzyloxycarbonyl derivatives of tripeptides and N-ethylbenzisoxazolium cation (7). Cyclization takes place upon removal of the protecting group by hydrogenation, a reaction which does not affect the intermediate active O-acyl-salicylamide esters. This approach gave relatively good yields (see Table 11), and the base-soluble by-product N-ethylsalicylamide was easily separated from the cyclic peptides.

TABLE 11

Cyclic Hexapeptides from (7)

Compound	Yield (%)
Cyclo(Gly-Gly-Gly)$_2$	70
Cyclo(Gly-Phe-Gly)$_2$	41
Cyclo(Gly-Pro-Gly)$_2$	44.4

C. Steroidal Peptides, Alkaloidal Peptides, Glycopeptides, and
    Nucleopeptides

The original isoxazolium salt reagent (6) has been one of the
major coupling agents used in recent work on the synthesis of
steroidal peptides. In 1965, Delépine reported the attachment of
N-protected amino acids and small peptide acids to 3α-amino-
5α-pregnanone-20 [169]. Pettit's group [170-173] has applied (6)
to the preparation of several arginine and proline derivatives of
17β-amino-5α-androstane and 17β-aminoandrost-5-ene. No difficulties
due to the 3β-hydroxyl group were mentioned; the reported yields were
about 70%. Tam obtained a 69% yield of the coupling product from
benzyloxycarbonyldiglycine and 3α-amino-5α-pregnanone-20 [174], while
acylations of other 6β-aminoandrostanes with (6) and benzyloxycarbonyl
histidine have been reported by DeFaye and Fetizon [175, 176].

The acylations of other types of amines of biologic interest
by enol esters from isoxazolium salts have not yet received much
attention. Pettit and Gupta [177] did not obtain favorable results
with (6) in attempts to make peptide derivatives of emetine. Some
use of (6) has been reported in the preparation of models for
alkaloid biosynthesis [178] and amino acid and peptide derivatives
of dihydrosphingosine [179]. Rao, Rebello, and Pogell [180] prepared a tyrosyl derivative of an amino nucleoside with (6) in 86%
yield, while a few N-acetylmuramyl peptides have been prepared
as analogs of bacterial cell wall glycopeptides [181].

D. Polypeptides

Bláha and Rudinger have also demonstrated the possibility of
using (6) for the preparation of polypeptides in the course of their
work on cyclic peptides [161]. Experiments with leucine (in methanol)

and phenylalanine (in dimethylformamide) revealed that enol esters
could be obtained from (6) and dipolar amino acids themselves. In
the latter instance, subsequent addition of base to the enol ester
solution resulted in the precipitation of poly(L-phenylalanine),
although only in small amounts.

E. Peptide and Protein Modification

    A number of investigators have found isoxazolium salts useful
for conducting reactions on acid or amine groups in peptide side-
chains. To obtain models for the tertiary structure of proteins,
Gill, Marfey, and Kunz [182, 183] cross-linked synthetic polypeptides
containing glutamic acid, lysine, and tyrosine by means of (6), as
well as with carbodiimides or 1,5-difluoro-2,4-dinitrobenzene.
Commercially, a variety of isoxazolium salts have been employed as
hardeners for gelatin in photographic emulsions [42, 184-187].
Conjugates of the polypeptide hormone angiotensin have been coupled
with the N-ethylbenzisoxazolium cation (7) to poly(L-lysine) [188].
    Patel and Price [189] reported the direct condensation of
carboxyl and amino side chains with retention of enzymatic activity
in their polymerization of chymotrypsin with (6). When excess
coupling agent was used, activity was diminished, probably as a
result of loss of tertiary structure. Similarly, enzymatic activity
was retained in the linking of α-chymotrypsin to carboxyl-containing
polymers by the use of (6) to form amide bonds [190]. The ease of
removal of the ionic ketoamide by-product was an advantageous feature
in the use of (6).
    An initial examination of the role of amino acid side chains
in the active sites of aminoacyl-RNA ligases has been conducted by
measuring the inhibition of the capacity to esterify transfer RNAs
with ^{14}C-labeled amino acids after using (6), or fluorodinitrofluoro-

benzene, to achieve chemical modification [191]. Bodlaender,
Feinstein, and Shaw [41, 192] found they could selectively modify
the essential carboxylic acid groups of the trypsin binding site
by reaction with isoxazolium salts. It was suggested that the
selectivity of the reaction with isoxazolium cations might result
from the preference of the active site for positively charged substrates or a relatively low pK_A of active site carboxyl side chains.
The fact that the zwitterion (6) reacted with lower specificity and
that the same effect was observed by increasing pH tends to confirm
these possibilities. Reactions with (6) were shown to be specific
for carboxyl groups below pH 4.75 and the intermediate enol esters
were sufficiently stable to permit isolation of the protein prior
to reactions with other nucleophiles. Very little cross-linking
was observed in these studies and the possibility of conducting the
activation and coupling steps separately under optimum conditions
for each reaction was recognized as an advantage, relative to the
use of water-soluble carbodiimides. Isoxazolium salts were not
considered promising for the estimation of the total carboxyl
content of proteins, however, since modification of more than five
groups resulted in precipitation.

Other instances of protein modification include the use of
isoxazolium salts as bridging agents to attach active proteins to
various carriers [193], to form reactive particles of insoluble
biological reagents [194], and to prepare ^{14}C-labeled antibiotic
peptides by acylation of free amine groups [195].

F. Preparation of Acylating Agents

Even though the enol ester acylating agents from isoxazolium
salts possess only a moderate level of reactivity, in two cases
they have proven to be sufficiently energetic to serve as reagents

for the relay synthesis of other acylating agents. Jones and Young
[196] found that the reagent (6) was valuable for the preparation
of optically pure N-protected peptide esters of 1-hydroxypiperidine
in small-scale work, even though some racemization was detected with
benzoyl-L-leucine. Finally, (6) gave N-thiocarboxyanhydrides (110)
of high optical purity from amino thio acids [197].

$$R-\underset{\underset{NHCOS^-,\ K^+}{|}}{C}HCOS^-,\ K^+ \quad \underset{\longrightarrow}{(6)}$$

(110)

REFERENCES

1. O. Mumm, Ph. D. Thesis, Kiel, 1902.

2. K. Meyer, Ph. D. Thesis, Kiel, 1903.

3. G. Münchmeyer, Ph.D. Thesis, Kiel, 1910.

4. C. Bergell, Ph. D. Thesis, Kiel, 1912.

5. A. Wirth, Ph. D. Thesis, Kiel, 1914.

6. W. Stülcken, Ph. D. Thesis, Kiel, 1935.

7. H. Hornhardt, Ph. D. Thesis, Kiel, 1937.

8. L. Claisen, Ber., 42, 59 (1909).

9. O. Mumm and G. Münchmeyer, Ber., 43, 3335 (1910).

10. O. Mumm and G. Münchmeyer, Ber., 43, 3345 (1910).

11. O. Mumm and C. Bergell, Ber., 45, 3040 (1912).

12. O. Mumm and C. Bergell, Ber., 45, 3149 (1912).

13. A. Knust and O. Mumm, Ber., 50, 563 (1917).

14. O. Mumm and H. Hornhardt, Ber., 70, 1930 (1937).

15. R. A. Olofson, Ph.D. Thesis, Harvard, Cambridge, 1961.

16. R. B. Woodward and R. A. Olofson, J. Am. Chem. Soc., 83, 1007 (1961).

17. R. B. Woodward and R. A. Olofson, Tetrahedron, Suppl., 7, 415 (1966).

18. R. B. Woodward, R. A. Olofson, and H. Mayer, J. Am. Chem. Soc., 83, 1010 (1961).

19. R. B. Woodward, R. A. Olofson, and H. Mayer, Tetrahedron, Suppl., 9, 321 (1966).

20. D. S. Kemp, Ph. D. Thesis, Harvard, Cambridge, 1964; Diss. Abstr., 25, 3845 (1965).

21. D. S. Kemp and R. B. Woodward, Tetrahedron, 21, 3019 (1965).

22. D. S. Kemp, Tetrahedron, 23, 2001 (1967).

23. D. J. Woodman, Ph. D. Thesis, Harvard, Cambridge, 1965.

24. R. B. Woodward and D. J. Woodman, J. Org. Chem., 31, 2039 (1966).

25. R. B. Woodward, D. J. Woodman, and Y. Kobayashi, J. Org. Chem., 32, 388 (1967).

26. R. B. Woodward and D. J. Woodman, J. Am. Chem. Soc., 88, 3169 (1966).

27. R. B. Woodward and D. J. Woodman, J. Am. Chem. Soc., 90, 1371 (1968).

28. D. J. Woodman and A. I. Davidson, J. Org. Chem., 35, 83 (1970).

29. D. J. Woodman and A. I. Davidson, J. Org. Chem., 38, 4288 (1973).

30. S. Rajappa and A. S. Akerkar, Chem. Commun., 826 (1966).

31. D. S. Kemp and S. W. Chien, J. Am. Chem. Soc., 89, 2743 (1967).

32. R. A. Olofson and Y. L. Marino, Tetrahedron, 26, 1779 (1970).

33. L. Claisen, Chem. Ber., 24, 3900 (1891).

34. N. K. Kochetov and S. D. Sokolov, Advan. Heterocycl. Chem., 2, 365 (1963).

35. A. Quilico, "Isoxazoles and Related Compounds," in The Chemistry of Heterocyclic Compounds (A. Weissberger, ed.), Vol. 17, Wiley (Interscience), New York, 1962, p. 1.

36. A. N. Nesmeyanov, M. I. Rybinskaya, and T. G. Kelekhsaeva, Zh. Org. Khim., 4, 921 (1968).

37. B. D. Wilson and D. M. Burness, J. Org. Chem., 31, 1565 (1966).

38. A. Conduché, Ann. Chim. [8], 13, 46 (1908).

39. H. Lindemann and H. Thiele, Ann., 449, 63 (1926).

40. H. Meerwein, E. Battenberg, H. Gold, E. Pfeil, and G. Wilfang, J. Prakt. Chem., 154, 83 (1939).

41. P. Bodlaender, G. Feinstein, and E. Shaw, Biochemistry, 8, 4941 (1969).

42. D. M. Burness and B. D. Wilson, British Pat. 1,030,882 (May 25, 1966); CA, 65, P20261d (1966).

43. C. H. Eugster, L. Leichner, and E. Jenny, Helv. Chim. Acta, 46, 543 (1963).

44. D. J. Woodman, J. Org. Chem., 33, 2397 (1968).

45. D. J. Woodman and Z. L. Murphy, J. Org. Chem., 34, 3451 (1969).

46. J. Thesing, A. Müler, and G. Michel, Chem. Ber., 88, 1027 (1955).

47. R. B. Woodward, D. J. Woodman, and Y. Kobayashi, unpublished results.

48. D. J. Woodman and Z. L. Murphy, J. Org. Chem., 34, 1468 (1969).

49. E. P. Kohler and A. H. Blatt, J. Am. Chem. Soc., 50, 1217 (1928).

50. E. P. Kohler and W. F. Bruce, J. Am. Chem. Soc., 53, 644 (1931).

51. D. J. Woodman and Z. L. Murphy, unpublished results.

52. M. Smith, J. G. Moffatt, and H. G. Khorana, J. Am. Chem. Soc., 80, 6204 (1958).

53. C. L. Stevens, R. C. Freeman, and K. Noll, J. Org. Chem., 30, 3718 (1965).

54. L. Claisen, Chem. Ber., 36, 3672 (1903).

55. D. F. DeTar and R. Silverstein, J. Am. Chem. Soc., 88, 1013 (1966).

56. C. L. Stevens and M. E. Munk, J. Am. Chem. Soc., 80, 4065, 4069 (1958).

57. J. C. Sheehan and G. P. Hess, J. Am. Chem. Soc., 77, 1067 (1955).

58. H. G. Khorana, Chem. Ind. (London), 1087 (1955).

59. R. B. Woodward and D. J. Woodman, J. Org. Chem., 34, 2742 (1969).

60. K. L. Shepard, W. Halczenko, and A. J. Cragoe, Jr., Tetrahedron Letters, 1969, 4757.

61. D. S. Kemp, J. M. Duclos, Z. Bernstein, and W. M. Welch, J. Org. Chem., 36, 157 (1971).

62. D. S. Kemp and S. W. Chien, J. Am. Chem. Soc., 89, 2745 (1967).

63. T. R. Govindachari, S. Rajappa, A. S. Akerkar, and V. S. Iyer, Indian J. Chem., 6, 557 (1968).

64. D. S. Kemp, in Peptides: Chemistry and Biochemistry, Proceedings of the First American Peptide Symposium, (B. Weinstein and S. Lande eds.), Dekker, New York, 1970, p. 33.

65. G. W. Anderson and F. M. Callahan, J. Am. Chem. Soc., 80, 2902 (1958).

66. M. Goodman and K. C. Steuben, J. Org. Chem., 27, 3409 (1962).

67. M. Goodman and L. Levine, J. Am. Chem. Soc., 86, 2918 (1964).

68. M. Goodman and W. J. McGahren, J. Am. Chem. Soc., 87, 3028 (1965).

69. M. W. Williams and G. T. Young, J. Chem. Soc., 1964, 3701.

70. I. Antonovics and G. T. Young, Chem. Commun., 1963, 398.

71. M. W. Williams and G. T. Young, J. Chem. Soc., 1963, 881.

72. F. Weygand, A. Prox, L. Schmidhammer, and W. König, Angew. Chem., 75, 282 (1963).

73. F. Weygand, A. Prox, and W. König, Chem. Ber., 99, 1451 (1966).

74. M. Bodanszky and L. E. Conklin, Chem. Commun., 1967, 773.

75. B. Halpern, L. F. Chew, and B. Weinstein, J. Am. Chem. Soc., 89, 5051 (1967).

76. B. Weinstein and A. H. Pritchard, J. Chem. Soc., 1972c, 1015.

77. N. Izumiya and M. Muraoka, J. Am. Chem. Soc., 91, 2391 (1969).

78. D. S. Kemp, S. W. Wang, G. Busby, III, and G. Hugel, J. Am. Chem. Soc., 92, 1043 (1970).

79. D. S. Kemp and J. Rebek, Jr., J. Am. Chem. Soc., 92, 5792 (1970).

80. M. Bodanszky and M. A. Ondetti, Peptide Synthesis, Wiley (Interscience), New York, 1966, pp. 146-155.

81. E. Schröder and K. Lübke, The Peptides, Academic, New York, 1965, p. 325.

82. M. Goodman and C. Glazer, in Peptides: Chemistry and Biochemistry, Proceedings of the First American Peptide Symposium (B. Weinstein and S. Lande, eds.), Dekker, New York, 1970, p. 267.

83. Chem. Sources, 1972, Directories Publishing Co., Flemington, New Jersey.

84. C. H. Li, J. Ramachandran, and D. Chung, J. Am. Chem. Soc., 85, 1895 (1963); 86, 2711 (1964).

85. J. Ramachandran, D. Chung, and C. H. Li, J. Am. Chem. Soc., 87, 2696 (1965).

86. H. Brockmann and H. Lackner, Chem. Ber., 101, 1312 (1968).

87. Ref. 78, p. 147.

88. B. Halpern, L. F. Chew, and J. W. Westley, Anal. Chem., 39, 399 (1967).

89. M. A. Ruttenberg, J. Am. Chem. Soc., 90, 5598 (1968).

90. A. A. Costopanagiotis, B. O. Handford, and B. Weinstein, J. Org. Chem., 33, 1261 (1968).

91. D. M. Theodoropoulos and J. S. Fruton, Biochemistry, 1, 933 (1962).

92. H. T. Chung and E. R. Blout, J. Org. Chem., 30, 315 (1965).

93. D. Theodoropoulos and I. Souchleris, J. Org. Chem., 31, 4009 (1966).

94. F. Weygand, W. König, E. Nintz, D. Hoffman, P. Huber, N. M. Khan, and W. Prinz, Z. Naturforsch., 21b, 325 (1966).

95. W. Oelofsen and C. H. Li, J. Org. Chem., 33, 1581 (1968).

96. W. Oelofsen and C. H. Li, J. Am. Chem. Soc., 88, 4254 (1966).

97. S. Lande, J. Org. Chem., 27, 4558 (1962).

98. P. G. Katsoyannis and K. Suzuki, J. Am. Chem. Soc., 85, 2659 (1963).

99. P. G. Katsoyannis, A. M. Tometsko, J. Z. Ginos, and M. A. Tilak, J. Am. Chem. Soc., 88, 164 (1966).

100. J. Ramachandran and C. H. Li, J. Org. Chem., 27, 4006 (1962).

101. N. C. Chaturvedi, W. K. Park, R. R. Smeby, and F. M. Bumpus, J. Med. Chem., 13, 177 (1970).

102. D. Theodoropoulos and I. Souchleris, Acta Chim. (Budapest), 44, 183 (1965).

103. J. Blake and C. H. Li, Biochim. Biophys. Acta, 147, 386 (1967).

104. M. E. Cox, H. G. Garg, J. Hollowood, J. M. Hugo, P. M. Scopes, and G. T. Young, J. Chem. Soc., 1965, 6806.

105. I. Photaki, J. Am. Chem. Soc., 88, 2292 (1966).

106. G. K. Garg and T. K. Virupaksha, Europ. J. Biochem., 17, 13 (1970).

107. V. K. Naithani, K. B. Mathur, and M. M. Dhar, Indian J. Biochem., 6, 10 (1969).

108. A. A. Costopanagiotis, J. Preston, and B. Weinstein, J. Org. Chem., 31, 3398 (1966).

109. J. Ramachandran and C. H. Li, J. Am. Chem. Soc., 87, 2691 (1965).

110. J. E. Shields, Biochemistry, 5, 1041 (1966).

111. R. G. Hiskey, G. W. Davis, M. E. Safdy, T. Invi, R. A. Upham, and W. C. Jones, Jr., J. Org. Chem., 35, 4149 (1970).

112. E. L. Gustus, J. Org. Chem., 32, 3425 (1967).

113. B. O. Handford, T. A. Hylton, K.-T. Wang, and B. Weinstein, J. Org. Chem., 33, 4251 (1968).

114. J. E. Folk and P. W. Cole, Biochim. Biophys. Acta, 122, 244 (1966).

115. M. Kikuchi, M. Hayashida, E. Nakano, and K. Sakaguchi, Biochemistry 10, 1222 (1971).

116. F. H. C. Stewart, Australian J. Chem., 18, 1095 (1965).

117. R. Camble, R. Purkayastha, and G. T. Young, J. Chem. Soc., 1968C, 1219.

118. H. T. Cheung, T. S. Murphy, and E. R. Blout, J. Am. Chem. Soc., 86, 4200 (1964).

119. E. Nicolaides, H. DeWald, R. Westland, M. Lipnik, and J. Poster, J. Med. Chem., 11, 74 (1968).

120. F. Weygand and W. König, Z. Naturforsch., 20b, 710 (1965).

121. C. H. Li, D. Chung, J. Ramachandran, and B. Gorup, J. Am. Chem. Soc., 84, 2460 (1962).

122. C. H. Li, B. Gorup, D. Chung, and J. Ramachandran, J. Org. Chem., 28, 178 (1963).

123. P. J. Thomas, M. Havranek, and J. Rudinger, Collect. Czech. Chem. Commun., 32, 1767 (1967).

124. D. Chung and C. H. Li, J. Am. Chem. Soc., 89, 4208 (1967).

125. C. H. Hassall and J. O. Thomas, J. Chem. Soc., 1968C, 1495.

126. G. R. Pettit and W. R. Jones, J. Org. Chem., 36, 870 (1971).

127. B. O. Handford, T. A. Hylton, J. Preston, and B. Weinstein, J. Org. Chem., 32, 1243 (1967).

128. D. L. Coleman and E. R. Blout, J. Am. Chem. Soc., 90, 2405 (1968).

129. E. L. Gustus, J. Biol. Chem., 239, 115 (1964).

130. S. Gerchakov and H. P. Schultz, J. Med. Chem., 12, 141 (1969).

131. H. Brockmann and H. Lackner, Tetrahedron Letters, 1964, 3523.

132. H. Brockmann and H. Lackner, Chem. Ber., 100, 353 (1967).

133. H. Brockmann and H. Lackner, Tetrahedron Letters, 1964, 3517.

134. H. Brockmann and H. Lackner, Chem. Ber., 101, 2231 (1968).

135. H. Brockmann and H. Lackner, Naturwissenschaften, 51, 384 (1964).

136. H. Brockmann and W. Schramm, Tetrahedron Letters, 1966, 2351.

137. H. Brockmann and F. Seela, Tetrahedron Letters, 1965, 4803.

138. H. Brockmann and F. Seela, Tetrahedron Letters, 1968, 161.

139. H. Takashima and V. du Vigneaud, J. Am. Chem. Soc., 92, 2501
     (1970).

140. R. G. Hiskey, J. T. Stables, and R. L. Smith, J. Org. Chem.,
     32, 2772 (1967).

141. P. G. Katsoyannis and M. Tilak, J. Am. Chem. Soc., 85, 4028
     (1963).

142. F. H. C. Stewart, Australian J. Chem., 18, 1877 (1965).

143. R. Schwyzer, J. P. Carrion, B. Gorup, H. Nolting, and
     A. Tun-Kyi, Helv. Chim. Acta, 47, 441 (1964).

144. S. M. Bloom, S. K. Dasgupta, R. P. Patel, and E. R. Blout,
     J. Am. Chem. Soc., 88, 2035 (1966).

145. P. LeFrancier and E. Bricas, Bull. Soc. Chim. Fr., 1969, 3561.

146. M. A. Ondetti, V. L. Narayanen, M. von Saltza, J. T. Sheehan,
     E. F. Sabo, and M. Bodanszky, J. Am. Chem. Soc., 90, 4711 (1968).

147. M. Okamoto, S. Kimoto, T. Oshima, Y. Kinomura, K. Kawasaki,
     and H. Yajima, Chem. Pharm. Bull., 15, 1618 (1967).

148. H. Yajima, K. Kawasaki, Y. Kinomura, T. Oshima, S. Kimoto, and
     M. Okamoto, Chem. Pharm. Bull., 16, 1342 (1968).

149. C. H. Li, J. Ramachandran, D. Chung, and B. Gorup, J. Am.
     Chem. Soc., 86, 2703 (1964).

150. H. Yajima, K. Kawasaki, Y. Okada, H. Minami, K. Kubo, and
     I. Yamashita, Chem. Pharm. Bull., 16, 919 (1968).

151. C. H. Li, D. Chung, and J. Ramachandran, J. Am. Chem. Soc.,
     96, 2715 (1964).

152. P. LeFrancier and E. Bricas, Bull. Soc. Chim. Biol., 49, 1257
     (1967).

153. M. C. Khosla, N. C. Chaturvedi, H. G. Garg, and N. Arnand, Indian J. Chem., 3, 111 (1965).

154. A. P. Fosker and H. D. Law, J. Chem. Soc., 1965, 4922.

155. R. W. Hanson and H. D. Law, J. Chem. Soc., 1965, 7285.

156. H. Kienhuis, J. P. J. van der Holst, and A. Verweij, Rec. Trav. Chim., 88, 592 (1969).

157. K. Kawamura, S. Kondo, K. Maeda, and H. Umezawa, Chem. Pharm. Bull., 17, 1902 (1969).

158. R. Schwyzer and H. Kappeler, Helv. Chim. Acta, 46, 1550 (1963).

159. M. Bodanszky, J. T. Sheehan, M. A. Ondetti, and S. Lande, J. Am. Chem. Soc., 85, 991 (1963).

160. G. S. Omenn and C. B. Anfinsen, J. Am. Chem. Soc., 90, 6571 (1968).

161. K. Blaha and J. Rudinger, Collect. Czech. Chem. Commun., 30, 3325 1965.

162. J. Rudinger and K. Jošt, Experientia, 20, 570 (1964).

163. K. Jŏst and J. Rudinger, Collect. Czech. Chem. Commun., 32, 1229 (1967).

164. K. Jŏst, Collect. Czech. Chem. Commun., 36, 218 (1971).

165. M. A. Ondetti and P. L. Thomas, J. Am. Chem. Soc., 87, 4373 (1965).

166. R. R. Crenshaw and L. Field, J. Org. Chem., 30, 175 (1965).

167. K. D. Kopple. M. Ohnishi, and A. Go, J. Am. Chem. Soc., 91, 4264 (1969).

168. S. Rajappa and A. S. Akerkar, Tetrahedron Letters, 1966, 2893.

169. M. M. Delépine, Compt. Rend., 260, 717 (1965).

170. G. R. Pettit, A. K. Das Gupta, and R. L. Smith, Can. J. Chem., 44, 2023 (1966).

171. G. R. Pettit, R. L. Smith, and H. Klinger, J. Med. Chem., 10, 145 (1967).

172. G. R. Pettit, R. L. Smith, A. K. Das Gupta, and J. L. Occolowitz, Can. J. Chem., 45, 501 (1967).

173. G. R. Pettit and A. K. Das Gupta, Can. J. Chem., 45, 567 (1967).

174. N. D. Tam, Bull. Soc. Chim. Fr., 1967, 3805.

175. G. DeFaye and M. Fetizon, Bull. Soc. Chim. Fr., 1969, 2835.

176. G. DeFaye, Bull. Acad. Pol. Sci., Ser. Sci. Chim., 19, 1 (1971).

177. G. R. Pettit and S. K. Das Gupta, Can. J. Chem., 45, 1561 (1967).

178. G. E. Krejcarek, B. W. Dominy, and R. G. Lawton, Chem. Commun., 1968, 1450.

179. B. Weiss and R. L. Stiller, J. Chem. Eng. Data, 13, 450 (1968).

180. M. M. Rao, P. F. Rebello, and B. M. Pogell, J. Biol. Chem., 244, 112 (1969).

181. N. C. Chaturvedi, M. C. Khosla, and N. Anand, J. Med. Chem., 9, 971 (1966).

182. T. J. Gill, III, P. S. Marfey, and H. W. Kunz, Biopolymers, 2, 395 (1964).

183. P. S. Marfey, T. J. Gill, III, and H. W. Kunz, Biopolymers, 3, 27 (1965).

184. J. H. Van Campen and J. L. Graham, U. S. Pat. 3,316,095 (April 25, 1967); CA, 67, P27587s (1967).

185. D. M. Burness and B. D. Wilson, U. S. Pat. 3,321,313 (May 23, 1967); CA, 68, P39612e (1968).

186. D. M. Burness and J. J. Looker, French Pat. 1,515,892 (March 8, 1968); CA, 71, P22126p (1969).

187. D. M. Burness and J. J. Looker, French Pat. 1,570,191 (June 6, 1969); CA, 72, P132709j (1970).

188. T. Goodfriend, G. Fasman, D. Kemp, and L. Levine, Immunochemistry, 3, 223 (1966).

189. R. P. Patel and S. Price, Biopolymers, 5, 583 (1967).

190. R. P. Patel, D. V. Lopiekes, S. P. Brown, and S. Price, Biopolymers, 5, 577 (1967).

191. J. A. Haines and P. C. Zamecnik, Biochim. Biophys. Acta, 146, 227 (1967).

192. G. Feinstein, P. Bodlaender, and E. Shaw, Biochemistry, 8, 4949 (1969).

193. S. Avrameas, G. Brown, E. Selegny, and D. Thomas, Ger. Offen. 1,915,970 (October 9, 1969); CA, 72, P63566w (1970).

194. L. L. Csizmas and V. Patel, Ger. Offen. 1,907,977 (September 25, 1969); CA, 72, P63564u (1970).

195. M. Havranek and K. Veres, Z. Naturforsch., 26b, 451 (1971).

196. J. H. Jones and G. T. Young, J. Chem. Soc., 1968C, 53.

197. R. S. Dewey, E. F. Schoenewaldt, H. Joshua, W. J. Paleveda, Jr.,
     H. Schwam, H. Barkemeyer, B. H. Arison, D. F. Verber, R. G.
     Stracham, J. Mikowski, R. G. Denkewalter, and R. Hirschmann,
     J. Org. Chem., 36, 49 (1971).

CHAPTER 5

SYNTHESES OF AMINO ACIDS AND
PEPTIDES UNDER POSSIBLE PREBIOTIC CONDITIONS[*]

Kaoru Harada

Institute for Molecular and Cellular Evolution
and Department of Chemistry
University of Miami
Coral Gables, Florida

I.   FORMATION OF AMINO ACIDS. 298
     A.   Introduction. 298
     B.   Electric Discharge. 299
     C.   Ultraviolet Rays. 305
     D.   Ionizing Radiations 307
     E.   Thermal Energy. 308
     F.   Other Energy Sources. 309
     G.   Amino Acids from Reactive Intermediates 311
II. FORMATION OF PEPTIDES AND POLYPEPTIDES. 316
     A.   Introduction. 316
     B.   Polyglycine Hypothesis. 317
     C.   Thermal Polycondensation of Amino Acids 321

_____

[*]Contribution No. 226 of the Institute for Molecular and
Cellular Evolution.

D. Formation of Peptide Bonds in Aqueous Solution. .   338
E. Formation of Peptide Bonds by the
    Use of Dehydrating Agents in Aqueous Solution . .   340
REFERENCES.   344

I. FORMATION OF AMINO ACIDS

A. Introduction

The formation of amino acids under inferred primitive Earth
conditions was first reported almost 60 years ago. However, because of the lack of good analytical techniques, the products were
not fully characterized. For example, Loeb [1] identified only
glycine from the product obtained by electric discharge in a carbon
monoxide, nitrogen, and water mixture. Lange [2] and Wipperman
[3] found glycine by hydrolysis of hydrogen cyanide oligomer; yet,
they did not detect any other amino acids. By contrast, many new
studies on the abiotic synthesis of amino acids under inferred
primitive terrestrial and cellestial conditions have been reported
in the past 20 years. The analysis of these complex reactions has
been greatly aided by the development of several elegant analytical
methods since the early 1950s. Some of these are: paper chromatography, electrophoresis, ion-exchange chromatography, gas chromatography, and mass spectrometry. These procedures and devices have
made it possible to identify very small amounts of standard and
unusual amino acids.

In general, amino acids have been synthesized from gas mixtures
or chemically reactive precursors by applying various types of energies, such as electric discharge, ultraviolet rays, ionizing radiation,
thermal energy, and other energy sources such as shock waves. The
facts indicate that the formation of amino acids occurs rather easily
under various prebiotic conditions.

B. Electric Discharge

Miller [4-7] synthesized amino acids and other organic compounds by the use of both spark and silent electric discharges from
a reducing gas mixture (hydrogen, 10 mm Hg; methane, 20 mm Hg; ammonia,
20 mm Hg; and water). The spark electric discharge apparatus is shown
in Fig. 1. Flask (A) was heated under boiling conditions and the gas
mixture was forced to circulate through the entire cyclic system. The
electric discharge took place in gas holder (B). The reaction products
passed through the condenser (C), and the condensate came back to flask
(A) through a U-tube (D). The reaction was continued for a few days to
a week. Gas analyses showed that carbon dioxide, carbon monoxide, nitrogen, and hydrogen cyanide were formed during the reaction. With the aid
of an ozonizer, the silent discharge was also used for amino acid syntheses. When nitrogen gas was substituted for ammonia in the spark
discharge apparatus, similar results were obtained. The results are
summarized in Table 1. The yield of glycine in the case of the spark
electric discharge was found to be 2.1% calculated from methane.

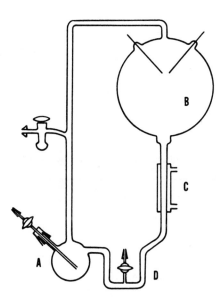

Fig. 1. Apparatus for spark electric discharge experiment.

TABLE 1

Organic Compounds Formed by Electric Discharge

Compound	Spark run 1 $\times 10^5$ mole	Silent run 3 $\times 10^5$ mole	Nitrogen run 6 $\times 10^5$ mole
Glycine	63 (2.1)[a]	80 (0.46)[a]	14.2 (0.48)[a]
Alanine	34	9	1.0
Sarcosine	5	86	1.5
β-Alanine	15	4	7.0
α-Aminobutyric acid	5	1	--
N-Methylalanine	1	12.5	--
Aspartic acid	0.4	0.2	0.3
Glutamic acid	0.6	0.5	0.5
Iminodiacetic acid	5.5	0.3	3.9
Iminoacetic-propionic acid	1.5	--	--
Formic acid	233	149	135
Acetic acid	15.2	135	41
Propionic acid	12.6	19	22
Glycolic acid	56	28	32
Lactic acid	31	4.3	1.5
α-Hydroxybutyric acid	5	1	--
Succinic acid	3.8	--	2
Urea	2	--	2
Methylurea	1.5	--	0.5
Sum of yields of compounds listed	15%	3%	8%

[a]Percentage yield of glycine, based on carbon placed in the apparatus as methane.

Aliquots of the reaction products were taken from the U-tube during the reaction and analyzed for ammonia, hydrogen cyanide, carbonyl compounds, and amino acids. Figure 2 summarizes the

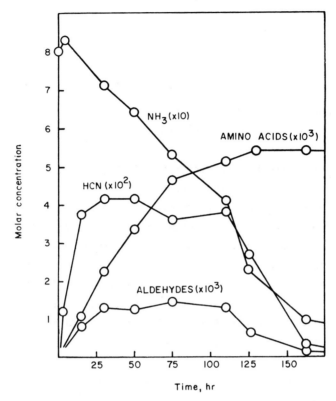

Fig. 2. Analyses of aqueous solution in the electric discharge experiment.

analytical results for the aqueous reaction mixture in the spark
discharge experiment. As is shown in Fig. 2, hydrogen cyanide is a
prime product and carbonyl compounds (which Miller believed were
aldehydes) were detected by 2,4-dinitrophenylhydrazine. Therefore
he thought that amino acid formation proceeded by a Strecker-type
reaction [8]. The synthesized aldehydes react with ammonia and
hydrogen cyanide to form aminonitriles that then hydrolyze to
amino acids.

     In order to prove this, a mixture of ammonia, hydrogen cyanide,
water, and formaldehyde, acetaldehyde, or propionaldehyde were heated
for a week. For the formaldehyde reaction, glycine, glycolic acid,

$$R-CHO + NH_3 + HCN \longrightarrow R-\underset{\underset{NH_2}{|}}{CH}-CN \xrightarrow{\text{H}_2\text{O}} R-\underset{\underset{NH_2}{|}}{CH}-COOH$$

$$\underline{(1)}$$

iminodiacetic acid, and iminoacetic acid-propionic acid were formed
in a 52% yield based on formaldehyde. In the acetaldehyde reaction,
alanine, lactic acid, and iminoacetic acid-propionic acid were found
in a 58% yield based on acetaldehyde. With propionaldehyde, α-aminobutyric acid and α-hydroxybutyric acid were obtained in a 36% yield
based on propionaldehyde. These results seem to support the Strecker-
type mechanism of amino acid formation in the electric discharge
process. In the electric discharge experiment, formation of β-alanine
was also observed. This could be explained by the addition of ammonia
to acrylonitrile, which formed from acetylene and hydrogen cyanide,
and subsequent hydrolysis of the amino nitrile. Similarly, succinic
acid could be synthesized from acrylonitrile by addition of hydrogen
cyanide and a later reaction with water.

$$CH \equiv CH + HCN \longrightarrow CH_2 = CH-CN \xrightarrow{\text{NH}_3} H_2N-CH_2-CH_2-CN \xrightarrow{\text{H}_2\text{O}}$$

$$H_2N-CH_2-CH_2-COOH$$

$$\underline{(2)}$$

Abelson [9] examined the effect of gas composition on amino
acid formation in the electric discharge. In addition to methane,
ammonia, hydrogen, and water, Abelson tried carbon monoxide, carbon
dioxide, nitrogen, and oxygen in his electric discharge experiments.
In any experiment when the over-all gas mixture is reducing, amino
acids (glycine, alanine, β-alanine, and sarcosine) are found among
the reaction products. However, if the over-all gas mixture is
oxidizing, no amino acids are obtained. Heyns and Walters [10] also
tested the effect of gas composition on the resulting amino acids
in a similar spark electric discharge experiment. They used several
combinations of methane, carbon dioxide, ammonia, nitrogen, hydrogen,

water and oxygen, and they found glycine, alanine, sarcosine, β-alanine, and α-aminobutyric acid. Even if some free oxygen was a component of the mixture, and sufficient methane, hydrogen, and ammonia were present, amino acids were synthesized. However, from the mixture of carbon dioxide, nitrogen, and water, ninhydrin positive substances were not detected in the products. When the reaction gas contained hydrogen sulfide, the products consisted of ammonium thiocyanate, thiourea, and thioacetamide; moreover, no sulfur-containing amino acid was identified here.

Pavlovskaya and Pasynskii [11] studied the effect of an electric discharge on a mixture of methane, ammonia, water, and carbon dioxide, which did not contain hydrogen gas. They reported the formation of glycine, alanine, aspartic acid, glutamic acid, β-alanine, and α-aminobutyric acid. It seems that the elimination of hydrogen increased the formation of amino acids. The strongly reducing condition may hinder the possible oxidation of methane to amino acids. According to thermodynamic calculations, the formation of alanine from methane, ammonia, and water under standard conditions requires a large input of free energy.

$$3 \ CH_4 + 2 \ H_2O + NH_3 \longrightarrow CH_3-\underset{\underset{NH_2}{|}}{CH} -COOH + 6 \ H_2$$

$$\Delta G = +62,040 \ cal/mole$$

$$\underline{(3)}$$

However, when water was replaced with carbon monoxide, the formation

$$CH_4 + 2 \ CO + NH_3 \longrightarrow CH_3-\underset{\underset{NH_2}{|}}{CH} -COOH$$

$$\Delta G = -5900 \ cal/mole$$

$$\underline{(4)}$$

of alanine takes place spontaneously. The existence or formation of carbon monoxide as a reactive intermediate under primitive terrestrial or celestial conditions could be considered very likely.

Frank [12] used isooctane and methanol as carbon sources for spark electric discharge experiments with ammonia and water. He observed that the formation of amino acids increased when methanol was used instead of methane. Oró [13] found that when ethane was added to a reaction gas containing methane, the formation of higher amino acids, such as proline, valine, and leucine, was observed in addition to the usual amino acids found in electric discharge experiments. Grossenbacher and Knight [14] detected aspartic acid, threonine, serine, glutamic acid, glycine, alanine, isoleucine, leucine, and lysine by spark electric discharge in a mixture of methane, ammonia, and water.

Matthews and Moser [15] applied a spark electric discharge to a gas mixture of methane and ammonia (1:1). After 60 hr, most of the methane was consumed. Then, the tarry brown material that stuck on the glass surface was hydrolyzed with hydrochloric acid. From the hydrolyzate, they observed peaks corresponding to lysine, histidine, aspartic acid, threonine, serine, glycine, alanine, and isoleucine. They also fractionated the unhydrolyzed brown product and claimed the existence of peptide bonding in the polymer. Yet, as indicated by the infrared absorption spectrum of the polymer, no positive evidence was found for the existence of peptide bonds.

Ponnamperuma and Flores [16] treated a mixture of methane, ammonia, and water with an electric discharge; they noted that 95% of the methane was converted after 24 hr. On hydrolysis of the reaction product, glycine, alanine, threonine, aspartic acid, glutamic acid, valine, leucine, isoleucine, and phenylalanine were observed both by amino acid analysis and by gas chromatography. Ponnamperuma and Woeller [17] passed an electric discharge through a mixture of methane and ammonia (1:1); the products were then examined by both gas chromatography and mass spectrometry. They noted the existence of α-aminoacetonitrile and α-aminopropionitrile among the volatile fractions of the reaction products. As mentioned earlier, Miller thought that these amino nitriles could be formed in his system. The electric discharge experiment without water there-

fore showed the possibility of direct formation of amino nitriles
without passing through the aldehyde stage.

Ishigami [18] and Yuasa et al [19] employed high-frequency
electric discharge (42.5 MC) for several combinations of methane,
ammonia, and water. They identified H and OH radicals from water
vapor, CN radicals from a mixture of methane and ammonia, and CO
radicals from a methane and water mixture. The product of electric
discharge of methane and ammonia (1:1) was condensed by dry ice
(-72°C) and liquid nitrogen (-195°C). The former material was
pale yellow and its infrared absorption spectrum was similar to that
of hydrogen cyanide polymer [20]. This compound contains carbon and
nitrogen in equal amounts. When a mixture of methane, ammonia, and
water (1:1:1) was applied to the high-frequency electric discharge,
a yellow aqueous solution was obtained. From this solution, a small
amount of glycine was detected; however, the yield of the amino acid
was increased considerably after hydrolysis, and glycine, alanine,
aspartic acid, and glutamic acid were detected, too. They found
dicyandiamide and hydrogen cyanide as primary products from a mixture
of methane and ammonia synthesis.

C. Ultraviolet Rays

Bahadur [21, 22] passed sunlight into an aqueous mixture of
paraformaldehyde, potassium nitrate, and ferric chloride. He
reported the formation of aspartic acid, serine, and lysine.
The fixation of molecular nitrogen using colloidal molybdenum
oxide was also claimed in a similar system. Groth and von
Wessenhoff [23, 24] treated mixtures of methane, ammonia, and
water and ethane, ammonia, and water with ultraviolet (uv) light
from a xenon lamp and observed the formation of glycine, alanine,
α-aminobutyric acid, and other organic acids. Ellenbogen [25]
reported the formation of amino acids by irradiation of uv
light to an aqueous mixture of ammonium chloride, iron sulfide,

and methane and subsequent hydrolysis. Terenin [26] applied
high-energy uv light (1000-2000 Å) from a hydrogen lamp to a
mixture of methane, carbon monoxide, ammonia, and water and
found alanine, another amino acid-like material, and aldehydes.
Reid [27] irradiated a mixture of methanol, formaldehyde,
ammonium salt, ammonia, and hydroxylamine containing zinc
oxide and cupric and ferric ions with uv light from a high-
pressure arc lamp. He found the presence of glycine, alanine,
and formic acid. Pavlovskaya and Pasynskii [28] applied uv
light to an aqueous solution of formaldehyde containing
ammonium nitrate and ammonium chloride. In this reaction
mixture they detected glycine, alanine, glutamic acid,
serine, and valine. Dodonova and Sidorova [29] irradiated
a gaseous mixture of methane, ammonia, carbon dioxide, and
water vapor with uv rays (1450-1800 Å) and noted glycine,
alanine, valine, leucine, ethylamine, hydrazine, urea, and
formaldehyde. Abelson [30] described the formation of aminoacetonitrile, a precursor of glycine, by irradiation with uv
light (2536 Å) of an aqueous mixture of ammonium formate,
ammonia, and sodium cyanide. The yield of glycine is up to
10% from ammonium formate. Abelson also treated an aqueous
solution of hydrogen cyanide with uv light from a mercury
lamp and observed the formation of glycine, alanine, serine,
and aspartic acid. Sagan and Khare [31] designed an elaborate
apparatus and irradiated a mixture of methane, ammonia, water,
and hydrogen sulfide with uv rays (2537 Å, 1849 Å). They
found glycine, alanine, cystine, serine, glutamic acid, and
aspartic acid; the formation of cystine is particularly
interesting here.

Although the following experiments were not started from postulated primitive gases, amino acids were synthesized by several types
of photochemical reactions. Deschreider [32] applied uv light
(2537 Å) to an aqueous mixture of a fatty acid and an ammonium salt
and found the formation of amino acids. Glycine and alanine were
formed from succinic acid, while glycine, alanine, and aspartic acid

were produced from maleic acid. Propionic acid formed only glycine, but not alanine. Cultera and Ferrari [33-37] have described a series of amino acid formations by similar photochemical reactions. Such amino acid formations also might have taken place on the primitive Earth.

D. Ionizing Radiations

Dose and Rajewsky [38] irradiated a gaseous mixture of ammonia, water, methane, hydrogen, carbon dioxide, and nitrogen with x-rays and found amines, neutral amino acids, and acidic amino acids in the reaction mixture after electrophoresis. However, they did not identify the individual amino acids. Hasselstrom et al. [39] irradiated an aqueous solution of ammonium acetate with β-rays and detected glycine, aspartic acid, and diaminosuccinic acids. Palm and Calvin [40] exposed a gaseous mixture of methane, ammonia, and water vapor to a 5-MeV electron beam and identified glycine and alanine in the reaction mixture. Oró [41] used a similar apparatus to irradiate a mixture of nitrogen, ethane, and water with electrons. Dose and Ponnamperuma [42] irradiated with γ-rays an aqueous solution of N-acetylglycine and ammonia and identified aspartic acid and threonine.

Paschke et al [43] reported the formation of glycine by irradiating ammonium carbonate with x-rays. Dose and Ettre [44] treated a solution of the ammonium salt of fatty acids with x-rays and observed amino acids that were formed by amination of

$$R-CH_2-COO^- + NH_3 \xrightarrow[X-ray]{} R\underset{\overset{|}{NH_2}}{-CH-COO^-}$$

$$R-CH_2-NH_2 + CO_3^{2-} \xrightarrow[X-ray]{} R\underset{\overset{|}{NH_2}}{-CH-COO^-}$$

(5)

the fatty acids. They also found that amino acids were formed
from amines and carbonate ions when irradiated with x-rays. The
latter reaction is a carboxylation of amines. Berger [45] performed
an experiment by using conditions thought to be present in interstellar
space. He bombarded a mixture of methane, ammonia, and water with
protons at 77°K. Urea, acetamide, and acetone were found; however,
the formation of amino acids was not reported. It was thought that
the reaction conditions were similar to those found in comets or
interstellar dust. Mehran and Pageau [46] prepared aspartic
acid by passing γ-rays from 60_{Co} through an aqueous solution of
alanine.

E. Thermal Energy

The form of energy in a spark electric discharge is qualitatively
composed of three elements: electron beams, ultraviolet light, and
heat. The use of electron beams (β-ray) and ultraviolet light for
syntheses of amino acids has been mentioned above. Thermal energy
is also a likely source of energy that was available on the primordial
Earth. Volcanic activity and outgassing could provide a considerable
amount of energy for the reaction of gases in the primitive atmosphere.
Some of the chemical reactions of primitive gases could be as follows:

$$CH_4 + H_2O \longrightarrow CO + 3 H_2$$
$$CO + NH_3 \longrightarrow HCN + H_2O$$
$$CH_4 + NH_3 \longrightarrow HCN + 3 H_2$$
$$CO + H_2 \longrightarrow hydrocarbon$$
$$CO + NH_3 + H_2 \longrightarrow nitrogen\text{-}containing\ organic\ compounds$$
$$CH_4 \longrightarrow CH_2 = CH_2,\ CH \equiv CH,\ aromatic\ hydrocarbons$$
$$Olefin + CO + H_2O \longrightarrow carbonyl\ compounds$$

(6)

Thus, once hydrogen cyanide, carbon monoxide, and carbonyl compounds were synthesized, various new organic compounds would be prepared by reacting with the coexisting ammonia, water, and hydrogen sulfide.

Harada and Fox [47, 48] heated a mixture of methane, ammonia, and water vapor over silica at 950-1050°C. The reaction gas was absorbed in aqueous ammonia. The ammoniacal solution was heated, then evaporated, and the residue was hydrolyzed. They detected many amino acids in the hydrolyzate by analysis on an amino acid analyzer. Among these were aspartic acid, threonine, serine, glutamic acid, proline, glycine alanine, valine, isoleucine, alloisoleucine, leucine, tyrosine, phenylalanine, α-aminobutyric acid, β-alanine, and lysine. Oró [49] similarly heated methane, ammonia, and water vapor at 1000°C and reported the formation of glycine, alanine, aspartic acid, glutamic acid, threonine, serine, leucine, isoleucine, tyrosine, phenylalanine, and β-alanine. Taube [50] studied thermal amino acid formation in a way similar to that used by Harada and Fox. He observed the formation of various amino acids, hydrocarbons, oxygen-containing organic compounds, and nitrogen-containing organic compounds. Taube et al. [51] heated gas mixture of acetylene and ammonia or acetylene, ammonia, and carbon dioxide at 600°C; the reaction gas was treated as described in their earlier study [50]. They found the amino acids shown in Table 2. When $^{14}CO_2$ was used, the radioactive carbon was incorporated into the synthesized amino acids.

F. Other Energy Sources

Other energy sources under primitive terrestrial conditions could be cosmic rays, kinetic energy of meteorites, and shock waves.

Cosmic rays are composed of about 80% protons, about 20% electrons, and less than 1% heavier charged particles. These particles have extremely high energies which average about 2×10^{10} eV.

TABLE 2

Amino Acids Formed by Thermal Reaction

Reacting gas	Amino acids formed
C_2H_2, NH_3 (1:1)	gly, ala, phe, ser, thr, β-ala, β-phe, try, asp
C_2H_2 NH_3, $^{14}CO_2$ (1:1:1)	gly, ala, phe, ser, thr, β-ala, β-phe, leu, val, lys, arg

However, the total energy of cosmic rays received on the surface of the Earth is only 0.0015 cal/cm^2/year, which corresponds to the light energy from stars at night. Therefore the role of cosmic rays in the chemical evolution of the Earth probably was not great, judging from the amount of available energy. There is no claim that amino acids were formed by utilizing the energy of cosmic rays. However, AMP and ATP formation [52] was reported by the use of thermal neutrons that could be derived from the fast neutrons of cosmic rays.

The velocity of heavenly bodies such as meteorites is very fast as compared with the speeds on the Earth. The speed of meteorites relative to the Earth has been estimated from 18-42 km/sec. Naturally when a meteorite enters the Earth's atmosphere, the surface of the meteorite has a very high temperature and many atmospheric gases are compressed in front of the meteorite. Hochstim [53] considered the chemical reactions accompanied by a high-speed flying body. Gilvarry and Hochstim [54] mentioned that if the oceans of the Earth were created by the collision of huge meteorites, a large amount of organic compounds would have been formed by the event. This type of energy source is very interesting in connection with the abiotic formation of amino acids. However, to try to recreate today those earlier conditions seems very difficult, especially if one seeks to avoid contamination. No positive report on the formation of amino acids using such energy sources has been recorded.

One other interesting energy source is shock waves; for example, thunder creates shock waves. Sagan et al. [55] reported the formation of amino acids from methane, ethane, ammonia, and water by the use of shock-wave energy. They applied shock-wave energy by the use of high-pressure helium (120 lb/in^2) to a mixture of gases (10-90 mm Hg/cm^2). They calculated that the temperature of the reaction gas rose to 1000-2000°K in 10^{-8} sec, and that when the shock wave hit the wall of the tube, the temperature rose to 2000-4000°K. As a result, the molecules were activated by heat and chemically reacted with each other. The reaction product was then cooled adiabatically without destruction of the products. They described the formation of glycine, alanine, valine, and leucine, and calculated that 36% of the ammonia was converted to amino acids.

G. Amino Acids from Reaction Intermediates

More than a century ago, Wolff [56] and Dessaignes [57] heated ammonium salts of fumaric acid, maleic acid, and malic acid and obtained aspartic acid by hydrolyzing the reaction products. Fox et al. [58] similarly heated ammonium salts of fumaric acid and malic acid at 160-200°C and obtained aspartic acid upon hydrolysis. They also observed the formation of a small amount of alanine and β-alanine among the hydrolyzates. Harada [59] proposed the following mechanism for these thermal reactions: the carboxylate (RCOO$^-$NH$_4^+$) group is converted to amide (-CONH$_2$) by heat; the resulting amide group attacks the double bond of other molecules to form an aspartic acid polymer that was hydrolyzed to aspartic acid. The structure of the aspartic acid polymer is discussed later.

Oró et al. [60] heated an aqueous solution of formaldehyde and hydroxylamine hydrochloride at varying temperatures. The reaction product consisted of glycine, alanine, β-alanine, serine, threonine, aspartic acid, formic acid, acetic acid, glycolic acid, and lactic acid. Oró postulated that glycine was formed by the usual Strecker mechanism with aminoacetonitrile as an intermediate.

When an aqueous mixture of hydrogen cyanide and ammonia
(ammonium cyanide solution) was allowed to stand at room temperature
for a few days, a brownish-colored polymer formed spontaneously.
This phenomenon has been recognized for one and a half centuries.
The material produced was named "azurmic acid" by Boullay [61]. The
formation of glycine from oligomers of hydrogen cyanide was already
mentioned [2, 3]. Oro and Kamat [62] heated an aqueous solution of
ammonium cyanide and analyzed for several amino acids (glycine,
alanine, aspartic acid, and serine) in the reaction products. The
carbon source for amino acid formation was hydrogen cyanide, which
contains only one carbon atom. Therefore some reaction yielding a
carbon-carbon bond must have taken place to produce glycine, alanine,
and aspartic acid. Lowe et al. [63] repeated the experiment of Oró
and noted the formation of glycine, alanine, aspartic acid, glutamic
acid, serine, threonine, leucine, isoleucine, β-alanine,
α, β-diaminopropionic acid, and α-aminobutyric acid as well as other
organic compounds.

Matthews and Moser [15, 64, 65] prepared polymers from anhydrous
hydrogen cyanide and ammonia. They observed histidine, aspartic acid,
threonine, serine, glycine, and isoleucine in the hydrolyzate. It
was speculated that diaminoacetonitrile was formed from hydrogen
cyanide and ammonia, that this material in turn became a biradical,
polymerized, and reacted with hydrogen cyanide to form polymers
containing $(C-C-N)_n$ residues, as shown in Reaction 7. The proposed
final product could be polyglycine. They thought that side chains
of the amino acid could be introduced at the polynitrile stage.
The idea is certainly related to Akabori's polyglycine hypothesis,
which is discussed later. However, the structure of the hydrogen
cyanide polymer has not been clarified and the existence of peptide
bonds in the polymer is not certain at this time.

Volker [20] has already proposed a structure for the hydrogen
cyanide polymer. On his mechanism, the dimer of hydrogen cyanide
polymerized to form polymers that constituted the carbon main chain,
which was cyclized to form azurmic acid in the presence of base, as

$$(7)$$

shown in Reaction 8. However, it is difficult to explain why several amino acids, such as glycine, alanine, and aspartic acid, were produced by hydrolysis of polymers I and II. Finally, we are uncertain

II   I

$$(8)$$

as to how the amino acids or their precursors are bonded to the
hydrogen cyanide polymer. The structure of the hydrogen cyanide
polymer and the nature of the precursors of amino acids remain as
future problems in this area.

Heyns and Pavel [66] heated glycine with silica powder at 260-
280°C in order to decompose it. Water, ammonia, hydrogen cyanide,
and carbon monoxide were generated during this procedure. In the
pyrolytic products, glycylglycine, alanine, aspartic acid,
α-aminobutyric acid, oxalic acid, succinic acid, and fumaric acid
were identified. Harada [67] thought that alanine, aspartic acid,
and α-aminobutyric acids could be formed by oligomerization of
hydrogen cyanide based on the existence of hydrogen cyanide in the
pyrolytic products. In order to prove this, he pyrolyzed formamide
at 230°C and then hydrolyzed the pyrolyzate. From the latter, glycine,
alanine, and aspartic acid were found to be the predominant amino acids.
These products were further confirmed by chromatography of several
derivatives. The detailed mechanism of formation is not known; however, the carbon skeleton of the amino acids could be formed by
oligomerization of hydrogen cyanide, as shown in Reaction 9. By
the hydrolytic procedure, these C_2, C_3, and C_4 compounds (or those
produced by hydrolysis) could be converted to glycine, alanine,
and aspartic acid.

Labadie et al. [68, 69] observed the formation of several amino
acids, urea, guanidine, and glycocyamine by hydrolysis of azurmic
acid. Labadie et al. [70] also treated azurmic acid with various
proteolytic enzymes (papain, trypsin, pepsin, and pronase) to detect
peptide bonding; however, azurmic acid was not affected by these
enzymes.

Thus several amino acids are easily formed from hydrogen
cyanide, which could be considered one of the simplest and most
important primitive compounds in the abiotic synthesis of organic
compounds.

Sanchez et al. [71] applied a spark electric discharge to a
mixture of methane and nitrogen and noted that cyanoacetylene was

$$
2 \text{ HCN} \longrightarrow \underset{\underset{C\equiv N}{|}}{H-C=NH} \longrightarrow \underset{\underset{C\equiv N}{|}}{\overset{C\equiv N}{|}{}}H-\overset{|}{C}-NH_2 \longrightarrow H-\underset{\underset{C\equiv N}{\overset{|}{C=NH}}}{\overset{\overset{C\equiv N}{|}}{C}-NH_2}
$$

$$
\underset{\underset{C\equiv N}{|}}{\overset{H-C=NH}{|}{}}C=NH
$$

gly,

glycolic
acid

ala, β-ala,
ser, α,β-diNH$_2$
propionic acid,
lactic acid

asp

(9)

the major constituent among the nitrogen-containing products. The
cyanoacetylene reacts with ammonia, hydrogen cyanide, and water to
form aspartic acid and cytosine according to Ferris et al. [72].
Cyanoacetylene in interstellar space was recently observed by Turner
[73]; as a result, it could play an important prebiotic role.

Friedmann and Miller [74] found the formation of valine (0.2%)
and isoleucine (0.01%) by the respective addition of acetone or
methyl ethyl ketone to an aqueous mixture of ammonium cyanide. They
explained the valine and isoleucine formation by the addition of the
ketones to α-aminoacetonitrile, which could be formed by condensation
of ammonia and hydrogen cyanide.

The formation of sulfur-containing amino acids has not been
reported to date. Hydrogen sulfide could be a constituent of the
materials from outgassing on the primitive Earth. Treatment of
methane, ammonia, water, and hydrogen sulfide with spark electric
discharge (Heyns et al. [10]) resulted in the formation of ammonium
thiocyanate, but not sulfur-containing amino acids, as described
earlier. Herrera [75] found that amino acids were formed by the
reaction of formaldehyde and ammonium thiocyanate; unfortunately,
the formation of sulfur-containing amino acids was not mentioned.

Steinman et al. [76] irradiated with ultraviolet rays an aqueous
solution (0.1 M) of ammonium thiocyanate labeled with ^{14}C. The
reaction mixture was hydrolyzed and then examined by two-dimensional
paper chromatography. They claimed the formation of methionine, as
detected by the autoradiography.

In some prebiotic syntheses of amino acids, phenylalanine was
found among the products. Harada [77] thought that the precursor
of phenylalanine might be phenylacetylene, which could arise from
the polymerization of methine radicals. Phenylacetylene was heated
with ammonia and hydrogen cyanide and the reaction products hydrolyzed
in the usual manner. The products included phenylalanine and β-phenyl-
β-alanine, in addition to glycine, alanine, and aspartic acid.
Friedmann and Miller [78] confirmed that phenylacetylene was obtained
from methane or ethane by treatment with high temperature or ultraviolet rays. It was noted that phenylacetaldehyde was formed from
phenylacetylene by heating with water at pH 7-10. The resulting
phenylacetaldehyde could react with hydrogen cyanide and ammonia
to form phenylalanine after hydrolysis. If hydrogen sulfide was
present, then these resulted in the formation of a small amount of
tyrosine.

II. FORMATION OF PEPTIDES AND POLYPEPTIDES

A. Introduction

The most important biopolymers that make up living structures
are protein, nucleic acid, and polysaccharide. In an evolutionary
context, these high polymers could be regarded as high level material
when compared with the lower-molecular weight bioorganic compounds.
Therefore these three classes are of great interest, for they can
form higher structures associating with other molecules. In order

to produce multimolecular systems, which could possibly develop
into primitive cells, the preexistence of these bioorganic macromolecules is absolutely necessary. Structurally speaking, proteins
are high-molecular-weight compounds that are formed by dehydration
of amino acids. Therefore most of the studies on the formation of
peptides and polypeptides under prebiotic conditions were carried out
by thermal polycondensation or by the use of appropriate dehydrating
agents. The free energy of peptide bond formation from two amino
acids is positive, and the peptide bond formation does not occur
spontaneously unless suitable energies are supplied to the system.

$$H_2N-CH-COOH + H_2N-CH-COOH \longrightarrow dipeptide + H_2O$$

with R below each CH.

$$\Delta G = 2 \sim 4 \text{ kcal}$$

(10)

B. Polyglycine Hypothesis

Akabori [79] started the so-called polyglycine hypothesis to
avoid this energetic difficulty and also to explain the side-chain
formation of amino acids in proteins. He proposed that on the
primitive Earth, the polymerization of aminoacetonitrile, which
formed from formaldehyde, ammonia, and hydrogen cyanide, yielded
a polymer that was partially hydrolyzed to polyglycine. This
polyglycine could react with aldehydes and unsaturated hydrocarbons

$$HCHO + NH_3 + HCN \longrightarrow H_2N-CH_2-CN$$

$$\longrightarrow H_2N-CH_2-\underset{\underset{H}{N}}{\overset{\|}{C}}-(NH-CH_2-\underset{\underset{H}{N}}{\overset{\|}{C}})_n -NH-CH_2-CN$$

$$\overset{H_2O}{\longrightarrow} H_2N-CH_2-\underset{O}{\overset{\|}{C}}-(NH-CH_2-\underset{O}{\overset{\|}{C}})_n -NH-CH_2HCN$$

(11)

to form various amino acid residues. For example, formaldehyde and acetaldehyde would condense at the methylene group of glycine residues to yield serine and threonine residues, respectively. Propene-1 and butene-2 could add to glycine residues to produce valine and isoleucine residues, likewise. The formation of aromatic side chains was explained by the reaction of glycine residues with aromatic aldehydes, followed by dehydration and reduction. The synthesis of aspartic acid residues involved the dehydration of seryl residues to dehydro-alanine residues, then the readdition of hydrogen cyanide to the dehydroalanine residues and subsequent hydrolysis. Thus Akabori thought that polyglycine must have been the most primitive protein on the Earth. The formation of some of the amino acid residues by the polyglycine hypothesis could be carried out, as shown in Reaction 12.

Akabori [80] initially prepared a kaoline-polyglycine complex using glycine N-carboxyanhydride. This kaoline-polyglycine complex was treated with formaldehyde and acetaldehyde under weak alkaline conditions. The reaction mixture was then hydrolyzed and the newly formed serine and threonine residues were measured by column chromatography of the 2,4-dinitrophenyl (DNP) derivatives. The yields of seryl and threonyl residues were 0.4∿3.1% and 1.4∿1.5%, respectively. The addition reactions took place only when the polyglycine was dispersed on kaoline. Akabori [80] also reported the formation of isoleucine by treating the polyglycine dispersed on the Japanese acid clay with an excess of butene-2 at 130°C and subsequent hydrolysis. Hanafusa and Akabori [81] heated aminoacetonitrile with kaoline and detected the existence of diglycine, triglycine, and tetraglycine in the reaction products. However, the formation of high-molecular-weight polyglycine has not been verified under these conditions. Furuyama et al. [82] reacted polyglycine with acetaldehyde in the presence of metallic sodium in liquid ammonia and obtained threonine (7∿9%) after hydrolysis. It was found that 65% of the threonine was in the threo form.

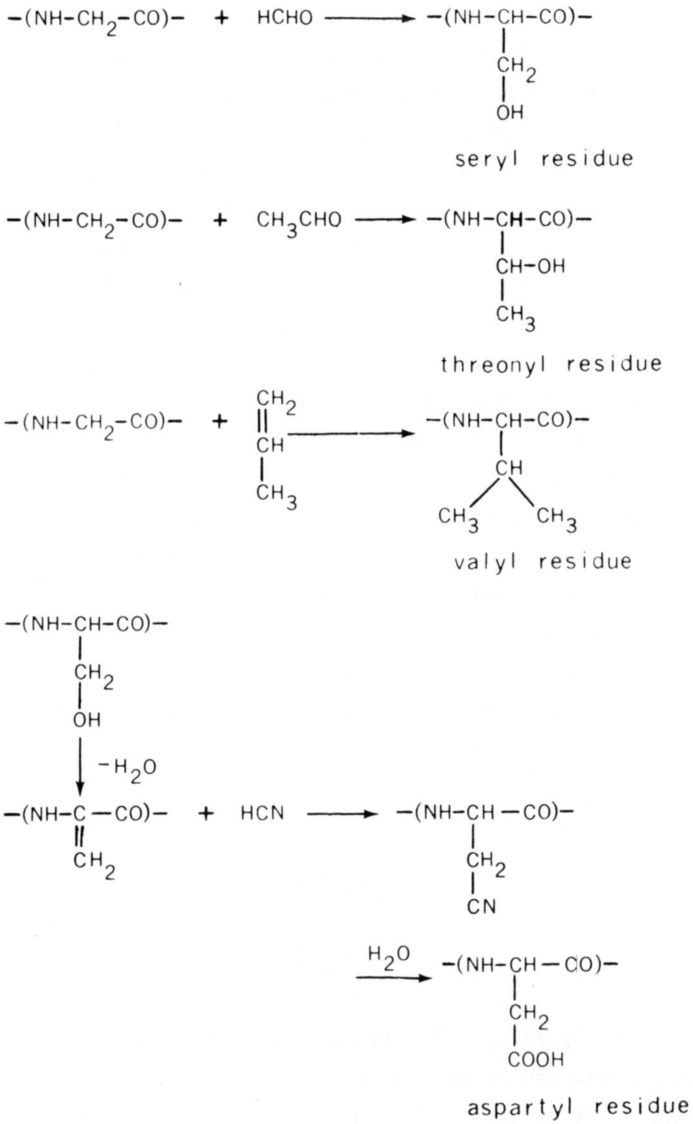

$-(NH-CH_2-CO)-$ + $HCHO$ $\longrightarrow$ $-(NH-CH-CO)-$
$$| \\ CH_2 \\ | \\ OH$$

seryl residue

$-(NH-CH_2-CO)-$ + CH_3CHO $\longrightarrow$ $-(NH-CH-CO)-$
$$| \\ CH-OH \\ | \\ CH_3$$

threonyl residue

$-(NH-CH_2-CO)-$ + $\begin{matrix} CH_2 \\ \parallel \\ CH \\ | \\ CH_3 \end{matrix}$ $\longrightarrow$ $-(NH-CH-CO)-$
$$| \\ CH \\ \diagup \ \diagdown \\ CH_3 \quad CH_3$$

valyl residue

$-(NH-CH-CO)-$
$$| \\ CH_2 \\ | \\ OH$$

$\downarrow$ $-H_2O$

$-(NH-C-CO)-$ + HCN $\longrightarrow$ $-(NH-CH-CO)-$
$$\parallel \qquad\qquad\qquad\qquad | \\ CH_2 \qquad\qquad\qquad\qquad CH_2 \\ \qquad\qquad\qquad\qquad\quad | \\ \qquad\qquad\qquad\qquad\quad CN$$

$\xrightarrow{H_2O}$ $-(NH-CH-CO)-$
$$| \\ CH_2 \\ | \\ COOH$$

aspartyl residue

(12)

Serine and cysteine have been converted to dehydroalanine by pyrolysis or alkali decomposition. When the serine residue in the polymer was dehydrated, the newly formed dehydroalanine residue was

found to react with several reagents. Sakakibara [83] synthesized
polydehydroalanine by the conventional organic techniques (~90% of
the residues were the pyrrolidone type; ~10% of the residues were
the dehydroalanine type) and the polymer was treated with hydrogen
cyanide [84]. The hydrolysis of this polymer showed aspartic acid
in addition to serine, glutamic acid, glycine, and alanine.

According to Akabori's hypothesis, the C-terminal residue of
the polyglycine might be aminoacetonitrile. The methylene group
of the aminonitrile residue would be more reactive compared with
other methylene groups of glycyl residues in the polyglycine.
Harada and Okawara [85] cyanoethylated various N-acylated amino-
acetonitriles in liquid ammonia; hydrolysis of the reaction products
yielded 10~35% of glutamic acid.

By using a method similar to Akabori's method, Losse and
Anders [86] polymerized α-aminopropionitrile in the presence of an
acid-clay at 80°C. This reaction product was partially hydrolyzed
with water (50°C, 50 hr) and there were obtained polymers composed
mostly of alanine with an estimated molecular weight of 20,000. It
was also reported that no polyalanine formation took place in this
reaction without acid clay.

$$n \; H_2N-CH-CN \longrightarrow (-HN-CH \; -C-)_n \xrightarrow[-nNH_3]{+nH_2O} (-HN-CH \; -C-)_n$$

$$\begin{array}{ccc} & & \\ CH_3 & CH_3 \; NH & CH_3 \; O \end{array}$$

polyalanine

(13)

Elad and Sperling [87] and Sperling [88] studied the photochemical alkylation of glycine residues with 1-butene or toluene in
the presence of acetone. The conversion of glycine residues to
either norleucine or phenylalanine ranged from 0.5-3%. When the
other amino acid residues in the peptides containing glycine were
optically active, the alkylated amino acid residues were found to
be partially optically active.

C. Thermal Polycondensation of Amino Acids

Several attempts to obtain protein-like material by heating amino acids were reported more than half a century ago. However, the results of the studies were not clear because of a lack of knowledge about polymer chemistry. Most of the amino acids were decomposed by heating, or were converted to corresponding diketopiperazines. Yet, according to the literature, glycine and aspartic acid formed horny materials by heating. In the following discussion, thermal polymers of glycine and aspartic acid are stressed. The experimental details for the thermal polycondensation of amino acids are described in a review by Fox and Harada [89].

1. Polyglycine

Balbiano and Trasciatti [90] heated glycine in glycerine and obtained a small amount of horny material and diketopiperazine. The material did not dissolve in ordinary organic solvents and gave glycine upon hydrolysis. Alanine did not give horn-like material when heated under similar conditions. Maillard [91, 92] also obtained horny material from glycine in a similar treatment. Abderhalden and Komm [93] placed glycine in a sealed tube at 106°C and found small amounts of polymeric material and diketopiperazine. Shibata [94] heated diketopiperazine in glycerine at 170°C and obtained a colloidal product that remained in a dialysis tube. Polyakova and Vereschagin [95] reacted diketopiperazine with water at 170°C under high pressure (200-400 atm). They found that most of the diketopiperazine was hydrolyzed to glycine; however, about 25% was converted to water-insoluble polyglycine. Meggy [96] studied polyglycine formation using diketopiperazine and noted that the yield of the polymer was dependent upon the heating time and the amount of water in the reaction system. The yield of polyglycine was the highest when the ratio of diketopiperazine and water was 2:1. From these results, Meggy described the equilibrium of the thermal polycondensation reaction as follows:

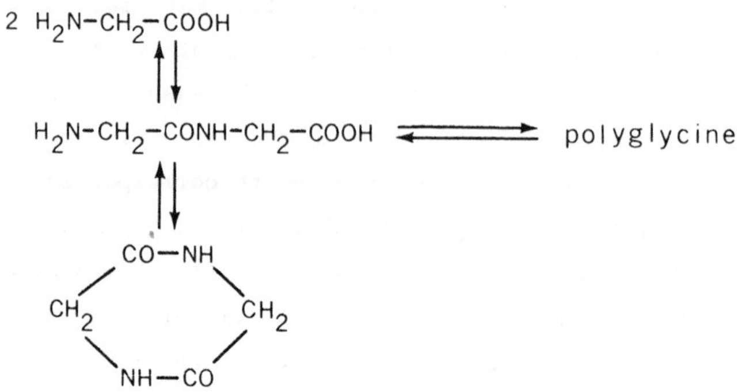

$$2 \ H_2N-CH_2-COOH$$

$$H_2N-CH_2-CONH-CH_2-COOH \rightleftharpoons polyglycine$$

$$
\begin{array}{c}
CO-NH \\
CH_2 \qquad\qquad CH_2 \\
NH-CO
\end{array}
$$

(14)

Meggy also measured the effect of acid and alkali on the thermal polycondensation of glycine. Watanabe and Kozai [97] also analyzed the effect of water, organic acid, and organic base on the thermal polycondensation of glycine. When water constituted one-tenth of the weight of the reactants, the yield of polyglycine was the highest.

2. Polyaspartic Acid

Schaal [98] heated aspartic acid and asparagine under a carbon dioxide atmosphere. Grimaux [99] and Schiff [100, 101] also heated aspartic acid and obtained material that was different from the original aspartic acid. They thought that these products were oligomers of aspartic acid. Schiff separated two components and believed them to be tetra- and octaaspartic acid anhydrides. The derivatives were converted to the acidic substances by heating in water and forming salts with alkali. The so-called octaaspartic acid was an amorphous glass and was easily soluble in water. The tetraaspartic acid was converted quantitatively to octaaspartic acid anhydride by heating at 190-200°C.

Kovacs et al. [102] prepared a polymer by heating DL-aspartic acid at 180-200°C. The polymer was dissolved in aqueous sodium hydroxide and dialyzed. The polyaspartic acid was water soluble and the molecular weight was more than 8000. On the other hand, the solubility of poly-α- DL-aspartic acid in water prepared by the N-carboxyanhydride method was very small. Therefore it was suspected that the peptide bonds of polyaspartic acid prepared by the heating of aspartic acid and subsequent hydrolysis was composed mainly of β-linkages. Kovacs and Konyves [103] converted the poly-aspartic acid to the corresponding methyl ester, and the ester was treated with liquid ammonia to form the amide, as shown in Reaction 15. The Hofmann reaction was then applied to the amide-type polymer. If the polyaspartic acid was a mixture of α- and β-linkages, then the α-linkage would result in the formation of α,β-diaminopropionic acid, and a β-linkage would yield acetaldehyde and carbon dioxide. Thus the ratio of α- and β-linkages could be determined by the ratio of these products from the Hofmann reaction. The results showed the polyaspartic acid was a mixture of α- and β-linkages and the ratio of α:β = 1:1.3. The polyaspartic acid was converted to the original anhydropolyaspartic acid by prolonged heating at 200°C or by heating in tetralin, according to Kovacs et al. [104]. This study of polyaspartic acid has been summarized by Kovacs et al. [105].

Vegotsky et al. [106] confirmed the structure of anhydropolyas-partic acid by the infrared absorption spectra, which had the characteristic bands of the five-membered imide structure at 1720 and 1780 cm^{-1}. It was noted that each anhydroaspartyl residue tightly bound water, and the polymer was actually anhydropolyaspartic acid polyhydrate.

As described earlier, aspartic acid was produced by the hydro-lysis of a polymer that was formed by heating ammonium fumarate and ammonium maleate [58, 59]. Harada [59] showed that the polymer produced by heating these ammonium salts had the anhydropolyaspartic structure as indicated by the infrared absorption spectra.

(15)

3. Thermal Copolycondensation of Amino Acids

The reason for the formation of polyglycine and anhydropolyaspartic acid could be that the reactive functional groups (-NH$_2$,
-COOH) constitute a larger part of these amino acids compared with
other amino acids. The reason for the failure of the thermal polycondensation of other amino acids may be a result of the reaction
conditions, i.e., a single amino acid was heated alone in the solid
state without solvent, or in glycerine. After the dipeptide
formation from amino acids, the reaction may have proceeded in two
ways: (a) to form the tripeptide or (b) to form the diketopiperazine.
The latter cyclization reaction is a first-order reaction; therefore
the use of excess solvent would increase formation of the diketopiperazine. In order to obtain high-molecular-weight polymers, the
amount of solvent must be minimized or, better yet, the amino acids
reacted under molten conditions. The melting of amino acids is
generally difficult because of their amphoteric properties. Yet
there are a few exceptions, such as glutamic acid and lysine, that
can be converted to the corresponding lactams, which melt because
they are not amphoteric.

Harada and Fox [107] heated glycine with molten pyroglutamic
acid at 170°C and the products were dialyzed to obtain a copolymer
of glutamic acid and glycine (see Table 3). The reasons for success
in this copolymerization could be as follows:

1. Reactive amino acids were used.

2. The melted pyroglutamic acid acted as a solvent in
the reaction.

3. The pyroglutamic acid was also a reactant.

4. The pyroglutamic acid acted as an acid catalyst,
as shown in Reaction 16.
Many other amino acids did not form peptides when they were heated
alone. However, when they were heated with glutamic acid, the
reaction mixtures showed positive biuret reactions [107].

$$R-COOH \ + \ H_2N-R' \ \xrightarrow{H^+} \ R-\overset{\overset{\displaystyle OH}{|}}{\underset{\underset{\displaystyle OH}{|}}{C}} \longleftarrow \overset{\overset{\displaystyle H}{|}}{\underset{\underset{\displaystyle H}{|}}{N}}{}^+\!-R'$$

$$\xrightarrow{-H_2O} \ R-\overset{+}{\underset{\underset{\displaystyle OH}{|}}{C}}-NH-R' \longrightarrow \ R-CO-NH-R' \ + \ H^+$$

$$(\underline{16})$$

Free lysine was converted to its lactam by heating. The lactam was then homopolymerized to polylysine by prolonged heating (Harada [108]). The polylysine had α- and ε-linkages; the ratio of linkages was determined by the chromatographic separation of α- and ε-DNP lysine after hydrolysis of the dinitrophenylated polylysine. Free lysine easily copolymerizes with other amino acids on heating.

Harada and Fox [109] copolymerized two acidic amino acids, glutamic acid and aspartic acid, at 160-200°C and 1∿6 hr. The molecular weight of the polymer was in the range of 5000-20,000.

TABLE 3

(Glutamic Acid, Glycine) Copolymer[a]

Heating (hr) (170°C)	Yield (g)	Glu (%)	Gly (%)	Mol wt
1/2	0.12	23	77	12,000
1	0.39	27	73	11,000
2	0.51	27	73	18,000
4	0.71	33	67	18,000

[a]L-Glutamic acid (0.01 mole) and glycine (0.025 mole) were heated at 170°C. The yields shown were after dialysis for 4 days.

$$H_2N-(CH_2)_4-\underset{\underset{NH_2}{|}}{CH}-COOH \xrightarrow[-H_2O]{\Delta}$$

(structure, ring with HN, NH₂, CO) $\xrightarrow{\Delta}$ polylysine

(<u>17</u>)

A similar copolymer of aspartic acid and glutamic acid was prepared
from glutamine and malic acid. In this reaction, glutamine was the
amino donor and malic acid was the amino acceptor. The infrared
absorption spectra showed that the aspartic acid residue in these
thermal polymers was a five-membered anhydroaspartyl structure.

glutamine + malic acid $\longrightarrow$ copolymer of glutamic acid

and aspartic acid

$$>N-CH-CO$$
(ring structure)
$$CH_2CO$$

(<u>18</u>)

The ring structure was hydrolyzed under alkaline conditions to form
the corresponding α- and β-aspartyl residues.

The effect of phosphoric acid on the thermal polycondensation
was examined by Fox and Harada [110]. Three-component thermal
copolymers made up of glutamic acid, aspartic acid, and one other
amino acid were prepared under similar conditions by Harada and
Fox [111].

In these latter polymers, there is a possibility that glutamic
acid and aspartic acid initially form the main chain of the polymer;
then neutral amino acids react with the carboxyl group of the main
chains. In order to examine this possibility, (aspartic acid,
alanine) copolymer, (glutamic acid, alanine) copolymer, and
(lysine, alanine) copolymer were prepared and partially hydrolyzed.
The hydrolyzates were then studied by two-dimensional paper chromatography. The alanylalanine portions of these chromatograms were

cut and eluted, and then analyzed with an automatic amino acid
analyzer. The results are shown in Fig. 3. Two diastereomers of
alanylalanine were observed in each case. The results suggest that
alanine also forms the main chain with aspartic acid and glutamic
acid.

Harada and Fox [112] polymerized amino acids by the use of
polyphosphoric acid, which was prepared from orthophosphoric acid

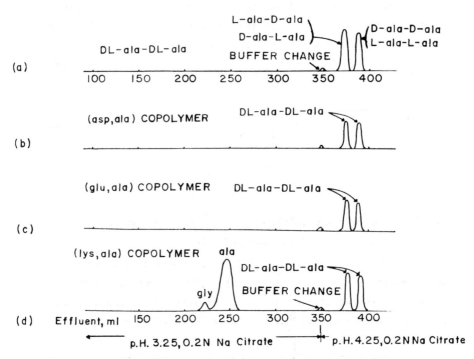

Fig. 3. Chromatograms of alanylalanine derived from
copolymers of (aspartic acid, alanine), (glutamic acid, alanine),
and (lysine, alanine). A through D indicate: standards of the
two diastereomers of alanylalanine; partial hydrolysis of copoly
(asp, ala) isolated on paper and chromatographed on amino acid
analyzer column; partial hydrolysis of copoly (glu, ala) isolated
on paper and chromatographed on amino acid analyzer column; partial
hydrolysis of copoly (lys, ala) put directly onto amino acid
analyzer column after partial hydrolysis.

by heating at 200°, 250°, 300°, and 350°C. The polycondensation
of amino acids using polyphosphoric acid proceeded smoothly at
100°C, and even at 65°C the polycondensation of aspartic acid took
place. The polyphosphoric acid could be regarded as a dehydrating
agent, a solvent, and also an acid catalyst. However, sulfuric
acid, which could be considered to have the same abilities, did not
work effectively for the thermal condensation of amino acids under
similar conditions. This fact might be interesting from the viewpoint of biochemistry.

In addition to the two- or three-component thermal polycondensation of amino acids, Fox and Harada [113, 114] copolymerized
16 other natural amino acids common to protein with glutamic acid
and aspartic acid. Glutamic acid was melted first, then aspartic
acid and the other 16 amino acids were added to the pyroglutamic
acid. The mixture was heated at 170°C for 3~7 hr under nitrogen
atmosphere. After dialysis and hydrolysis, many amino acids were
observed by paper chromatography and by analysis on the amino acid
analyzer. The protein-like material (proteinoid) showed many color
tests common to natural proteins. The proteinoid salted-out in
concentrated salt solutions and salted-in with diluted salt solution.
The average molecular weight by N-terminal assay ranged from 3000-
5000. The infrared absorption spectra showed characteristic peptide
bands at 3300, 3080, 1650 (amide I), and 1550 cm^{-1} (amide II), in
addition to the strong bands at 1720 and 1780 cm^{-1} exhibited by the
five-membered anhydroaspartyl residue.

The nutritive property was measured by the growth of L. arabinosus as measured by lactic acid production. The 1:1:1 acidic
proteinoid* was significantly more utilizable than the 2:2:1 acidic
proteinoid. The result is in accord with the higher content in the

---

*1:1:1 acidic proteinoid means that proteinoid prepared by
thermal polycondensation from amino acids in the following weight
ratio: aspartic acid, 1; glutamic acid, 1; and all other equal
molar neutral and basic amino acids, 1.

2:2:1 acidic proteinoid was prepared from amino acids in the
following weight ratio: aspartic acid, 2; glutamic acid, 2; and
all other equal molar amino acids, 1.

1:1:1 proteinoid of neutral and basic amino acids that are essential
for the growth of L. arabinosus. The nutritive property of the proteinoid was, however, less than that of peptone, and the most active
proteinoid yet tested was calculated roughly to be significantly
over one-half as nutritionally utilizable as peptone by L. arabinosus.
The electrophoretic mobility and proteolyzability of the acidic proteinoid were also measured at the same time.

The amino acid compositions of proteinoid are shown in Table 4
and Fig. 4 [115]. In Table 4, "unpurified" means that the polymer
was isolated from the reaction mixture and was dialytically washed
in cellophane tubing. "Purified" means that the unpurified proteinoid
was dissolved in hot water and the soluble part was separated by filtration while it was hot. The purified proteinoid in the filtrate
precipitated by standing at room temperature. The resulting suspension
was dialyzed for a few hours and was lyophilized. "Repurified" means
that the purification process was repeated. The amino acid composition of unpurified, purified, and repurified proteinoids were almost
the same. High aspartic acid and glutamic acid contents were observed,
and the recovery values increased depending on the purification. The
amino acid composition of the proteinoid and the composition of N-
terminal position are compared in Table 5. For the total amino acid
composition of 1:1:1 acidic proteinoid, the ratio of aspartic acid,
glutamic acid, and basic and neutral amino acids was 55:15:30. However, for the N-terminal composition, the ratio of aspartic acid,
glutamic acid, and basic and neutral amino acids ranged from
$8\sim10:20\sim42:50\sim71$. Similarly, the C-terminal amino acid composition analyzed by hydrazinolysis [116-118] was found to be
different from total amino acid composition (see Table 6). For
the C-terminal composition, the content of acidic amino acids
was only $10\sim15\%$.

The difference of N- and C- terminal amino acid compositions
from the total amino acid compositions of the polymer is interesting
in connection with the mechanism of polymer formation as well as
with the amino acid sequence in the polymer. If the sequence of

TABLE 4

Amino Acid Composition of Proteinoid $(2:2:3)^a$

Amino acid	Unpurified (%)	Purified (%)	Repurified (%)
Lysine	5.1	5.4	5.4
Histidine	1.8	2.0	2.0
Ammonia	8.6	8.1	6.9
Arginine	2.0	2.3	2.4
Aspartic acid	51.7	50.2	51.5
Glutamic acid	10.7	11.6	12.0
Proline	0.7	0.6	0.6
Glycine	2.7	3.1	2.8
Alanine	4.0	4.3	5.5
1/2 Cystine	4.5	3.5	3.4
Valine	1.2	1.2	1.2
Methionine	1.8	1.9	1.7
Isoleucine	1.2	1.3	0.9
Leucine	1.3	1.2	1.1
Tyrosine	2.0	1.9	1.7
Phenylalanine	1.8	1.7	1.5
Recovery	81.8	97.5	100.0

^{a}Proteinoid was hydrolyzed at 110°C for 4 days under reduced pressure in a sealed tube.

amino acids in the polymer was completely random, the amino acid compositions of N- and C-terminal positions would be the same as in the total amino acid composition. Therefore the reason for the wide difference of N- and C-terminal amino acid compositions from the total amino acid compositions suggests that the proteinoid was not completely random.

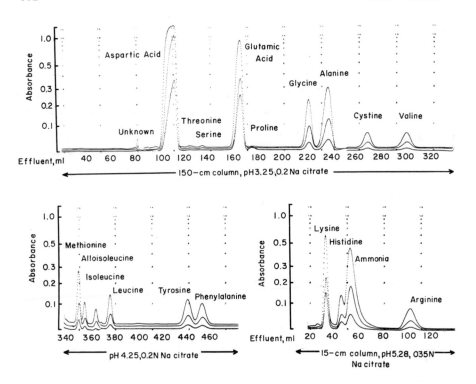

Fig. 4. Chromatogram of the hydrolyzate of (2:2:3) proteinoid
as obtained on the automatic amino acid analyzer.

TABLE 5

Comparison of Total Amino Acid Composition and

N-Terminal Amino Acid Composition of (1:1:1) Proteinoid

Reaction temp. (°C)	Mol wt	Total amino acid (%)			N-Terminal amino acid (%)		
		Asp	Glu	Other amino acid	Asp	Glu	Other amino acid
160	3600	56	14	30	8	42	50
170	3800	55	13	32	7	30	63
180	4000	55	14	31	10	20	71

TABLE 6

C-Terminal Amino Acid Composition of Proteinoid

C-Terminal amino acid	2:2:1 Proteinoid (%)	2:2:3 Proteinoid (%)
Lys	14.5	18.5
His	5.2	5.4
Asp	4.7	9.2
Glu	10.1	1.1
Gly	8.0	6.9
Ala	18.1	11.8
1/2 Cys	7.6	11.0
Val	4.3	3.4
Met	6.0	6.0
Ilu	5.1	4.4
Leu	5.0	3.8
Tyr	6.0	6.6
Phe	5.4	5.0
Total	7.5 μmole/10 mg	7.6 μmole/10 mg

Fox and Nakashima [119] treated proteinoid with liquid ammonia and fractionated the amide-type proteinoid on a DEAE-cellulose column. They obtained six peaks (see Fig. 5). After further purification, fractions 3, 4, and 5 were found to be very similar in amino acid composition and appeared to be rather homogeneous by high-voltage electrophoresis, ultracentrifugation, and gel filtration. The word "homogeneous" used here does not have the same meaning as in contemporary protein chemistry, but these polymers must have some order in their amino acid sequence. Such an arrangement in a thermal proteinoid without control by nucleic

acids may be regarded as the starting point for the evolution
of protein on the primordial Earth. Chemically speaking, the
reactivity of amino acids, the stability of peptides, and the
steric factors would control the amino acid sequence in the
polymer. The fact that polymer formation takes place by the

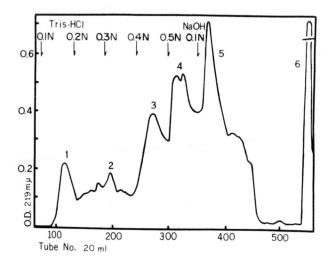

Fig. 5. Elution pattern of amidated (1:1:1) proteinoid from
DEAE cellulose.

polycondensation mechanism should not be overlooked; splitting of
peptide bonds and also transpeptidation reactions would constantly
occur during the elongation of the peptide chain in order to
stabilize the sequence of amino acids under the reaction conditions.

A large part of the aspartyl residues in the thermally prepared
polyamino acids had the anhydroaspartyl structure. The anhydroaspartyl
residues were hydrolyzed easily in basic and even in neutral conditions.
Therefore it was difficult to analyze the amount of the free carboxyl
group in the polymer by titration. If a nonaqueous solvent (dimethylformamide, for example) was used, then the carboxyl groups of the
polymer were titrated easily with sodium methoxide without hydrolyzing

the anhydroaspartyl residues. For another analysis, the anhydroaspartyl residues were converted to asparginyl residues (α-and β-amides) by treatment with liquid ammonia. The amide-type polyamino acid was purified and the amide content (imide content) was analyzed by the Kjedahl method by Fox and Harada [120].

$$
\underset{\underset{CH_2CO}{|}}{\overset{\text{>N-CH-CO}}{|}}\text{N-} \quad \xrightarrow{NH_3} \quad
\left[
\begin{array}{c}
\overset{H}{\underset{}{-N-CH-CO-}} \\[2pt]
\overset{|}{CH_2} \\[2pt]
\overset{|}{CONH_2} \\[12pt]
\overset{H}{\underset{}{-N-CH-CH-CO-}} \\[2pt]
\underset{CONH_2}{|}
\end{array}
\right]
\quad \xrightarrow{NaOH} \quad NH_3
$$

(19)

Based on these determinations, as well as on the results of total amino acid composition, the contents of the carboxylic group and the anhydroaspartyl residue for the thermal polyamino acid were calculated, as shown in Table 7 [121].

The thermal polyamino acids mentioned above are very rich in acidic amino acids or basic amino acids. Fox and Waehneldt [122] also synthesized polymers thermally from an equimolar mixture of amino acids. The yield of polymer was rather low; however, the polymer contained all of the amino acids in the starting mixture, especially glycine, alanine, and glutamic acid. This composition could indicate the reactivity of these amino acids. On the other hand, glycine, alanine, and glutamic acid are generally the highest constituents of protein and, as we discussed earlier, these amino acids were also commonly formed in the various prebiotic syntheses of amino acids (electric discharge, thermal, hydrogen cyanide, etc.).

Some interesting properties of thermal proteinoids have been studied in recent years. These include: (a) microsphere formation in aqueous solution; (b) catalytic activity; and (c) hormonal activity.

TABLE 7

Functional Group Analyses of Proteinoids

Functional groups in the polymer	Anhydropoly-aspartic acid (μmole/10 mg)	2:2:1 Proteinoid (μmole/10 mg)	2:2:3 Proteinoid (μmole/10 mg)
-COOH	19.5	17.9	15.8
-CONH$_2$	2.0	2.3	2.6

-CH-CO
      \N-
    /
CH$_2$-CO
	69.2	51.9	36.6

C-Terminal amino acid	4.9	7.9	7.6
Aspartyl residue Glutamyl residue	14.6	10.0	8.3
Amino acid residue in 10 mg	87.0[a]	84.3[b]	84.6[b]

[a]Calculated as anhydropolyaspartic acid polyhydrate.

[b]Calculated from amino acid analysis.

When an acidic proteinoid was heated with water and allowed to cool at room temperature, the solution became turbid. Fox et al. [123] found that numerous microspheres had formed. They called these spherules "proteinoid microspheres" (see Fig. 6). The microspheres were of rather uniform size (1∿10 μ) and their size varied depending upon the proteinoid, salt solution, and rate of cooling. Microspheres were also formed by cooling the temperature of the solution from 25° to 0°C, according to Young [124]. Since a proteinoid could be regarded as a possible primitive protein on the primordial Earth, proteinoid microspheres that form upon simple heating with water could also be regarded as primitive multimolecular systems. Gram staining [125], various forms [126], electron micrographs [127], dynamic budding phenomena [128],

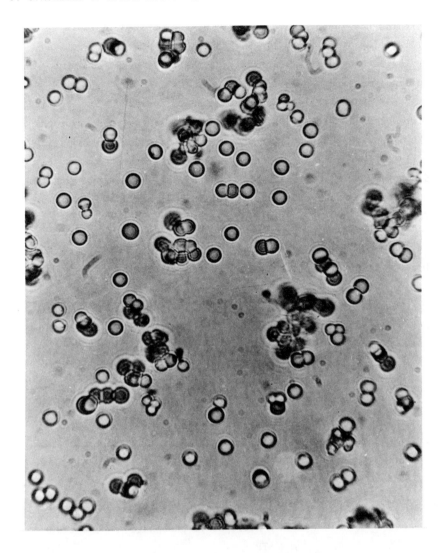

Fig. 6. Proteinoid microspheres. Average diameter 1.9 μ.

and interaction of proteinoid and polynucleotides [129, 130] and other related studies have been described for the microspheres.

From the viewpoint of chemical evolution, a primitive protein might have some weak catalytic activities that would develop into early enzymes. Several catalytic activities of the thermal pro-

teinoids have been studied. These are summarized in Table 8
[131-141]. Along these lines, Rohlfing and Fox [142] wrote a
review on the catalytic activity of thermally prepared amino
acid polymers.

Fox and Wang [143] prepared a thermal polymer from arginine,
glutamic acid, glycine, histidine, phenylalanine, and tryptophan.
They found that the polymer showed clear melanocyte-stimulating
activity (MSH activity) (see Fig. 7). The mixture of amino acids
did not show the MSH activity. The polymer was not fractionated;
therefore it may be possible to obtain by fractionation polymers
that have higher MSH activity.

D. Formation of Peptide Bonds in Aqueous Solution

When Oró and Guidry [144] dissolved glycine in a N ammonia in
water and heated the solution at 140°C, polyglycine was obtained.

TABLE 8

Catalytic Activities of Proteinoids and
Other Thermal Polyamino Acids

Type of reaction	Substrate	Reference
Hydrolysis	p-Nitrophenyl acetate	131-134
	p-Nitrophenyl phosphate	135
	ATP	136
Decarboxylation	Glucoronic acid	137
	Pyruvic acid	138, 139
	Oxaloacetic acid	140
Amination	α-Ketoglutaric acid	141

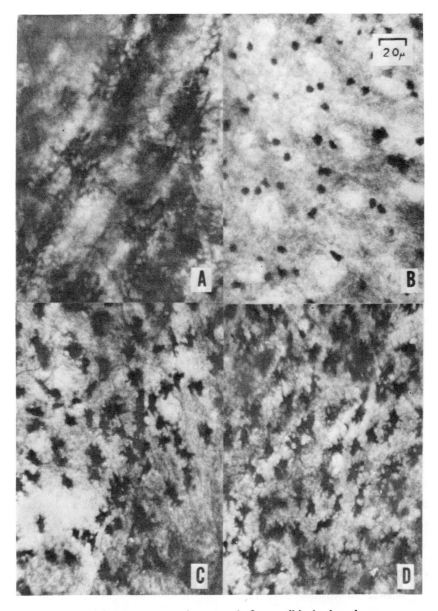

Fig. 7. (a) Melanocyte in normal frog; (b) in hypohysec-
tomized frog; (c) in hypophysectomized frog treated with native
MSH; (d) in hypophysectomized frog treated with thermal copolymer
of six amino acids.

The degree of polymerization was up to 18 residues. The infrared
spectra of the polymer agree with that of polyglycine. When the
temperature was raised above 140°C, the yield of polymer decreased;
however, prolonged heating at a lower temperature increased the yield
of polyglycine (40%). When the glycine was heated with other amino
acids in the ammoniacal solution, there were obtained copolymers that
contained other amino acids, and which showed a positive biuret
reaction. Oró and Guidry [145] reported the polycondensation of
glycinamide by heating in ammoniacal solution. The polycondensation
reaction of glycinamide proceeded faster than that of glycine and
the reaction takes place slowly, even at room temperature.

 Kovacs and Nagy [146] heated a solution of asparagine and
obtained polyaspartic acid. The polymer was very soluble in water,
implying a mixture of α- and β-linkages in peptide bonds. The
formation of polyaspartic acid, could be explained by transpeptidation
from the amide bond of asparagine to the α-amino group of a second
molecule of asparagine.

$$\text{asparagine} \xrightarrow[\text{heat}]{\text{H}_2\text{O}} \text{polyaspartic acid}$$

$$\underline{(20)}$$

 As mentioned earlier, peptide bond formation from amino acids
generally requires energy, and the equilibrium of the reaction is
inclined to the amino acid side. Since peptide or polyamino acid
formation from glycine and asparagine takes place, the differences
of free energies of these reactions could be considered to be small
under these conditions.

E. Formation of Peptide Bonds by the Use of Dehydrating Agents in Aqueous Solution

 Several investigators have studied the formation of peptide
bonds in aqueous solution by coupling this reaction to another
chemical reaction.

Calvin and Calvin [147] proposed that hydrogen cyanide could act as a dehydrating agent in peptide bond formation, as shown in Reaction 21. In this mechanism, hydrogen cyanide was converted to

(21)

formamide during peptide bond formation. Such an idea led to the investigation of several nitrile derivatives as dehydrating agents for peptide bond formation in aqueous solution.

Carbodiimide-type compounds have been commonly used in peptide syntheses. For example, dicyclohexylcarbodiimide was first employed in peptide formation by Sheehan et al. [148, 149], while the formation of phosphoric acid esters by the use of carbodimide was developed by Khorana [150]. However, dicyclohexylcarbodiimide may not be relevant to peptide formation on the primitive Earth, although cyanamide may have been such a dehydrating agent. Cyanamide could be considered as a tautomer of carbodiimide. Several cyanamide-type compounds have been evaluated for peptide bond formation in aqueous solution, which includes cyanamide (CA), dicyandiamide (DCDA), and dicyanamide (DCA). The structures of these compounds are shown in Table 9.

TABLE 9

Structures of Dehydrating Agents

Name	Structure	Abbreviation
Cyanamide	H_2N-CN	CA
Dicyandiamide (cyanoguanidine, cyanamide dimer)	$H_2N-\underset{\overset{\|}{NH}}{C}-NH-CN$	DCDA
Dicyanamide (Dicyanamide Na salt)	NC-NH-CN	DCA (NaDCA)

Steinman et al. [151, 152] reported alanylalanine and alanyla-lanylalanine formation from alanine in aqueous solution by the use of DCDA (yield, 1-3%). Using these dilute conditions, DCDA was found to be more effective in peptide formation than CA,* while peptide formation was favored in acidic conditions (pH 2).

Ponnamperuma and Peterson [153] reacted glycine-^{14}C and leucine-^{14}C in aqueous solution in the presence of CA under irradiation by uv rays (1800 Å). They found the formation of glycylglycine, glycylleucine, leucylglycine, leucylleucine, glycylglycylglycine, and leucylglycylglycine. Steinman et al. [154] reported the effect of DCA in peptide formation from alanine. They found the formation of alanylalanine (1.6%) in the reaction mixture. Steinman et al. [155] noted glycine peptide formation from glycine-^{14}C by the use of NaDCA in aqueous solution and detected the formation of glycylglycine, triglycine, and tetraglycine. Again the condensation went best under acidic conditions (pH 2). The dehydration mechanism is probably similar to the mechanism involving dicyclohexylcarbodiimide,

$$H_2N-C{\equiv}N \rightleftarrows HN{=}C{=}NH$$

$$\underline{(22)}$$

*The commercial product called cyanoguanidine is a dimer of cyanamide, mp 209°C.

as shown in Reaction 23. The dehydrating agent DCA was converted
to cyanourea during peptide formation. N,N,N'-Tricyanoguanidine and
N-cyano-N'-glycylurea were observed as byproducts of the peptide
formation reaction.

Finally, such dehydrating agents must be formed continuously,
since these compounds were consumed in the peptide formation or in
other dehydration reactions. Schimpl et al. [156] irradiated an
aqueous mixture of $H^{14}CN$ and ammonia with ultraviolet rays and noted
the formation of DCDA (2-3%).

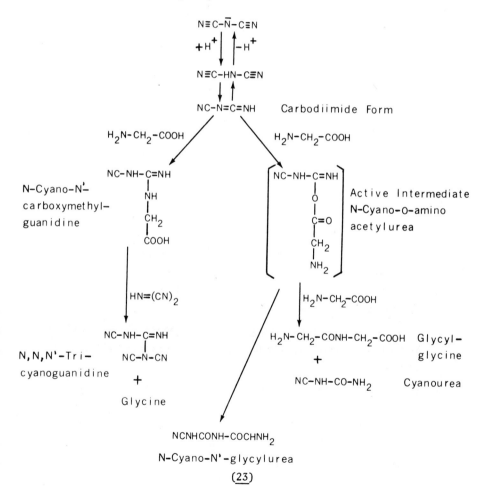

(23)

ACKNOWLEDGMENT

This work was supported by grant no. NGR 10-117-052 from the U.S. National Aeronautics and Space Administration. The author thanks Dr. Sidney W. Fox, director of the Institute, for his valuable discussions, encouragement, and help.

REFERENCES

1. W. Loeb, Ber., 46, 690 (1913).

2. O. Lange, Ber., 6, 99 (1872).

3. R. Wippermann, Ber., 7, 767 (1874).

4. S. L. Miller, Science, 117, 528 (1953).

5. S. L. Miller, J. Am. Chem. Soc., 77, 2351 (1955).

6. S. L. Miller, Ann. N.Y. Acad. Sci., 69, 260 (1957).

7. S. L. Miller, Biochim. Biophys. Acta, 23, 480 (1957).

8. A. Strecker, Ann. Chem., 75, 27 (1950).

9. P. H. Abelson, Carnegie Inst. Wash. Yearbook, 53, 97 (1953-54).

10. K. Heyns, W. Walters, and E. Meyer, Naturwissenschaffen, 44 385 (1957).

11. T. E. Pavlovskaya and A. G. Pasynskii, The Origin of Life on the Earth, I.U.B. Symposium Series I, Pergamon Press, New York 1959, p. 151.

12. B. Frank, Chem. Ber., 93, 446 (1960).

13. J. Oró, Nature, 197, 862 (1963).

14. K. A. Grossenbacher and C. A. Knight, in The Origins of Pre-
     biological Systems and of Their Molecular Matrices (S. W. Fox,
     ed.), Academic, New York, 1965, p. 173.

15. C. N. Matthews and R. E. Moser, Proc. Nat. Acad. Sci., 56, 1087
     (1966).

16. C. Ponnamperuma and J. Flores, American Chemical Society 152nd
     Natl. Meeting, 1966.

17. C. Ponnamperuma and F. Woeller, Currents in Modern Biology, 1
     156 (1967).

18. M. Ishigami, Proc. Japan Acad., 45, 3519 (1969).

19. S. Yuasa, M. Ishigami, Y. Honda, and K. Imahori, Sci. Rept.,
     Osaka Univ., 19, No. 1, 33; No. 2, 7 (1970).

20. T. Völker, Agnew. Chem., 72, 379 (1960).

21. K. Bahadur, Nature, 173, 1141 (1954).

22. K. Bahadur et al., Nature, 182, 1668 (1958).

23. W. Groth and H. von Wessenhoff, Naturwissenschaffen, 44, 510(1957).

24. W. Groth and H. von Wessenhoff, Planetary Space Sci., 2, 79
     (1960).

25. E. Ellenbogen, Abstr. Amer. Chem. Soc. Meeting, Chicago,
     1958, p. 47c.

26. A. N. Terenin, The Origin of Life on the Earth, I.U.B.
     Symposium Series I, Pergamon Press, New York, 1959, p. 136.

27. C. Reid, The Origin of Life on the Earth, I.U.B. Symposium
     Series I, Pergamon Press, New York, 1959, p. 183.

28. T. E. Pavlovskaya and A. G. Pasynskii, The Origin of Life on the
     Earth, I.U.B. Symposium Series I, Pergamon Press, New York,
     1959, p. 151.

29. N. Y. Dodonova and A. L. Sidorova, Biofizika, 6, 14 (1961).

30. P. H. Abelson, Proc. Nat. Acad. Sci., 55, 1365 (1966).

31. C. Sagan and B. N. Khare, Science, 173, 417 (1971).

32. A. R. Deschreider, Nature, 182, 528 (1958).

33. R. Cultera and G. Ferrari, Ann. Chim. (Rome), 47, 1321 (1957);
     47, 1331 (1957); 48, 1410 (1958); 48, 1419 (1958); 49, 176
     (1959); 49, 1639 (1959).

34. G. Ferrari, Ann. Chim (Rome), 49, 2017 (1959).

35. R. Cultera and G. Ferrari, Agrochimica, 5, 108 (1961).

36. G. Ferrari and R. Cultera, Nature, 190, 326 (1961).

37. G. Ferrari, C. Passera, and A. Pedrotti, Gazz. Chim. Ital., 94, 223 (1964).

38. K. Dose and B. Rajewsky, Biochim. Biophys. Acta, 25, 225 (1957).

39. T. Hasselstrom, M. C. Henry, and B. Murr, Science, 125, 350 (1957).

40. C. Palm and M. Calvin, J. Am. Chem. Soc., 84, 2115 (1962).

41. J. Oró, Nature, 197, 971 (1963).

42. K. Dose and C. Ponnamperuma, Radiation Res., 31, 650 (1967).

43. R. Paschke, R. W. H. Chang, and D. Young, Science, 125, 881 (1957).

44. K. Dose and K. Ettre, Z. Naturforsch., 13b, 784 (1958).

45. R. Berger, Proc. Nat. Acad. Sci., 47, 1434 (1961).

46. A. R. Mehran and R. Pageau, Can. J. Biochem., 43, 1359 (1965).

47. K. Harada and S. W. Fox, Nature, 210, 335 (1964).

48. K. Harada and S. W. Fox, The Origins of Prebiological Systems and of Their Molecular Matrices (S. W. Fox. ed.), Academic, New York, 1965, p. 187.

49. J. Oró, The Origins of Prebiological Systems and of Their Molecular Matrices (S. W. Fox, ed.), Academic, New York, 1965 p. 137.

50. M. Taube, Z. Zdrojewski, K. Samochocka, and K. Jerierska, Angew. Chem., 79, 239 (1967); Angew. Chem. Intern. Ed., Engl. 6, 247 (1967).

51. K. Samochocka, A. L. Lawcznski, and M. Taube, Angew. Chem., 80, 396 (1968).

52. M. Akaboshi, K. Kawai, and A. Waki, Biochim. Biophys. Acta, 238, 5 (1971).

53. A. R. Hochstim, Proc. Nat. Acad. Sci., 50, 200 (1963).

54. J. J. Gilvarry and A. R. Hochstim, Nature, 197, 624 (1963).

55. A. Bar-Nun, N. Bar-Nun, S. H. Bauer, and C. Sagan, Science, 168, 470 (1970).

56. J. Wolff, Ann. Chem., 75, 294 (1850).

57. V. Dessaignes, Compt. Rend., 30, 324; 31, 432 (1850).

58. S. W. Fox. J. E. Johnson, and M. Middlebrook, J. Am. Chem. Soc., 77, 1048 (1955).

59. K. Harada, J. Org. Chem., 24, 1662 (1959).

60. J. Oró, A. P. Kimball, R. Fritz, and F. Master, Arch. Biochem. Biophys., 85, 115 (1959).

61. P. Boullay, Ann. Chim. Phys., 43, 273 (1830).

62. J. Oró and S. S. Kamat, Nature, 190, 442 (1961).

63. C. U. Lowe, M. W. Rees, and R. Markham, Nature, 199, 219 (1963).

64. C. N. Matthews and R. E. Moser, Nature, 215, 1230 (1967).

65. R. E. Moser, A. R. Claggett, and C. N. Matthews, Tetrahedron Letters, 1968, 1605.

66. K. Heyns and K. Pavel, Z. Naturforsch., 12b, 97, 109 (1957).

67. K. Harada, Nature, 214, 479 (1967).

68. M. Labadie, R. Jensen, and E. Neuzil, Bull. Soc. Chim. Biol., 49, 673 (1967).

69. M. Labadie, R. Jensen, and E. Neuzil, Biochim. Biophys. Acta, 165, 525 (1968).

70. M. Labadie, S. Ducastaing, and J. C. Breton, Bull. Soc. Pharm. Bordeaux, 107, 61 (1968).

71. R. A. Sanchez, J. P. Ferris, and L. E. Orgel, Science, 154, 784 (1966).

72. J. P. Ferris, R. A. Sanchez, and L. E. Orgel, J. Mol. Biol., 33, 693 (1968).

73. B. Turner, Astrophys. J., 163, L35 (1971).

74. N. Friedmann and S. L. Miller, Nature, 221, 1152 (1969).

75. A. L. Herrera, Science, 96, 14 (1942).

76. G. D. Steinman, A. E. Smith, and J. J. Silver, Science, 159, 1108 (1968).

77. K. Harada, Fourth Annual Report of Institute of Molecular Evolution, University of Miami, 1968, p. 32.

78. N. Friedmann and S. L. Miller, Science, 166, 766 (1969).

79. S. Akabori, Kagaku (Tokyo), 25, 54 (1955).

80. S. Akabori, K. Okawa, and M. Sato, Bull. Chem. Soc. Japan, 29, 608 (1956).

81. H. Hanafusa and S. Akabori, Bull. Chem. Soc. Japan, 32, 626
     (1959).

82. T. Furuyama, F. Sakiyama, and K. Narita, Bull. Chem. Soc.
     Japan, 36, 903 (1963).

83. S. Sakakibara, Bull. Chem. Soc. Japan, 33, 814 (1960).

84. S. Sakakibara, Bull. Chem. Soc. Japan, 34, 205 (1961).

85. K. Harada and T. Okawara, in Molecular Evolution: Prebiological
     and Biological (D. L. Rohlfing and A. I. Oparin, eds.),
     Plenum, New York, 1972.

86. G. Losse and K. Anders, Z. Physiol. Chem., 323, 111 (1961).

87. D. Elad and J. Sperling, Chem. Commun., 1968, 655.

88. J. Sperling, J. Am. Chem. Soc., 91, 5389 (1969).

89. S. W. Fox and K. Harada, "Thermal Polymerization of α-Amino
     Acids," in A Laboratory Manual of Analytical Methods of Protein
     Chemistry (P. Alexander and H. P. Lundgren, eds.), Vol. 4,
     Pergamon, New York, 1966, p. 129.

90. L. Balbiano and D. Trasciatti, Ber., 33, 2323 (1900).

91. L. C. Maillard, Compt. Rend., 153, 1078 (1911).

92. L. C. Maillard, Ann. Chim., 1, 519 (1914).

93. E. Abdelhalden and E. Komm, Z. Physiol. Chem., 139, 147 (1924).

94. K. Shibata, Acta Phytochim., 2, 193 (1925).

95. A. L. Polyakova and M. F. Vereschagin, Dok. Akad. Nauk S.S.S.R.,
     64, 607 (1949).

96. A. B. Meggy, J. Chem. Soc., 1953, 851; 1956, 1444.

97. H. Watanabe and Y. Kozai, Nippon Kagaku Zasshi, 84, 744 (1963).

98. E. Schaal, Ann. Chem., 151, 26 (1871).

99. E. Grimaux, Bull. Soc. Chim., 38, 64 (1882).

100. H. Schiff, Ber., 30, 2449 (1897).

101. H. Schiff, Ann. Chem., 307, 231 (1899).

102. J. Kovacs, J. Konyves, and A. Pusztai, Experientia, 9, 459 (1953).

103. J. Kovacs and J. Konyves, Naturwissenschaften, 41, 333 (1954).

104. J. Kovacs, J. Konyves, and J. Casszar, Naturwissenschaften, 41,
     575 (1954).

105. J. Kovacs, H. Nagy, J. Konyves, J. Casszar, T. Vajda, and H. Mix,
     J. Org. Chem., 26, 1084 (1961).

106. A. Vegotsky, K. Harada, and S. W. Fox. J. Am. Chem. Soc., 80, 3361 (1958).

107. K. Harada and S. W. Fox. J. Am. Chem. Soc., 80, 2694 (1958).

108. K. Harada, Bull. Chem. Soc. Japan, 32, 1008 (1959).

109. K. Harada and S. W. Fox, Arch. Biochem. Biophys., 86, 274 (1960).

110. S. W. Fox and K. Harada, Arch. Biochem. Biophys., 86, 281 (1960).

111. K. Harada and S. W. Fox, Arch. Biochem. Biophys., 109, 49 (1965).

112. K. Harada and S. W. Fox, in The Origins of Prebiological Systems and of Their Molecular Matrices (S. W. Fox, ed.), Academic, New York, 1965, p. 289.

113. S. W. Fox and K. Harada, Science, 128, 1214 (1958).

114. S. W. Fox and K. Harada, J. Am. Chem. Soc., 82, 3745 (1960).

115. S. W. Fox, K. Harada, K. R. Woods, and C. R. Windsor, Arch. Biochem. Biophys., 102, 439 (1963).

116. S. Akabori, K. Ohno, and K. Narita, Bull. Chem. Soc. Japan, 25, 214 (1952).

117. S. Akabori, K. Ohno, T. Ikenaka, A. Nagata, and I. Haruna, Proc. Japan Acad., 29, 561 (1953).

118. J. H. Bradbury, Nature, 178, 912 (1956).

119. S. W. Fox and T. Nakashima, Biochim. Biophys. Acta, 140, 155 (1967).

120. S. W. Fox and K. Harada, Fed. Amer. Soc. Exptl. Biol., 22, 479 (1963).

121. K. Harada and S. W. Fox, unpublished results.

122. S. W. Fox and T. V. Waehneldt, Biochim. Biophys. Acta, 160, 246 (1968).

123. S. W. Fox, K. Harada, and J. Kendrick, Science, 129, 1221 (1959).

124. R. S. Young, in The Origins of Prebiological Systems and of Their Molecular Matrices (S. W. Fox, ed.), Academic, New York, 1965, p. 347.

125. S. W. Fox and S. Yuyama, J. Bacteriol., 85, 279 (1963).

126. S. W. Fox and S. Yuyama, Ann. N.Y. Acad. Sci, 108, 487 (1965).

127. S. W. Fox and T. Fukushima, in Problems of Evolutionary and Industrial Biochemistry (V. C. Kretovich, T. E. Pavlovskaya, and G. A. Deborin, eds.), U.S.S.R. Publishing House, Moscow, 1964, p. 93.

128. S. W. Fox, R. J. McCauley, and A. Wood, Comp. Biochem. Physiol., 20, 773 (1967).

129. T. V. Waehneldt and S. W. Fox, Biochim. Biophys. Acta, 160, 239 (1968).

130. A. Yuki and S. W. Fox, Biochem. Biophys. Res. Commun., 36, 657 (1969).

131. S. W. Fox, K. Harada, and D. L. Rohlfing, in Polyamino Acids, Polypeptides, and Proteins (M. Stahmann, ed.), University of Wisconsin Press, Madison, 1962, p. 47.

132. J. Noguchi and T. Saito, in Polyamino Acids, Polypeptides, and Proteins (M. Stahmann, ed.), University of Wisconsin Press, Madison, 1962, p. 313.

133. V. R. Usdin, M. A. Mitz, and P. J. Killos, Arch. Biochem. Biophys., 122, 258 (1967).

134. D. L. Rohlfing and S. W. Fox, Arch. Biochem. Biophys., 118, 122, 127 (1967).

135. T. Oshima, Arch. Biochem. Biophys., 126, 478 (1968).

136. S. W. Fox, in The Origins of Prebiological Systems and of Their Molecular Matrices (S. W. Fox, ed.), Academic, New York, 1965, p. 371.

137. S. W. Fox and G. Krampitz, Nature, 203, 1362 (1964).

138. G. Krampitz and H. Hardebeck, Naturwissenschaften, 53, 81 (1966).

139. D. H. Durant and S. W. Fox, Fed. Amer. Soc. Exptl. Biol., 25, 342 (1966).

140. D. L. Rohlfing, Arch. Biochem. Biophys., 118, 468 (1967).

141. G. Krampitz, S. Diehl, and T. Nakashima, Naturwissenschaften, 19, 516 (1967).

142. D. L. Rohlfing and S. W. Fox, Advan. Catalysis, 20, 373 (1969).

143. S. W. Fox and C.-T. Wang, Science, 160, 547 (1968).

144. J. Oró and C. L. Guidry, Arch. Biochem. Biophys., 93, 166 (1961).

145. J. Oró and C. L. Guidry, Nature, 186, 156 (1960).

146. J. Kovacs and H. Nagy, Nature, 190, 531 (1961).

147. M. Calvin and G. J. Calvin, Am. Scientist, 52, 163 (1964).

148. J. C. Sheehan and G. P. Hess, J. Am. Chem. Soc., 77, 1067 (1955).

149. M. Goodman and G. W. Kenner, Advan. Protein Chem., 12, 488 (1957).

150. H. G. Khorana, Some Recent Developments in the Chemistry of Phosphate Esters of Biological Interest, Wiley, New York, 1961.

151. G. Steinman, R. M. Lemmon, and M. Calvin, Proc. Nat. Acad. Sci. U.S., 52, 27 (1964).

152. G. D. Steinman, R. M. Lemmon, and M. Calvin, Science, 147, 1574 (1965).

153. C. Ponnamperuma and E. Peterson, Science, 147, 1572 (1965).

154. G. D. Steinman, D. H. Kenyon, and M. Calvin, Nature, 206, 707 (1965).

155. G. D. Steinman, D. H. Kenyon, and M. Calvin, Biochim. Biophys. Acta, 124, 339 (1966).

156. A. Schimpl, R. M. Lemmon, and M. Calvin, Science, 147, 149 (1965).

AUTHOR INDEX

Numbers in brackets are reference numbers and indicate that an author's work is referred to although his name is not cited in the text. Underlined numbers give the page on which the complete reference is listed.

A

Abbott, N. B., 23, 25, 36
Abdelhalden, E., 159[25], 195
   [25], 202, 321, 348
Abdullaev, N. D., 23[86], 26
   [86], 27[86], 31[115], 36, 38
Abe, O., 15[55,56,59], 18[55,
   59], 21[74], 34, 35
Abelson, P. H., 302, 306, 344,
   345
Adkins, H., 175[51], 203
Akabori, S., 317, 318, 330[116,
   117], 347, 348, 349
Akaboshi, M., 310[52], 346
Akerkar, A. S., 210, 236[30],
   238, 239[63], 275[30,63], 283
   288, 290, 294
Akhrem, A. A., 150[12], 201
Ambrose, E. J., 23, 25, 36
Anders, K., 320, 348
Anderson, G. W., 242, 290
Anderson, R. N., 48[78], 112
   [78], 128[268,269], 133, 142
Andreatta, R., 70[123,124], 115
   [124], 135
Andreoli, T. E., 31[117], 38
Anfinsen, C. B., 10[41], 33,
   103, 140, 281, 294
Antonovics, I., 242[70], 290
Aoyagi, H., 15[53,57,59,60,61,
   62], 18[53,57,59,60,62], 21,
   34
Arison, B. H., 287[197], 296
Arkhipova, S. F., 31[115], 38
Arnand, N., 269[153], 272[153],
   284[181], 294, 295

A (continued)

Astwood, E. B., 40, 41[20], 44,
   48[76], 129, 130, 131, 133
Avrameas, S., 286[193], 295

B

Bagnall, K. W., 154[18], 201
Bahadur, K., 305, 345
Bajusz, S., 88, 89, 90[185,186],
   91[184,186,187,188,189,190],
   110[186], 111(184,190), 138,
   139
Baker, B. R., 172[40], 203
Balasubramanian, D., 23[87], 24
   [87], 27[87], 36
Balbiano, L., 321, 348
Bankert, R. A., 191[95], 199
   [95], 206
Barkemeyer, H., 287[197], 296
Bar-Nun, A., 311[55], 346
Bar-Nun, N., 311(55), 346
Barrett, J. F., 119[239], 141
Barrett, R. J., 48[76], 133
Barth, A., 92[195], 139
Barthe, P., 42[30], 43[30], 79
   [148], 121[148], 130, 136
Battenberg, E., 212[40], 214
   [40], 289
Battersby, A. R., 2[7], 32
Bauer, S. H., 311[55], 346
Baxst, W., 48[77], 133
Bayer, E., 100, 139
Bell, P. H., 44, 45, 48[52], 49,
   63[54], 76[54,55], 108, 131, 132
Bell, R. P., 161[29], 202
Bellamy, D., 40, 129
Belleau, B., 104, 140

Belozerskii, A. N., 2[13], 28, 32, 37
Benisck, W. F., 62[108], 134
Benoit, J., 128[276], 143
Benoiton, N. L., 155[23], 198 [23], 202
Berde, B., 128[272], 142
Bergell, C., 209[4,11,12], 287
Berger, A., 12[48], 34
Berger, R., 308, 346
Bergmann, M., 177[66], 196[66], 204
Berlinguet, L., 184[89], 185 [89], 198[89], 199[89], 205
Bernstein, Z., 236[61], 289
Berse, C., 148[7], 159[26], 160 [26], 165[34], 168[26,34,38, 41], 169[39], 171[26,41], 172 [49], 174[34,41,53], 176[53], 177[60], 186[26,38,49], 188 [49], 189[26,53,60],192[26,38, 41], 193[26,34,38], 194[26,34, 38,39,49], 195[34,38,53], 196 [60], 198[26], 199[26], 200, 202, 203, 204
Berson, S. A., 43, 44[44], 131
Berthet, L., 126[263], 142
Bewley, T., 120, 141
Bhagavan, N. V., 5[31], 6[31], 33
Bichowski-Slomnitzki, L., 12 [47,48], 28[47], 34
Bickers, E. J., 184[89], 185 [89], 198[89], 199[89], 205
Birk, Y., 48[78], 112[78], 133
Bláha, K., 281, 284, 294
Blake, J., 62[103], 100, 115 [103], 119[103], 122[103], 134, 139, 255[103], 258[103], 259[103], 291
Blatt, A. H., 218[49], 289
Bloom, S. M., 265[144], 269 [144], 270[144], 293
Blout, E. R., 252[92], 254[92], 258[92], 259[118], 260[118], 262[118,128], 266[118,144], 270[144], 271[144], 272[118], 273[118], 291, 292, 293
Bodanszky, M., 243, 249[80], 266 [146], 267[146], 275[159], 290, 293, 294

Bodlaender, P., 214[41], 286, 289, 295
Bohn, H., 70[123,124], 115[124], 135
Boissonnas, R. A., 50, 51[85], 52 [84], 55, 111[84], 115[87], 121 [86,87], 133
Borovas, D., 80[151], 136
Boskin, M. J., 153[16], 201
Bossert, H., 80[159], 137
Boullay, P., 312, 347
Bowers, C. Y., 119[239], 128 [271], 141, 142
Boyland, E., 174[52], 204
Bradbury, E. M., 24, 36
Bradbury, J. H., 330[118], 349
Braz, G. I., 176[58], 195[58],204
Brazhnikova, M. G., 2, 3[4], 32
Bredesen, J. E., 5[28], 33
Breton, J. C., 314[70], 347
Bricas, E., 265[145], 269[152], 293
Brockmann, H., 251[86], 252, 263 [86,131,132,133,134,135,136, 137,138], 268[86], 291, 293
Brown, G., 286[193], 295
Brown, H. C., 147[1], 200
Brown, R. A., 120[250], 141
Brown, S. P., 285[190], 295
Bruce, W. F., 218[50], 289
Bruckner, V., 91[187,188,189], 138
Brugger, M., 79[148], 80[158], 121[148], 136, 137
Brüning, W., 100[206], 139
Buchi, G., 183[85], 198[85], 199 [85], 205
Bulavian, L. G., 174[54], 195[54], 204
Bulgakova, V. G., 29, 30[112], 37
Bumpus, F. M., 255[101], 291
Bunding, I. M., 49[81], 133
Burness, D. M., 211, 212, 213, 214[37,42], 215[37], 285[42,185, 186,187], 288, 289, 295
Busby, G., III, 244[78], 248[78], 249[78], 251[87], 290, 291
Butcher, R. W., 126[261,265], 142
Bystrov, V. F., 23[86], 26[86], 27 [86], 31[115], 36, 38

C

Cacciola, A. R., 44[52], 48[52], 131

Cairns, T. L., 177[74], k96[74], 204

Callahan, F. M., 243[65], 290

Callen, J. E., 177[68], 196[68], 204

Calvin, G. J., 341, 351

Calvin, M., 307, 341, 342[151, 152, 154, 155], 343[156], 346, 351

Camble, R., 258[117], 292

Camiletti, C., 26, 37

Capp, L. B., 177[64], 196[64], 204

Cark, J., 182[80], 199[80], 205

Carpenter, B. G., 24[91], 36

Carrion, J. P., 266[143], 293

Carstensen, H., 49[83], 133

Carter, S. K., 162[33], 202

Casszar, J., 323[104,105], 348

Chang, R. W. J., 307[43], 346

Chaturvedi, N. C., 255[101], 270[153], 273[153], 284[181], 291, 294, 295

Chauvet, J., 128[276], 143

Chawla, R. K., 100[216], 140

Cherng, C. J., 29[110], 37

Cheung, H. T., 259[118], 260 [118], 262[118], 266[118], 272 [118], 273[118], 292

Chew, L. F., 244, 251, 290, 291

Chien, S. W., 210, 225[31], 238 [62], 239[31], 245[31], 246 [62], 248, 249[31], 278[31], 280[31], 288, 289

Child, R. G., 44[52], 48[52], 131

Claggett, A. R., 312[65], 347

Claisen, L., 209, 211[33], 213 [8,33], 222[54], 287, 288, 289

Clayton, G. W., 41[28], 130

Chou, F. C. H., 100[216], 140

Chung, D., 45[70,71,72], 56[93, 94], 58[93,94], 59[96], 61[96], 62[97,98,99,102,104,105], 75 [94], 108[97,98,99], 110[98, 99], 111[96,99], 115[102], 122[104,105], 123[105], 132,

134, 251[84,85], 255[84,85], 260[121,122,124], 267[85,149], 268[84,85,151], 269[149], 270 [85], 290, 291, 292, 293

Chung, H. T., 252[92], 254[92], 258[92], 291

Cohn, P., 172[40], 203

Cole, P. W., 257[114], 265[114], 292

Cole, R. D., 45[69], 62[108], 132, 134, 177[63], 196[63], 204

Coleman, D. L., 262[128], 292

Coleman, G. H., 177[68], 196 [68], 204

Conduché, A., 212[38], 288

Conklin, L. E., 243, 290

Consden, R., 2[8], 32

Conti, F., 24[89,94], 36

Cook, P., 31[117], 38

Cooke, J., 184[89], 185[89], 198 [89], 199[89], 206

Corey, R. B., 8, 33, 159[24], 202

Costopanagiotis, A. A., 79[144], 110[144], 128[279], 136, 143, 252, 256[108], 259[108], 261 [90], 291, 292

Coulombe, R., 159[26], 160[26], 168[26], 171[26], 186[26], 189 [26], 192[26], 193[26], 194 [26], 198[26], 199[26], 202

Cox, H. R., 120[250], 141

Cox, M. E., 255[104], 256[104], 272[104], 291

Cragoe, A. J., Jr., 232[60], 235. 239[60], 289

Craig, L. C., 2[7], 3[20], 23 [85], 24[92], 25, 26, 32, 36, 37

Cram, D. J., 150[10], 201

Crane-Robinson, C., 24[91], 36

Crenshaw, R. R., 283[166], 294

Csizmas, L. L., 286[194], 295

Cullinane, N. M., 154[18], 201

Cultera, R., 307, 345, 346

Culvenor, C. C. J., 153[16], 201

D

Daigle, J. Y., 194[98], 195[98], 196[98], 206

D'Angeli, F., 183[84], 198[84], 199[84], 205
Das Gupta, S. K., 265[144], 269 [144], 270[144], 284, 293, 294, 295
David, R., 177[72], 196[72], 204
Davidson, A. I., 210, 232[28, 29], 233[29], 235[28,29], 241 [29], 246[29], 276[28], 277 [29], 288
Davies, M. C., 44[52], 48[52], 120[250], 131, 141
Davies, W., 153[16], 201
Davis, D. S., 45[55], 76[55], 132
Davis, G. W., 257[111], 292
Davis, R. E., 153[16], 198[94], 199[94], 201, 206
Davis, S. B., 45[55], 76[55], 132
Dedman, M. L., 116[234,237], 140, 141
DeFaye, G., 284, 294
de Garilhe, M. P., 128, 142, 143
Delépine, M. M., 284[169], 294
Denkewalter, R. G., 287[197], 296
Denney, D. B., 153[16], 201
Dermer, O. C., 182[79], 199[79], 205
De Santis, P., 23[83], 24[83], 26[104], 36, 37
Desaulles, P. A., 42[30], 43[30], 79[146,148], 120[146], 121[146, 148], 130, 136
Deschreider, A. R., 306, 345
Dessaignes, V., 311, 346
DeTar, D. F., 227[55], 289
DeWald, H., 259[119], 292
Dewey, R. S., 287[197], 296
Dhar, M. M., 256[107], 273[107], 291
Dickens, F., 184[89], 185[89], 198[89], 199[89], 206
Diehl, S., 338[141], 350
Dixon, H. B. F., 46, 112, 116, 133, 140
Dixon, J. S., 45[66,67,69,71, 72], 132
Dodonova, N. Y., 306, 345
Doepfner, W., 128, 142
Dominy, B. W., 284[178], 295

Dose, K., 307, 346
Doty, P., 26[103], 37
Drake, M. P., 45[58], 132
Dualszky, S., 91[189], 138
Dubos, R. J., 2[2], 31
Dubrovskii, V. A., 150[12], 201
Ducastaing, S., 314[70], 347
Duclos, J. M., 236[61], 289
Dunn, M., 154[21], 202
Durant, D. H., 338[139], 350
du Vigneaud, V., 55, 133, 264, [139], 293

E

Eckstein, H., 100[206], 139
Efremov, E. S., 23[86], 26[86], 27[86], 31[115], 36, 38
Eigner, E. A., 45[55], 76[55], 132
Eisenbeiss, F., 19, 20[66], 35
Elad, D., 320, 348
Elderfield, R. C., 172[40], 203
Eliel, E. L., 147[4], 200
Ellenbogen, E., 305, 345
Elpiner, J. E., 112[228], 140
Emmons, W. D., 183[81], 198[81], 199[81], 205
Engel, F. L., 40, 41[3,19], 116 [232], 119[3], 129, 130, 140
Englert, M., 120[250], 141
English, J. P., 44[52], 48[52], 131
Erlanger, B. F., 12[44,45], 33
Etienne, Y., 183[81], 198[81], 199[81], 205
Ettre, K., 307, 346
Eugster, C. H., 214, 219, 289
Euler, V., 159[24], 202
Evans, H. M., 44[47], 131
Evstratov, A. V., 31[115], 38

F

Fairburn, E. I., 172[40], 203
Fanta, P. E., 148[6], 182[79], 199[79], 200, 205
Farmer, T. H., 116[234,236,237], 140, 141
Fasman, G. D., 24[93], 36, 285 [188], 295
Feinstein, G., 214[41], 286,

289, 295
Feinstein, M., 42[33], 131
Felber, J. P., 43, 131
Feldstein, A., 150[13],, 201
Fenton, J. S., 148[6], 200
Ferrari, G., 307, 345, 346
Ferris, J. P., 314[71], 315, 347
Fetizon, M., 284, 294
Fiedorek, F. T., 191[95], 199
    [95], 206
Field, L., 283[166], 294
Fields, D. L., 182[79], 199[79],
    205
Fierce, W. L., 45[57], 49[57],
    132
Filira, F., 183[84], 198[84],
    199[84], 205
Finn, B. M., 44[52], 45[54], 48
    [52], 49[54], 63[54], 76[54],
    108[54], 131, 132
Finn, F. M., 127[266], 142
Fischer, N., 183[81], 198[81],
    199[81], 205
Fisher, J. D., 41[25], 49[81],
    130, 133
Fishman, J., 44[42], 131
Fittkau, S., 107[226], 140
Fles, D., 184[91], 198[91], 199
    [91], 206
Fletcher, R. S., 147[1], 200
Flores, J., 304, 345
Folk, J. E., 257[114], 264[114],
    292
Fontana, A., 15, 34
Forsham, P. H., 44[45], 131
Fosker, A. P., 270[154], 294
Fox, S. W., 309, 311, 321, 323,
    [58,106], 325, 326, 327, 328,
    329, 330[115], 333, 335, 336,
    337[129,130], 338, 346, 347,
    348, 349, 350, 351
Frank, B., 304, 344
Freeman, R. C., 221[53], 289
Frhølm, L. D., 4[24], 32
Friedlander, P., 172[40], 203
Friedmann, N., 315, 316, 347
Friesen, H., 48[76], 133
Fritz, R., 311[60], 347
Frobensus, M., 177[71], 196[71],
    204
Fromageot, C., 128[276], 143

Frøyshov, Ø. 5[27], 33
Fruton, J. S., 177[66], 196[66],
    204, 252, 261[91], 274[91],
    291
Fuchs, R., 150[13], 201
Fujikawa, K., 3[19], 22[19], 32
Fujino, M., 92, 93, 94, 97[200],
    99[199,200,201,202,203,204],
    111[199,201,202], 112, 114
    [197,201], 115[203,204], 116,
    117[202], 118[233], 119[233],
    129, 139, 140, 143
Fujita, Y., 21[72], 35
Fukagawa, T., 172[40], 203
Fukui, K., 184[87,89], 185[87,
    89], 198[87,89], 199[87,89],
    205, 206
Fukushima, T., 336[127], 350
Fürst, A., 150, 201
Furuyama, T., 318, 348

                          G

Gandry, R., 184[89], 185[89],
    198[89], 199[89], 205
Garg, G. K., 256[106], 257[106],
    291
Garg, H. G., 255[104], 256[104],
    269[153], 271[104], 272[153],
    291, 294
Garnuchot, B., 128[274], 142
Gause, G. F., 2, 3[4], 32
Gauthier, R., 174[53], 176[53],
    189[53], 195[53], 204
Gazis, E., 80[151], 136
Geiger, R., 69[161,162], 80, 81
    [161,162], 82, 83, 84, 85, 111
    [160,166], 112, 114[163,166],
    115[163], 116[166], 135, 136,137
Gerchakov, S., 263[130], 292
Geschwind, I. I., 45[65,66,67,
    69], 119[241], 120[245], 132
    141
Gevers, W., 4[26], 5[26,29,30],
    33
Ghirardelli, R., 148[6], 200
Gibbons, W. A., 23[85], 25[85,
    99], 26[85], 36, 37
Gilhuus-Moe, C. C., 5[28], 33
Gill, T. J., III, 285, 295
Gilliom, R. D., 161[29], 202

Gilman, A., 177[70], 196[70], 204
Gilman, H., 172[40], 177[65], 196[65], 203, 204
Gilvarry, J. J., 310, 346
Ginos, J. Z., 255[99], 291
Giordano, N. D., 42[36], 43[36], 131
Gish, D. T., 55[89], 133
Glazer, C., 249[82], 290
Gō, A., 283[167], 294
Godovaribai, S., 184[88], 185 [88], 198[88], 199[88], 205
Gold, H., 212[40], 214[40], 289
Goldman, L., 172[40], 203
Goodall, M. C., 31[116], 38
Goode, L., 12[44,45], 33
Goodfriend, T., 285[188], 295
Goodman, M., 242[66,67,68], 249 [82], 290, 341[149], 351
Gordon, A. H., 2[8], 32
Gordon, S., 55[88], 133
Gorup, B., 62[97,98], 108[97,98], 110[98], 134, 260[121,122], 265 [143], 266[149], 269[149], 292 293
Gould, E. S., 150[10], 201
Govindachari, T. R., 239[63], 275, 276[63], 290
Grady, A. B., 44[49], 131
Graham, J. L., 285[184], 295
Gregory, V. P., 183[81], 198 [81], 199[81], 205
Gresham, T. L., 191[95], 199 [95], 206
Grieshaber, T., 80[150], 136
Grimaux, E., 322, 348
Grodsky, G. M., 44[45], 131
Gros, C., 128[274,275,276,277, 278,279], 142, 143
Gross, E., 15, 34
Grossenbacher, K. A., 304, 345
Groth, W., 305, 345
Guidry, C. L., 338, 340, 351
Guillemin, R., 41, 45[53], 128, 130, 131, 142
Gustus, E. L., 257[112], 262 [112,129], 292
Gutte, B., 100, 139
Guttmann, St., 50[84], 51[85], 52[84], 55[86,87], 80[159],

111[84], 115[87], 121[86,87], 133, 137

H

Hagdahl, L., 49[83], 133
Hägele, K., 100[206], 139
Hagenmaier, H., 100[206,209], 139
Haines, J. A., 286[191], 295
Halczenko, W., 232[60], 235, 239 [60], 289
Halkerston, I. D. K., 42[33], 131
Hall, J. B., 5[31], 6[31], 33
Hallsworth, A. S., 150[11], 201
Halpern, B., 244, 251, 290, 291
Halstrøm, J., 20, 35
Ham, G. E., 182[79], 199[79], 205
Hamaguchi, K., 120[253], 141
Hamalidis, C., 80[152], 136
Hammond, G. S., 150[10], 201
Hanafusa, H., 318, 348
Handford, B. O., 252, 257[113], 261[90,127], 291, 292
Haning, R., 42[35], 131
Hano, K., 120[247], 141
Hanson, R. W., 270[155], 294
Hanze, A. R., 172[40], 203
Harada, K., 309, 311, 314, 316, 320, 321, 323, 325, 326, 328, 330[115], 335, 336[123], 228 [131], 346, 347, 348, 349, 350
Hardebeck, H., 338[138], 350
Harris, J. I., 11[42,43], 27, 33, 41[24], 45, 130, 132
Haruna, I., 333[117], 349
Harvey, R., 180[77], 205
Hase, S., 20[68], 35
Hashimoto, Y., 29[10], 37
Hassall, C. H., 269[125], 292
Hasselstrom, T., 307, 346
Hatanaka, C., 92, 93, 94[198,199], 97[200], 99[199,201,202,203, 204], 111[199,201,202], 112 [202], 114[197,201], 115[203, 204], 117[202], 139
Havranek, M., 260[123], 286[195], 292, 296
Hayashida, M., 257[115], 292

Haynes, R. C., 126[263], 142
Hays, E. E., 40, 49[81], 129, 133
Hays, H. R., 153[16], 201
Heath, N. S., 153[16], 201
Heine, H. W., 154[20], 201
Hellerman, L., 177[69], 179[69],
  196[69], 204
Henderson, R. B., 147[5], 200
Henebest, H. B., 150[11], 154
  [19], 201
Henry, M. C., 307[39], 346
Herrera, A. L., 315, 347
Hess, G. P., 229[57], 289, 341
  [148], 351
Heyns, K., 302, 314, 315, 344,
  347
Hiramatsu, T., 29, 37
Hirschmann, R., 287[197], 296
Hiskey, R. G., 257[111], 264
  [140], 292, 293
Hochstim, A. R., 310, 346
Hodgkin, D. C., 2[11,12], 23, 32
Hoffman, D., 254[94], 257[94],
  259[94], 260[94], 261[94], 291
Hofmann, K., 40, 41[21], 62
  [109], 63, 64[113,116], 65,
  68[109,117,118,119,120], 69
  [121,122,123,124,125], 78,
  79, 80, 81, 107, 110[109,
  118], 111[111,120], 114[122],
  115[111,118,120,124],117
  [110,111], 118, 127, 129,
  130, 134, 135, 142
Hollenberg, C. H., 41[20], 130
Hollowood, J., 255[104], 256
  [104], 271[104], 291
Holm, H., 4[24], 5[27], 32, 33
Honda, Y., 305[19], 345
Hornhardt, H., 209[7,14], 287
Hotchkiss, R. D., 2[2], 31
Hoving, H., 24[93], 36
Howard, K. S., 44[52], 45[54,
  55], 48[52], 49[54], 63[54],
  76[54,55], 108[54], 131, 132
Howell, T., 172[50], 203
Huber, P., 254[94], 257[94], 259
  [94], 260[94], 261[94], 291
Hugel, G., 244[78], 248[78], 249
  [78], 251[87], 290, 291
Hugo, J. M., 255[104], 256[104],
  271[104], 291

Humes, J. L., 69[119,120], 111
  [120], 115[120], 135
Hunger, K., 104[224], 140
Hylton, T. A., 257[113], 261
  [127], 292

                    I

Ikeda, K., 120, 141
Ikenaka, T., 333[117], 349
Imahori, K., 305[19], 345
Imura, H., 44[45], 131
Ingold, C. K., 161[30], 202
Inouye, H., 63[113], 64[113],
  134
Inouye, K., 84, 85[175], 86
  [175], 87[168,176,177,178,179,
  180,181,182], 108[168,177],
  110[168,176], 111[175], 114
  [176,177,179,180], 115[178],
  119[238], 121[178], 123[178,
  181,182], 124[178,181,182],
  137, 138, 141
Invi, T., 257[111], 292
Isaacs, N.S., 147[2], 149[9],
  153[2], 200
Iselin, B., 6[32], 33, 71[140],
  80[153,154,155], 115[134], 136
Ishigami, M., 305, 345
Island, D. P., 40[4], 42[32],
  129, 130
Ivanov, V. T., 20[69], 23[86],
  26[86], 27[86], 31[115], 35,
  36, 38
Iyer, V. S., 239[63], 275[63],
  290
Izumiya, N., 9, 10[40], 12[46],
  15, 18[52,53,55,57,58,59,60,
  62,63], 20, 21, 22, 23[79,80],
  24[36], 27[105], 29[64,105],
  33, 34, 35, 36, 37, 100[211],
  139, 244, 290

                    J

Jackman, L. M., 171[48], 172
  [48], 203
Jankowski, K., 148[7], 150[15],
  154[18,20], 159[26], 160[26,27],
  162[31], 165[34], 168[26,27,34,
  38,41], 169[27,39], 171[26,27,

41,45,46,47], 172[27,49], 174
[27,34,41,53], 176[53], 177[60],
178[15], 180[15,77], 186[15,26],
187[26,38,49], 188[26,49], 189
[53,60], 192[26,27,38,41], 193
[15,26,27,34,38], 194[26,27,34,
38,39,49], 195[15,27,34,38,53],
196[15,60], 197[15], 198[15,26],
199[15,26], 200, 201, 202, 203,
204, 205
Janowski, R., 160, 194, 198, 199
[28], 180[75], 190[75], 202, 204
Jansen, J. E., 191[95], 199[95],
206
Janusch, V. B., 45[73,74], 88[73,
74], 120[244], 133, 141
Jaquenoud, P. A., 50[84], 51[85],
52[84], 80[159], 111[84], 133,
137
Jenny, E., 214, 219, 289
Jensen, R., 314[68,69], 347
Jerierska, K., 309[50], 346
Johannesen, R. B., 147[1], 200
Johnson, J. E., 311[58], 323[58],
347
Johnson, L. F., 25[99], 37
Johnson, M. K., 184[88], 185[88,
92], 198[88,92], 199[88,92],
205, 206
Jones, J. B., 162[32], 202
Jones, J. H., 100[215], 140,
287, 296
Jones, W. C., Jr., 257[111], 292
Jones, W. R., 261[126], 292
Jono, Y., 84[169], 137
Joshua, H., 287[197], 296
Jost, K., 281, 282, 294
Jukes, T. H., 155[22], 202
Jung, G., 100[209], 139
Junk, G. A., 171[44], 203

K

Kadish, A. F., 172[40], 203
Kagiya, T., 184[87,89], 185[87,
89], 198[87,89], 199[87,89],
205, 206
Kamat, S. S., 312, 347
Kamber, B., 80[158], 137
Kamernitskii, A. V., 150[12],
201
Kanayama, M., 84[168,170,171],

87[168,181,182], 108[168],
110[168], 123[181,182], 124
[181,182], 137, 138
Kappeler, H., 70[137], 71[139,
141], 72[135,136], 75[139], 79
[146], 80, 84[139], 111[136],
115[134], 120, 121[146], 128
[273], 136, 142, 273[158], 294
Katchalski, E., 12, 28[47,106],
34, 37
Katjár, M., 91[187], 138
Kato, T., 15[53,55,57,59,60,61],
18[53,55,57,58,60], 21[60,72,
73], 22[76], 23[80], 27[105],
29[105], 34, 35, 36, 37,
100[211], 139
Katsoyannis, P.G., 40[10], 55
[88,89], 129, 133, 255[98,99],
265[141], 291, 293
Kauzmann, W., 24, 36
Kawai, K., 310[52], 346
Kawamura, K., 273[157], 294
Kawasaki, K., 68[126,129,130],
135, 266[147,148], 267[150],
293
Kawatani, H., 68[128,129,130],
102, 104[221,223], 105, 106
[217], 135, 140
Kelekhsaeva, T. G., 211[36], 288
Kemp, D. S., 210, 212[20,21],
213, 219, 222[20,21], 225[31],
227[20,22], 229[20], 231[20],
236, 238[20,62], 239[20,31,
64], 244, 245[20,22,31], 246
[62,79], 248, 249[31,78], 251
[87], 275[20], 278[31,64], 279
[31,64], 285[188], 288, 289,
290, 291, 295
Kendrick, J., 336[123], 349
Kenner, G. W., 341[149], 351
Kenyon, D. H., 342[154,155], 351
Khan, N. M., 254[94], 257[94],
259[94], 260[94], 261[94], 291
Khare, B. N., 306, 345
Khorana, H. G., 221[52], 229[58],
277[58], 289, 341, 351
Khosla, M. C., 269[153], 272
[153], 284[181], 294, 295
Kibler, R. F., 100[216], 140
Kienhuis, H., 271[156], 272[156],
294
Kikuchi, M., 257[115], 292

Kil'disheva, O. V., 184[90], 198
 [90], 199[90], 206
Killos, P. J., 338[133], 350
Kimball, A. P., 311[60], 347
Kimoto, S., 266[147,148], 293
King, T. P., 3[16,17], 32
Kinomura, Y., 68[127,128], 87
 [177], 135, 137, 266[147,148],
 293
Kiprianov, A. I., 168[35], 171
 [42], 192[35], 193[35], 194
 [35], 203
Kiprianov, G. I., 168[35], 171
 [42], 192[35], 193[35], 194
 [35], 203
Kiryushkin, A. A., 21[71], 35
Kisfaludy, L., 91[187,188,189,
 191], 129, 138, 143
Kishida, Y., 91[193], 138
Kiso, Y., 68[130], 104[225],
 106, 135, 140
Kleinkauf, H., 4[26], 5[26,29,
 30], 33
Klinger, H., 284[171], 294
Kloppenborg, P. W. C., 42[32],
 130
Klostermeyer, H., 10, 20, 33, 35
Knight, C. A., 304, 345
Knorr, E., 192[96], 193[96], 206
Knorr, L. A., 177[67], 192[67],
 196[67], 204
Knunyants, I. L., 184[90], 198
 [90], 199[90], 206
Knust, A., 209[13], 287
Kobayashi, Y., 210, 216[25,47],
 234[25], 288, 289
Kochetov, N. K., 211[34], 288
Kocsis, J. J., 49[81], 133
Kohler, E. P., 218, 289
Koida, M., 120[247], 141
Komm, E., 321, 348
Kondo, M., 15[52,53,54,57,60,
 61], 18[52,53,57,60]m 21[60,
 72,73,74], 34, 35
Kondo, S., 273[157], 294
Konig, W., 83, 137, 273[72,73],
 254[97], 257[94], 259[94,120],
 260[94], 261[94], 290, 291,
 292
König, W. A., 100[206], 140
Konyves, J., 323, 348

Kopple, K. D., 283[167], 294
Korosi, I., 184[86], 190[86],
 196[86], 198[86], 199[86], 205
Kovacs, A. L., 23[83], 24[83],
 36
Kovacs, J., 323, 340, 348, 351
Kozai, Y., 322, 348
Kozhevnikova, I. V., 21[71], 35
Krampitz, G., 338[137,138,141],
 350
Krassusky, K., 148, 200
Krejcarek, G. E., 284[178], 295
Krishnamistly, K., 184[88], 185
 [88], 198[88], 199[88], 205
Kristensen, T., 5[28], 33
Kubo, K., 41[23], 68[126], 120,
 130, 135, 141, 267[150], 293
Kunz, H. W., 285, 295
Kurahashi, K., 3[19], 4[23], 22,
 32, 33,
Kurihara, M., 100, 139
Kuromizu, K., 10[40], 21[74],
 22[77,78], 23[80], 33, 35, 36
Kurtz, J., 12[48], 34

L

Labadie, M., 314, 347
Lackner, H., 251[86], 252, 263
 [86,131,132,133,134,135], 268
 [86], 291, 293
Laiken, S. L., 24[92], 25[100],
 27[92], 36, 37
Laland, S. G., 4[24], 5[28], 32
Lande, S., 48[77], 63[110],
 64 [116], 65[110], 68
 [117], 69[118], 78[110],
 79[110], 80[110], 81
 [117], 110[118], 115
 [118], 117[110], 118
 [110], 133, 292, 294
Landgrebe, F. W., 119[240], 141
Landmann, W. A., 45[58,60,62,63],
 132
Landolt, R., 41[21], 130
Lang, Z., 88[183,184], 89[183,
 184], 91[184], 111[184], 138
Lange, O., 298, 344
Larkin, J., 154[21], 202
Law, H. D., 270[154,155], 294
Lawczmslo. A. L., 309[51], 346

Lawton, R. G., 284[178], 295
Lázár, T., 91[190], 111[190],
  138
Leach, S. J., 23[84], 25[84], 36
Lebovitz, H. E., 40, 116[232],
  129, 140
Lee, S., 15[58], 18[58], 34
Lee, T. H., 45, 88, 120[244],
  133, 141
Lefkowitz, R. J., 124[258,259,
  260], 126[258], 127, 142
LeFrancier, P., 211[145], 265
  [145, 269[152], 293
Leichner, L., 214, 219, 289
Leman, J. D., 162[32], 202
Lemmon, R. M., 342[151,1522, 343
  [156], 351
Lenard, J., 103[219], 140
Lenard, K., 91[188], 139
Lengyel, I., 177[61,62], 179
  [62], 184[91], 196[61,62],
  198[91], 204, 206
Lenormant, H., 154[21], 159[24],
  202
Leonis, J., 45[70],120[251],
  132, 141
Lerner, A. B., 41[24], 45[73,
  74], 88, 130, 133
Lesh, J. B., 49[81], 133
Levine, L., 242[67],285[188],
  290, 295
Levy, A. L., 45[65,66,67], 132
Li, C. H., 40, 41[14,22], 44,
  45, 48, 49[82,83], 55, 56, 58,
  59, 61, 62[97,98,99,100,101,
  102,103,104,105], 75[94], 100,
  108, 110[98,99], 111[96,99,
  100], 112[78], 113, 114[100],
  115[102,103], 119, 120, 122,
  123[105], 129, 130, 131, 132,
  133, 134, 139, 140, 141, 143,
  251[84,85], 255[84,85,95,96,
  100,103], 256[109], 258[103],
  259[103], 260[121,122,124],
  267[85,149], 268[84,85,95,96,
  151], 269[149], 270[85,109],
  275[109], 290, 291, 292, 293
Liddle, G. W., 40, 42[32], 129
  130
Lindemann, H., 212[39], 289
Lindner, E. B., 128[277], 143

Lin'kova, M. G., 184[90], 198
  [90], 199[90], 206
Lipkind, G. M., 31[115], 38
Lipmann, F., 4, 5[26,29,30], 33
Lipnik, M., 259[119], 292
Lipscomb, H. S., 41[28],43, 128
  [268,269,270], 130, 131, 142
Liquori, A. M., 23[83], 24, 36
Liu, T. Y., 63[110,111], 65[110,
  111], 68[117], 69[119,120], 78
  [110,111], 79[110,111], 80
  [110,111], 81[117], 111[111,
  120], 115[111,120], 117[110,
  111], 118[110,111,119,120],
  134, 135
Livingston, R. B., 162[33], 202
Lo, T. B., 56[93,94], 58[93,94],
  75[94], 134
Loeb, W., 298, 344
Lohmar, P., 48[78], 112[78], 133
Long, C. N. H., 44[48], 131
Long, F. A., 182[80], 199[80],
  205
Long, J. M., 128[268,269],142
Looker, J. J., 285[186,187], 295
Lopiekes, D. V., 285[190], 295
Losse, G., 10[39], 33, 80[150],
  92, 136, 139, 320, 348
Löw, M., 88[184], 89[184], 91
  [184,187,191], 111[184], 138
Lowe, C. U., 312, 347
Lozach, Y., 128[274], 142
Lübke, K., 40, 130, 249[81], 290
Lucas, H. J., 148[6], 200
Ludescher, U., 25, 36
Lukas, G., 183[85], 198[85], 199
  [85], 205
Lumbroso, H., 183[81], 198[81],
  199[81], 205
Lutz, E. F., 153[16], 201
Lyons, W. R., 44, 131

                    M

Mach, B., 3[18], 32
Machell, G., 183[83], 198[83],
  199[83], 205
Mackie, J. B., 124[257], 126
  [257], 141
Maeda, K., 273[157], 294

Maekawa, K., 168[36], 192[36], 193[36], 203
Magerlin, B. J., 172[40], 203
Maier, R., 79[146], 120[146], 121[146], 136
Maillard, L. C., 321, 348
Majumder, S. K., 184[88], 185 [88], 198[88], 199[88], 205
Makisumi, S., 12, 15[55,57,59, 60,61,63], 18[55,57,59,60,63], 21[60], 22[76], 23[79], 34, 35
Malek, G., 104, 140
Mannich, C., 194[97], 206
Marckwald, H. G. B., 177[71], 196[71], 204
Marfey, P. S., 285, 295
Marglin, A., 100, 139
Marino, Y. L., 210, 221[32], 227 [32], 230[32], 233[32], 237, 246[32], 278[32], 288
Markham, R., 312[63], 347
Markovac-Prpic, A., 184[91], 198 [91], 199[91], 206
Marquardt, I., 107[226], 140
Marshall, F. J., 172[40], 203
Martell, A. E., 155[23], 199 [23], 202
Martin, A. J. P., 2[8,9], 32
Massie, S. P., 172[40], 203
Master, F., 311[60], 347
Mathur, K. B., 256[107], 273 [107], 291
Matsuura, S., 9, 12[46], 15[63], 18[63], 22[75], 23[79], 27 [105], 33, 34, 35, 37
Matsuyama, H., 44[43], 131
Matthews, C. N., 304, 312, 345 347
Mayer, H., 209[18,19], 234[18, 19], 245[18,19]. 250[18], 252 [18,19], 256[19],257[19] 258 [19], 259[19], 260[19], 268 [19]. 269[19], 275[19], 288
Maynard, J. T., 172[40], 203
Mazzarella, L., 23[83],24[83], 36
McCauley, R. J., 336[128], 350
McDonald, R. K., 48[77], 133
McGahren, W. J., 242[68], 290
Meador, C. K., 40[4], 129
Medzihradszky, K., 88, 89[184,

186,187,188,189], 107, 110[186], 111[184], 138, 140
Meerwein, H., 212[40], 214[40]. 289
Meggy, A. B., 321, 348
Mehran, A. R., 308, 346
Meienhofer, J., 56[93,94,95], 58[93,94], 75[94], 134
Meisenhelder, J. H., 44[52], 45 [54], 48[52], 49[54], 63[54], 76[54], 108[54], 131, 132
Merrifield, R. B., 100, 102 [207], 139
Meyer, E., 302[10], 315[10], 344
Meyer, K., 209[2], 287
Meyers, D. R., 172[40], 203
Michel, G., 215[46], 289
Michelakis, A. M., 42[32], 130
Middlebrook, M., 311[58], 323 [58], 347
Miki, Y., 29, 30[111], 37
Mikowski, J., 287[197], 296
Miller, S. L., 299, 315, 316, 344, 347
Mims, R. B., 44[43], 131
Minami, H., 68[126,127,129,130], 135, 267[150], 293
Miroshnikov, A. I., 23[86], 26 [86], 27[86], 36
Mitsuyasu, N., 23[79], 35, 100 [211], 139
Mittleman, R., 2[9], 32
Mitz, M. A., 338[133], 350
Mix, H., 323[105], 348
Mizokami, N., 68[128,129,130], 135
Moffatt, J. G., 221[52], 289
Moiseenkov, A. M., 150[12], 201
Molnar, S. P., 198[94], 199[94], 206
Momany, F. A., 25[96], 36
Moore, J. A., 183[81], 198[81], 199[81], 205
Moret, V., 112, 140
Moroder, L., 70[124,125], 107 [125], 115[124], 135
Morris, C. J. O. R., 116[234, 236,237], 140, 141
Moser, R. E., 304, 312, 345, 347
Moyer, A. W., 44[52], 48[52],131

Müler, A., 215[46], 289
Mumm, O., 209, 224[1], 287
Münchmeyer, G., 209[3,9,10], 287
Munk, M.E., 229[56], 234[56],
  289
Muraoka, M., 244, 290
Murphy, A. L., 215[45], 216[48].
  221[45,51]. 289
Murphy, T. S., 259[118]. 260
  [118]. 262[118], 265[118].
  271[118], 272[118], 292
Murr, B., 307[39], 346
Muthu, M., 184[88],185[88],
  198[88], 199[88], 205

                      N

Nagata, A., 333[117], 349
Nagata, R., 15[57], 18[57], 34
Nagy, H., 323[105], 340, 348
  351
Naithani, V. K., 256[107], 273
  [107], 291
Nakamura, M., 43, 109, 110[37],
  131
Nakano, E., 257[115], 292
Nakashima, T., 333, 338[141],
  349, 350
Namba, K., 87[180], 114[180],
  137
Narayanen, V. L., 265[146], 267
  [146], 293
Narita, K., 318[82], 333[116],
  348, 349
Neelakahtan, K., 149[9], 200
Nehring, R., 198[94], 199[94],
  206
Nelson, D. H., 43, 44[43], 131
Némethy, G., 23[84], 25[84], 36
Nesmeyanov, A. N., 211[36], 288
Neuberger, A., 155[22], 202
Neubert, K., 10[39], 33
Neuzil, E., 314[68,69], 347
Ney, R. L., 126[265], 142
Nicholson, W. E., 42[32], 130
Nicolai, F., 174[56], 177[56],
  195[56], 204
Nicolaides, E., 259[119], 292
Nintz, E., 254[94], 257[94],
  259[94], 260[94], 261[94], 291
Nichimura, O., 92[196], 93, 94
  [198,199], 97[200], 99[199,

201,202,203,204], 111[199,201,
  202], 112[202], 114[201],
  115[203,204], 117[202], 129,
  139, 143
Noda, K., 100, 139
Noguchi, J., 338[132], 350
Noll, K., 221[53], 289
Nolting, H., 265[143], 293

                      O

Occolowitz, J. L., 284[172], 294
Oelofsen, W., 62[100], 113[100],
  114[100], 134, 255[95,96], 268
  [95,96], 291
Ogawa, H., 10[40], 33
Ohnishi, M., 25, 37, 283[167],
  294
Ohno, K., 330[116,117], 349
Ohno, M., 10, 15[60], 18[60], 21
  [60,72,73], 22[76], 23[79],
  33, 34, 35
Okada, M., 91[193], 138
Okada, Y., 68[126,127,128,129],
  135, 267[50], 293
Okamoto, K., 23[80], 36
Okamoto, M., 267[147,148], 293
Okawa, K., 20, 35, 318[80], 347
Okawa, R., 15[55,58,59], 18[55,
  58,59], 34
Okawara, T., 320, 348
Olafson, R. A., 209, 210, 212
  [15,16,17], 213, 214[15,16,
  17], 217[15,16,17], 218[15,16,
  17], 221[32], 222[15,16,17],
  224[15,16,17], 225[15,16,17],
  226[15,16,17], 227[15,16,17,
  32], 229[15,16,17], 230[32],
  233[32], 234[15,16,17,18,19],
  237, 238[15,16,17], 245[15,18,
  19], 246[32], 250[18], 252[15,
  18,19], 256[19], 257[19], 258
  [19], 259[19], 260[19, 268
  [19], 269[19], 278[32],
  287, 288
Omenn, G. S., 103, 140, 281, 294
Ondetti, M. A., 249[80], 266
  [146], 267[146], 275[159],
  282, 290, 293, 294
Orgel, L. E., 314[71], 315[72],
  347
Orlander, V., 159[24], 202

Oró, J., 304, 307, 309, 311. 312, 338, 340, 344, 346, 347, 351
Oshima, T., 120[247], 141, 266 [147,148], 293, 338[135], 350
Otani, S., 4[22], 32
Otsuka, H., 84, 85, 86, 87[168, 176,177,178,179,180,181,182], 108, 110[168,176], 111[175], 114[176,177,179,180], 115[178], 119[238], 121[178], 123[178, 181,182], 124[178,181,182], 137, 138, 141
Oughton, B. M., 2[11,12], 23[12], 32
Ovchinnikov, Yu. A., 20[69], 21 [71], 23[86], 26, 27[86], 31 [114,115], 35, 36, 37, 38
Oye, I., 126[261], 142

P

Pageau, R., 308, 346
Paladini, A., 3[15], 32
Paleveda, W. J., Jr., 287[197], 296
Palm, C., 307, 346
Pan, W., 100[206], 139
Paquette, L. A., 150[14], 154 [17], 176[59], 195[59], 201, 204
Park, W. K., 255[101], 291
Parker, R. E., 147[2], 153[2], 200
Pascal, M. L., 168[37,41], 171 [41], 174[41], 186[37], 192 [37,41], 203
Paschke, R., 307, 346
Paskhina, T. S., 2[13], 32
Passera, C., 307[37], 346
Pastan, I., 124, 126[254,255, 256,257,258], 127[260], 141, 142
Pasynskii, A. G., 303, 306, 344, 345
Patchornik, A., 62[106,107], 134
Patel, R. P., 265[144], 269 [144], 270[144], 285, 293, 295
Patel, V., 286[194], 295
Paulay, Z., 88[183,184], 89[183, 184], 91[184], 111[184], 138
Pauling, L., 8, 33

Pavel, K., 314, 347
Pavlovskaya, T. E., 303, 306, 344, 345
Payne, R. W., 40[1], 44[49,50], 129, 131
Pederson, K. O., 2[6], 32, 49 [82,83], 133
Pedrotti, A., 307[37], 346
Pervin, D. D., 182[80], 199 [80], 205
Peters, R. L., 48, 133
Peterson, E., 342, 351
Petrova, E. Ya., 184[90], 198 [90], 199[90], 206
Pettit, G. R., 261[126], 284, 292, 294, 295
Pfeil, E., 212[40], 214[40], 289
Philips, F. S., 177[70], 196 [70], 204
Phillips, J. G., 40, 129
Photaki, I., 256[105], 291
Pickering, B. T., 41[22], 48, 112, 119, 120[243], 130, 133, 141
Plattner, Pl. A., 150, 201
Pless, J., 55[86,87], 80[159], 115[87], 121[86,87], 133, 137
Plummer, C. A. J., 154[18], 201
Pogell, B. M., 284, 295
Polin, A. N., 28[108], 29[112], 30[112], 37
Pollard, L. W., 5, 6[31], 12 [31], 33
Polyakova, A. L., 321, 348
Pon, N. G., 45[66], 132
Ponnamperuma, C., 304, 307, 342, 345, 346, 351
Popov, E. M., 31[115], 38
Porath, J. O., 45[66], 119[241], 128[277], 132, 141, 143
Poremski, J., 159[24], 202
Poster, J., 259[119], 292
Pratt, Y. T., 159[24], 202
Prelog, V., 147[1], 200
Preston, J., 256[108], 259[108], 261[127], 292
Price, C. C., 172[40], 203
Price, S., 285, 295
Pricer, W., 124[257,259], 126 [257], 141
Printz, M. P., 24[92], 25[100], 27[92], 36, 37

Prinz, W., 254[94], 257[94], 259
  [94], 260[94], 261[94], 291
Pritchard, A. H., 244[76], 290
Pritchard, J. G., 182[80], 199
  [80], 205
Prox, A., 242[72,73], 290
Purdie, J. E., 155[23], 199[23],
  202
Purkayastha, R., 258[117], 292
Pusztai, A., 323[102], 348
Pysh, E. S., 26, 37

Q

Quadrifoglio, F., 24[90], 36
Querry, M. V., 172[40], 203
Quilico, A., 211[35], 288

R

Raacke, I. D., 45[69], 132
Raben, M. S., 40[1], 41[20,21],
  44[49,50], 129, 130, 131
Raftery, M. A., 62[108], 134
  177[63,72], 196[63,72], 204
Rajappa, S., 210, 236[30], 238,
  239[63], 275[30,63], 283, 288,
  290, 294
Rajewsky, B., 307, 346
Rall, T. W., 126[262], 142
Ramachandran, J., 55,56, 58[93,
  94], 59[96], 61[96], 62[97,98,
  99,101,104,105], 75[94], 108
  [97,98,99[, 110[98,99], 111
  96,99,101], 122[104,105], 123
  [105], 133, 134, 251[84,85],
  255[84,85,100], 256[109], 260
  [121,122], 266[85,149[, 267
  [84,85,151], 268[149], 269[85,
  109], 274[109], 290, 291, 292,
  293
Rao, M. M., 284, 295
Rashkovan, A., 171[43], 203
Rattle, H. W. E., 24[91], 36
Rebek, J., Jr., 246[79], 290
Rebello, P. F., 284, 295
Redding, T. W., 119[239], 141
Redgate, E. S., 128[267], 142
Rees, A. G., 154[18], 201
Rees, M. W., 312[63], 347
Reid, C., 306, 345
Reinhardt, W. O., 119[241], 141

Reinhardt, W. O., 119[241], 141
Reitsema, R. H., 172[40], 203
Reppe, W., 174[56], 177[56],
  195[56], 204
Ressler, C., 55[88], 133
Reynolds, D. D., 182[79], 199
  [79], 205
Reynolds, R. J. W., 183[83], 184
  [89], 185[89], 198[83,89], 199
  [83,89], 205
Riniker, B., 70[137], 79[146,
  149], 80[158], 120[146], 121
  [146,149], 136, 137
Rittel, W., 69, 71[133,138], 79
  [144,146,147,149], 80[158], 81,
  110[144], 115[134], 120[146,
  149], 121[146,147], 136, 137
Rizzo, R., 26[104], 37
Roberts, C. W., 55[88], 133
Robinson, A. B., 103[219], 140
Rockbuck, A., 174[51], 203
Rockland, L., 154[21], 202
Rohlfing, D. L., 338, 350
Rolfson, S. J., 172[40], 203
Rosenthaler, J., 70[121,122],
  114[122], 135
Roskoski, R., Jr., 5[29,30], 33
Rosowsky, A., 182[79], 199[79],
  205
Roth, J., 124[255,256,258,259,
  260], 126[254,255,256,258],
  127[260], 141, 142
Rothe, M., 19, 20[66], 35
Royce, P. C., 128[267], 142
Rudinger, J., 260[123], 281,
  281, 285, 292, 294
Ruhman, W., 44[43], 131
Ruttenberg, M. A., 3[17,18], 32
  252, 256[89], 291
Ryabova, I. D., 20[69], 35
Rybinskaya, M. I., 211[36], 288

S

Sabo, E. F., 265[146], 265[146],
  293
Safdy, M. E., 257[111], 292
Saffran, M., 42, 130
Sagan, C., 306, 311, 345, 346
Saito, T., 29[110], 37, 338[132],
  350
Saito, Y., 4[22], 32

Sakaguchi, K., 257[115], <u>292</u>
Sakakibara, S., 91, <u>138</u>, <u>320</u>, <u>348</u>
Sakamoto, Y., 3[19], 22[19], <u>32</u>
Sakiyama, F., 318[82], <u>348</u>
Samochocka, K., 309[50,<u>51</u>], <u>346</u>
Sanborn, I. H., 159[24], <u>202</u>
Sanchez, R. A., 314, 315[<u>72</u>], <u>347</u>
Sander, M., 153[16], 174[57], 182[79], 195[57], 199[79], <u>201</u>, <u>204</u>, <u>205</u>
Sandrin, E., 51[85], 80[159], <u>133</u>, <u>137</u>
Sanger, F., 2[10], <u>32</u>
Sannier, H., 154[21], 159[24], <u>202</u>
Sano, S., 100, <u>139</u>
Sarges, R., 3[21], 15, <u>32</u>, <u>34</u>
Sarid, S., 62[106,107], <u>134</u>
Sato, G. H., 126[264], <u>142</u>
Sato, M., 318[80], <u>347</u>
Sato, T., 184[87,89], 185[87, 89], 198[87,89], 199[87,89], <u>205</u>, <u>206</u>
Sayers, G., 41[26], 42, 43, 44, 128[267], <u>130</u>, <u>131</u>, <u>142</u>
Sayers, M. A., 41, <u>130</u>
Schaal, E., 322, <u>348</u>
Schally, A. V., 42, 119[239], 128, <u>130</u>, <u>141</u>, <u>142</u>
Schär, B., 42[30], 43[30], 79 [146], 120[146], 121[146], <u>130</u>, <u>136</u>
Schatz, K., 92[195], <u>139</u>
Scheraga, H. A., 23[84], 25, <u>36</u>
Schiff, H., 322, <u>348</u>
Schiller, P., 42[31], <u>130</u>
Schimmer, B. P., 126, <u>142</u>
Schimpl, A., 343, <u>351</u>
Schmidhammer, L., 242[72], <u>290</u>
Schmidt, C. L. A., 155[22], <u>202</u>
Schmidt, G. M. J., 2[11], <u>32</u>
Schnabel, E., 55, 56, 58[93,94], 75[94], 120, <u>133</u>, <u>134</u>, <u>141</u>
Schoenewaldt, E. F., 287[197], <u>296</u>
Scholer, F., 154[20], <u>201</u>
Schott, H., 154[21], <u>202</u>
Schramm, W., 252[136], 263[136], <u>293</u>

Schröder, E., 40, <u>130</u>, 249[81], <u>290</u>
Schröder, H. G., 83[163], 114 [163], 115[163], <u>137</u>
Schultz, H. P., 263[130], <u>292</u>
Schwam, H., 287[197], <u>296</u>
Schwartz, E. T., 63[112,113,114, 115], 64[113], <u>134</u>, <u>135</u>
Schwyzer, R., 3[14], 6[32], 8, 10, 15, 18, 20[67], 24[34], 25, <u>32</u>, <u>33</u>, <u>34</u>, <u>35</u>, <u>36</u>, 40, 42[31], 55, 70[134,137], 71 [133,139,140], 72, 75[139], 76, 77, 79[144,145], 80, 81, 84, 88, 110[144], 111[136, 143], 115[134,145], 128[273, 279], 129[142,143], 130, 133, <u>136</u>, <u>142</u>, <u>143</u>, 265[143], 273 [158], <u>293</u>, <u>294</u>
Scoffone, E., 183[84], 198[84], 199[84], <u>205</u>
Scopes, P. M., 255[104], 256 [104], 271[104], <u>291</u>
Scotoni, R., 150[11], <u>201</u>
Scott, R. A., 23[84], 25[84,95], <u>36</u>
Searle, C. E., 184[88], 185[88], 198[88], 199[88], <u>205</u>
Searles, S., 153[16], 183[81,82], 198[81,82], 199[81,82], <u>201</u>, <u>205</u>
Seela, F., 252[137,138], 263 [137,138], <u>293</u>
Seelig, S., 42[31], <u>130</u>
Segel, I. H., 155[23], 199[23], <u>202</u>
Sela, M., 28[106], <u>37</u>
Selegny, E., 286[193], <u>295</u>
Senyavina, L. B., 23[86], 26[86], 27[86], <u>36</u>
Shakespear, N. E., 45[55], 76 [55], <u>132</u>
Shames, P. M., 25[95], <u>36</u>
Shapira, K., 100[216], <u>140</u>
Shaver, F. W., 191[95], 199[95], <u>206</u>
Shaw, E., 214[41], 286, <u>289</u>, <u>295</u>
Sheehan, J. C., 177[61,62,], 198[91], 199[91], <u>204</u>, <u>206</u>, 229[57], <u>289</u>, 341, <u>351</u>

Sheehan, J. T., 265[146], 266
[146]. 274[159], 293, 294
Shekina, V. V., 184[90], 198
[90], 199[90], 206
Shemyakin, M. M., 20, 21, 31
[115], 35, 38
Shepard, K. L., 232[60], 235,
239[60], 289
Shepel, E. N., 23[86], 26[86],
27[86], 36
Shepherd, R. G., 44[52], 45[54,
55], 48, 49[54], 63[54], 76
[54,55], 108[54], 131, 132
Sherman, C. S., 172[40], 203
Shibata, K., 321, 348
Shields, J. E., 256[110], 258
[110], 269[110], 271[110],
273[110], 292
Shimonishi, Y., 91, 138
Shin, M., 87[177,178], 108[177],
114[177], 115[178], 121[178],
123[178], 124[178], 137, 138
Shinagawa, S., 99[204], 115
[204], 139
Shinozaki, F., 84[168,179,171],
87[168,181,182], 108[168],
110[168], 123[181,182], 124
[181,182], 137, 138
Sidorova, A. L., 30%, 345
Sieber, P., 3[14], 6, 8, 10, 15,
32, 33, 34, 76, 77, 80[153,
154,155,158], 81, 88, 111[143],
129, 136, 137
Siedel, W., 68[131,132], 80[160],
81[161,162], 82[160], 83[163,
166], 84[162], 85[160], 111
[160,166], 112[166], 114[163,
166], 115[163], 116[166], 135,
136, 137
Silaev, A. B., 28[108,109], 30
[109], 37
Silver, J. J., 316[76], 347
Silverstein, R., 227[55], 289
Simpson, M. E., 44[47], 131
Smeby, R. R., 255[101], 291
Smith, A. E., 316[76], 347
Smith, D. G. G., 126[265], 142
Smith, F. A., 41[21], 130
Smith, J. D., 41[28], 130
Smith, M., 221[52], 289
Smith, R. L., 264[140], 284

[170,171,172], 293, 294
Snyder, H. R., 174[55], 177[55],
195[55], 204
Sogn, J. A., 25[99], 37
Sokolov, S. D., 211[34], 288
Souchleris, I., 254[93], 255[93,
102], 256[93,102], 257[93,
102], 258[93], 262[93,102], 291
Soulas, R., 183[81], 198[81], 199
[81], 205
Sparks, L. L., 44[45], 131
Spencer, T. A., 172[50], 203
Sperling, J., 320, 348
Spühler, G., 63[112], 134
Squire, P. G., 120, 141
Sribney, M., 48[77], 133
Stables, J. T., 264[140], 293
Stack-Dunne, M. P., 46, 116[75],
133
Staehelin, M., 42[30], 43[30],
79[146], 120[146], 121[146],
130, 136
Stein, W. H., 177[66], 196[66],
204
Steinman, G. D., 316, 342, 347,
351
Stekolnikov, L. I., 112[228], 140
Stepanov, V. M., 28, 30[109], 37
Stern, A., 23[85], 25[85,99], 26
[85], 36, 37
Steuben, K. C., 242[66], 290
Stevens, C. L., 221[53], 229[56],
234[56], 289
Stewart, C. A., 147[3], 153[3],
200
Stewart, F. H. C., 257[116], 258
[116], 261[116], 265[142],271
[116], 272[116],273[116], 292,
293
Stewart, J. M., 174[55], 177[55],
195[55], 204
Stiller, R. L., 284[179], 295
Stoll, E., 5[27], 33
Stracham, R. G., 287[197], 296
Strecker, A., 301, 344
Stülcken, W., 209[6], 215[6], 287
Sturm, K., 80[160], 81[161,162],
82[160], 83[166], 84[162], 85
[160], 111[160,166], 112[166],
114[166], 116[166], 137
Stürmer, E., 128[272], 142

Stutz, E., 63[112], 134
Suba, L. A., 198[94], 199[94], 206
Sugihara, H., 91[193], 138
Sutherland, E. W., 126[261,262, 265], 142
Suzuki, K., 255[98], 291
Suzuki, T., 3[19], 22[19], 32
Svec, H. J., 171[44], 203
Swallow, R. L., 42[34,36], 43 [36], 131
Swan, J. M., 55[88], 133
Synge, R. L. M., 2[5,6,8], 32

T

Tait, J. F., 42[35], 131
Tait, S. A. S., 42[35], 131
Takashima, H., 264[139], 293
Talbot, G., 184[89], 185[89], 198[89], 199[89], 205
Tam, N. D., 284[174], 294
Tamura, F., 68[130], 104[225], 105, 106, 135, 140
Tanaka, A., 41[22,23], 43, 87 [179], 109, 110[37], 114[179], 130, 131, 138
Taube, M., 309, 346
Taunton, O. D., 124[255,256], 126, 141, 142
Terada, S., 100[211], 139
Terenin, A. N., 306, 345
Tesser, G. I., 79[145], 115 [145], 136
Theodoropoulos, D. M.,252, 254 [93], 255[93,102], 256[93, 102], 257[93,102], 258[93], 261[91], 262[93,102], 274 [91], 292
Thesing, J., 215[46], 289
Thiele, E., 194[97], 206
Thiele, H., 212[39], 289
Thomas, D., 286[193], 295
Thomas, J. O., 269[125], 292
Thomas, P. J., 260[123], 292
Thomas, P. L., 282, 294
Thompson, R. B., 185[93], 199 [93], 206
Thompson, T. A., 63[113], 64 [113], 134
Tieffenberg, M., 31[117], 38

Tilak, M. A., 102, 140, 255[99], 264[141], 291, 293
Timasheff, S. N., 24[93], 36
Tiselius, A., 49[83], 133
Tolman, L., 172[40], 203
Tomasic, V., 184[91], 198[91], 199[91], 206
Tometsko, A. M., 255[99], 291
Tomino, S., 4[23], 32
Tosteson, D. C., 31[117], 38
Trasciatti, D., 321, 348
Tsumura, S., 168[36], 192[36], 193[36], 203
Tun-Kyi, A., 265[143], 293
Turner, B., 315, 347
Tuttle, R. W., 25[95,96], 36

U

Ueda, K., 126[264], 142
Uemura, I., 29, 37
Umezawa, H., 273[157], 294
Uno, A., 20[68], 35
Upham, R. A., 257[111], 292
Urano, W., 120[253], 141
Urnes, P. J., 26[103], 37
Urry, D. W., 24[90], 25, 36, 37
Usdin, V. R., 338[133], 350

V

Vajda, T., 323[105], 348
Valcani, E., 12[47], 34
Van Campen, J. H., 285[184], 295
van der Holst, J. P. J., 271 [156], 272[156], 294
Vanderkooi, G., 23[84], 25[84, 95,96], 36
van der Scheer, 44[52], 48[52], 131
VanderWerf, C. A., 147[3], 150 [13], 153[3], 200, 201
VanTamelen, E. E., 180[76], 205
Varfalvy, L., 171[45,46,47], 203
Vargha, H. S., 107[226], 140
Vegotsky, A., 323, 349
Verber, D. F., 287[197], 296
Veres, K., 286[195], 296
Vereschagin, M. F., 321, 348
Verweij, A., 271[156], 272[156], 294

Vickers, E. J., 183[83], 198[83], 199[83], 205
Virupaksha, T. K., 256[106], 257 [106], 291
Vogel, G., 83[166], 111[166], 112 [166], 114[166], 116[166], 137
Volker, T., 305[20], 312, 345
Voloss, C. M., 128[276], 143
von Saltza, M., 265[146], 266 [146], 293
von Wessenhoff, H., 305, 345

W

Waehneldt, T. V., 335, 337[129], 349, 350
Waki, A., 310[52], 346
Waki, M., 9, 12[46], 15[36,55, 57,59,60,63], 18[55,57,59,60, 63], 20, 21[60], 23[79,80], 24 [36], 27[105], 29[64,105], 33, 34, 35, 36, 37, 100[211], 139
Walaszek, L. J., 49[81], 133
Waller, J. P., 50[84], 51[85], 52[84], 111[84], 116[229], 133, 140
Walters, W., 302, 315[10], 344
Wang, C.-T., 338, 351
Wang, K.-T., 257[113], 292
Wang, S. W., 244[78], 248[78], 249[78], 251[87], 290, 291
Waring, H., 119[240], 141
Warner, D. T., 23[82], 24, 36
Watanabe, H., 102[217], 106[217], 140, 322, 348
Watanabe, K., 85[175], 86[175], 87[180], 111[175], 114[180], 137, 138
Weinstein, B., 244, 252, 256 [108], 257[113], 259[108], 261 [90,127], 290, 291, 292
Weisblat, D. I., 172[40], 203
Weiss, B., 284[179], 295
Weitkamp, L. R., 116, 140
Welch, W. M., 236[61], 289
Wells, R. D., 70[121,122], 114 [122], 135
Wendlberger, G., 80[157], 137
Westland, R., 259[119], 292
Westley, J. W., 251, 291
Weygand, F., 104[224], 140, 243,

254[94], 257[94], 259[94,120], 260[94], 261[94], 290, 291, 292
White, A., 44[48], 131
White, J. E., 41[19], 130
White, W. F., 40, 45, 48, 49[57, 81], 112, 116[59], 129, 132, 133
Wilchek, M., 62[106,107], 134
Wilfang, G., 212[40], 214[40], 289
Willems, H., 80[159], 137
Williams, K., 174[52], 204
Williams, M. W., 242, 290
Williamson, K. L., 172[50], 203
Wilson, B. D., 211, 212, 213, 214[37,42], 215[37], 285[42, 85], 288, 289, 295
Wilson, G. L., 161[29], 202
Wilson, S. D., 45[55], 76[55], 132
Windsor, C. R., 330[115], 349
Wingender, W., 127[266], 142
Winstein, S., 147[5], 200
Wippermann, R., 298, 344
Wirth, A., 209[5], 287
Witkop, B., 3[21], 15, 32, 34
Woeller, F., 304, 345
Wolff, J., 311, 346
Wood, A., 336[128], 350
Woodbury, L. A. 41[26], 130
Woodman, D. J., 210, 214, 215 [23,24,44,45], 216[23,25,47, 48], 219, 220[23,26], 221[44, 45,51], 222[23], 227[23], 229 [23], 230[23], 231[23,59], 232 [28,29], 234, 235[23,27,28,29], 239[23], 241[29], 245[23,59], 246[29], 247[59], 276[28], 277 [27,29], 288, 289
Woods, K. R., 330[115], 349
Woodward, R. B., 209, 210, 212 [16,17,21], 213, 214, 215[24], 216[25,47], 217[16,17], 218 [16,17], 219, 220[26], 222[16, 17,21], 224[16,17], 225[16,17], 226[16,17], 227[16,17], 229[16, 17], 231[59], 233[29], 234. 235 [27], 236, 238[16,17], 245[18, 19,59], 247[59], 250[18], 252 [18,19], 256[19], 257[19], 258

[19], 259[19], 260[19], 268
[19], 269[19], 274[19], 277
[27], 288, 299
Works, T. S., 11[42,43], 27, 33
Wrigley, T. I., 154[19], 201
Wünsch, E., 80[156,157], 137

Y

Yajima, H., 40, 41[21,23], 62
[109], 63[109,110,111,112,113,
114,115], 64[113], 65, 68[109,
117,118,119,120], 69[121,122,
126,127,128,129,130], 70[121,
122], 78[110,111], 79[110,111],
80[110,111], 81[117], 102,
104[221,223,225], 105, 106,
110[109,118], 111[111,120],
114[122], 115[111,118,120],
117[110,111], 118[110,111,118,
119,120], 120, 129, 130, 134,
135, 140, 141, 266[147,148],
267[150], 293
Yalow, R. S., 43, 44[44], 131
Yamada, M., 4, 33
Yamanoi, T., 4[22], 32
Yamashiro, D., 100, 129, 140, 143
Yamashita, I., 68[126], 135, 267
[150], 293

Yanaihara, C., 69[119,120], 111
[120], 115[120], 135
Yanaihara, N., 63[110,111], 65
[110,111], 68[117], 69[118,119,
120], 78[110,111], 79[110,111],
80[110,111], 81[117], 110[118],
111[111,120], 115[111,118,120],
117[110,111], 118[110,111,118,
119,120[, 134, 135
Yoshida, N., 100[211], 139
Young, D., 307[43], 346
Young, G. T., 242, 255[104], 256
[104], 258[117], 271[104],287,
290, 291, 292, 293, 296
Young, R. S., 336, 349
Yuasa, S., 305, 345
Yuki, A., 337[130], 350
Yuyama, S., 336[125,126], 349

Z

Zamecnik, P. C., 286[191], 295
Zdrojewski, Z., 309[50], 346
Zeidler, D., 80[150], 136
Zervas, L., 80[151,152], 136
Ziegler, J. B., 174[55], 177
[55], 195[55], 204
Zimmer, T. L., 5[27,28], 33
Znamenskaya, M. P., 28, 37

SUBJECT INDEX

A

Abnormal opening, 148
N-Acetylamino acids in isoxazolium salt couplings, 252, 262
Acid catalyst, 325
Acidic proteinoids
  1:1:1, 330, 331
  2:2:1, 330
  electrophoretic mobility of,
    330
  nutritive property of as meas-
    ured by growth of L.
    arabinosus, 330
  proteolyzability of, 330
ACTH amino acid composition of,
    45
  heterogeneity of, 46
  isolation of, 44
  structure of
    bovine, 45, 47
    human, 45, 47
    ovine, 45, 47, 48
    porcine, 45, 47
ACTH assay
  ascorbic acid depleting test,
    41
  in vitro steroidogenetic
    assay, 41
    improved assay, 42
  in vivo steroidogenetic
    assay, 41
  radioimmunoassay, 43
ACTH, effect of, 40
  adipokinetic activity of, 41
  extra activity of, 41
  MSH activity of, 41
ACTH modification
  acetylation of, 116
  alkali treatment of, 119
  hydrogen peroxide oxidation
    of, 116

leucine aminopeptidase diges-
    tion of, 112, 113
  peptic digestion of, 49
  periodic acid oxidation of, 116
ACTH receptor, 124
ACTH synthesis
  active fragments, 108
  analogs, 113
  des-chain peptides, 122
  human, 47, 88, 129
  porcine, 76
  solid phase synthesis of, 100,
    129
  stereoisomers, 119
Activated esters from isoxazolium
    salts; see Enol esters
Activation of the carboxyl group
    with isoxazolium salts,
    217-233
Acylketenimines, 218-221
O → N Acyl shift, 233-237
Acyl-type protecting groups in
    isoxazolium salt couplings,
    251
Akabori's polyglycine hypothesis,
    312, 317
$[\beta\text{-Ala}^{4,4'}]$-gramicidin S, 21, 30
$[\text{D-Ala}^{4,4'}]$-gramicidin S, 30
$[\beta\text{-Ala}^5]$-semigramicidin S, 30
Alanylalanine, diastereomers of,
    328
Albumines, 183
Aliphatic epoxides opening, 168
Alkaloidal peptides, synthesis
    with isoxazolium salts,
    284, 285
ω-Amides, dehydration; see
    Dehydration
Amino acids
  amphoteric properties of, 325
  anhydrides; see Diketopipera-
    zines

basicity of, 155 - 156
bromo, 181, 185
composition of ACTH, 45; see
    also ACTH, amino acid
    composition of
composition of N and C termi-
    nals in proteinoids, 330
cysteine type, 163, 185
dimerization of, 155, 158-159
esters, opening with, 155-157
as opening reagents, 154
pK_a, 155-157
salts, metallic, opening with,
    168
sequence in proteinoids, 330
substituted, N-, 147
    α-, 148
zwitterion, 154 - 155
Aminolysis of enol esters, 238-
    239
Anhydroaspartyl structure, 327
Anhydropolyaspartic acid, 323
Anhydropolyaspartic acid poly-
    hydrate, 323
Arginine, activation with isox-
    azolium salts, 255, 280
Aromatic epoxides, opening, 154,
    169, 172, 174
N-Arylisoxazolium salts, 210,
    215, 216, 234
    synthesis, 215, 216
Asparagine, activation with isox-
    azolium salts, 252, 255
    270, 275-276, 278-280
Aspartyl residues, α- and β-,
    327
Azetidines, 182
    opening of, 175
Aziridines, 175, 177
    opening of, 175
Azlactone side reaction with
    isoxazolium salts, 229-
    231, 242-250, 251, 252,
    275-280
    hippuric acids spectral test,
    246
Azurmic acid, 312

B

β-Rays, 307

N-Benzoylamino acids, in isox-
    azolium salt couplings,
    252, 262, 263
N-Benzyloxycarbonylamino acids,
    in isoxazolium salt
    couplings, 251, 275-278
Bio-organic macromolecules, 317
Bis-homogramicidin S, 15
Bis-salicyl-aldiminato-
    gramicidin S, 26
Biuret reactions, 325
Blocking agents, 161, 181-184
Branched quaternizing groups for
    isoxazolium salts, 214
tert-Butyloxycarbonyllysine, 69
N-t-Butylisoxazolium salts
    preparation, 214
    stability of derived enol
    esters, 235, 276-277
N-t-Butyl-5-methylisoxazolium
    perchlorate
    activation of the carboxyl
    group, 226, 230-233
    azlactone formation, 230-232,
    245, 247, 277-278
    peptide synthesis with, 277-
    278
    polyfunctional amino acids,
    276-277
    racemization; see Azlactone
    formation
    side reactions, 231-233, 240,
    241, 245, 277-278
    synthesis, 24
    use with peptide acids, 277
N-t-Butyloxycarbonylamino acids,
    in isoxazolium salt coup-
    lings, 251
t-Butyl pentachlorophenyl
    carbonate, 92

C

N-Carbobenzoxyamino acids; see
    N-Benzyloxycarbonylamino
    acids
N,N-Carbonylgramicidin S, 27
Carboxymethylcellulose column
    chromatography, 16
Catalytic activity of protein-
    oids, 335, 338

Cationic detergent, 28, 30
N-Chloroacetylamino acids, in
    isoxazolium salt
    couplings, 252, 262
Cobalt-60, 308
Copolymerization, 325
Corticotropin
    A, 45, 46
    β, 44, 45, 48, 49
    A₁, 46
    A₂, 46
    B, 49
Cosmic rays, 309
Cyanamide (CA), 341, 342
    dimer of, 342
Cyanoguanidine, 342
Cyclic peptides, syntheses; see
    Cyclization of peptides
Cyclization of peptides, with
    isoxazolium salts, 281-284
Cyclobutanes, 184 - 185
Cyclopropanes, 181
    openings of, 191
Cysteines, opening with, 159,
    174, 181, 186, 187

                D

[Dab²,²']-gramicidin S, 30
DEAE-cellulose, 333
Dehydrating agents, 317, 340
Dehydration of ω-amides, 252,
    276-278
N,N'-Diacetylgramicidin S, 26,
    27
N,N-Diacylamides; see Imides
Dicyanamide (DCA), 341, 342
    sodium salt of, 342
Dicyandiamide (DCDA), 341, 342
Dicyclohexylcarbodiimide, 341
Digramicidin S, 21
2,5-Diketopiperazines, 321, 325
    formation of, 157-159
    opening with, 157, 159
Dimerization of amino acids,
    157-159
Dimerizing cyclization with isox-
    azolium salts, 281-284
N,N'-Diphthaloylgramicidin S,
    25
Di-p-nitrophenyl sulfite, 6
Ditosylgramicidin S, 8

Double opening, 152

                E

Electric discharge, 298, 299,
    300
    high-frequency, 305
    silent, 299
Electron beams, 307, 308
Electrophoresis, high voltage,
    333
Electrophoretic mobility, of
    acidic proteinoids, 330
Enneatin, 31
Enol esters from isoxazolium
    salts
    azlactone formation, 229, 242-
    250
    formation, 217-233, 250, 275-
    279
    hydrolysis, 278-279
    of arginine, 279-280
    of asparagine, 252, 275-276,
    279
    of glutamine, 252, 275-276,
    279
    of histidine, 279-280
    of hydroxyl-containing amino
    acids, 252, 275-276, 279-
    280
    of hydroxyproline, 252
    of serine, 252, 276, 280
    of tryptophan, 252
    of tyrosine, 252
    of unprotected amino acids
    and peptides, 284-285
    racemization, 242-250
    reaction with hydrazides, 252
    rearrangement, 233-237
    salt coupling, 278-279
Enzyme modification with isoxa-
    zolium salts, 286
Epihalohydrines, opening, 170
Episelenates; see Seleniranes
Episulfides
    opening of, 189
    reaction with amino acids,
    174, 176, 183
Epoxides
    abnormal opening, 148 - 149
    as blocking agents, 161
    cyclic openings, 171-172

gaseous, opening, 189
liquid, opening, 188
reaction of, with amino acids,
    148-149
solid, opening, 186
N-Ethoxycarbonyl-2-ethoxy-1,2-
    dihydroquinoline, 104
N-Ethylbenzisoxazolium cation
    activation of the carboxyl
        group, 227, 229, 230
    azlactone formation, 229-230
        245-248, 275-277
    peptide synthesis, 276
    racemization; see Azlactone
        formation
    synthesis, 212, 214
    use for cyclization of
        peptides, 284
    peptide acids, 276
    peptide modification, 286
    use with peptide acids, 276
N-Ethyl-7-hydroxybenzisoxazolium
    cation
    activation of the carboxyl
        group, 225, 226
    azlactone formation, 246, 247-
        250, 279-281
    peptide synthesis with 279-281
    racemization; see Azlactone
        formation
    salt couplings, 280
    use with peptide acids, 280,
        281
N-Ethyl-5-phenylisoxazolium-3'-
    sulfonate
    activation of the carboxyl
        group, 217, 224-226, 229,
        250-253
    with N-acylamino acids, 252
        hydrazides, 252
        peptide acids, 252
        polyfunctional amino acids,
            252
    azlactone formation, 242-245,
        249, 252
    peptide synthesis with, 250-
        253
    racemization; see Azlactone
        formation in solid phase
        peptide syntheses, 280
    synthesis, 211-214
    for alkaloidal peptides,
        284-285

cyclization of peptides, 281-282
glycopeptides, 284-285
nucleopeptides, 284-285
peptide and protein modifi-
    cation, 285-287
polypeptides, 284-285
preparation of acylating
    agents, 287
steroidal peptides, 284
Extrusion reaction, 153-154,
    176

F

First-order reaction, 325
N-Formylamino acids, unisox-
    azolium salt couplings,
    252, 262
Formyllysine, 63
Four-member rings
    opening with amino acids, 182
    reactions of, 182, 198-199
        azetidenes, 182
        cyclobutanes, 184-185
        lactams, 183
        lactones, 183
        oxetanes, 182
        propiolactone, 183
        selenetenes, 182
        thietanes, 182
        thiolactones, 183
    reactivity of, 147
Free energy, 303, 317, 340
Fürst-Plattner principle, 150

G

γ-Rays, 307, 308
Gelatin, hardening with isoxazo-
    lium salts, 285
Gel filtration, 334
Glutamine, activation with isox-
    azolium salts, 252, 257
    270, 276-279
Glycine esters, 157-158
Glycines, opening with, 148,
    157-159
Glycopeptides, synthesis with
    isoxazolium salts, 285
[Gly$^{4,4'}$]-gramicidin S, 30
[Gly$^{5,5'}$]-gramicidin S, 20, 21
[Gly^6]-tyrocidine A, 23

[Gly⁷]-tyrocidine A, 23
[Gly¹⁰]-tyrocidine A, 23
[Gly⁶,⁷,¹⁰]-tyrocidine, 23
Gramicidin S, 1
  analogs, 15
  biosynthesis, 4
  chemical synthesis, 6
  circular dichroism, 24, 27
  conformation, 23
  deuterium exchange experiment,
    25
  differential dialysis, 25
  energy minimization calcula-
    tion, 25
  linear peptides related to, 11
  molecular models, 23, 24, 25
  nuclear magnetic resonance,
    24, 25
  optical calculation, 26
  optical rotatory dispersion,
    24, 25, 27, 29
  possible precursor, 5
  relationship between structure
    and activity, 27
  solid-phase synthesis, 10
  structure, 2, 3

H

α-Helix, 24
Heterocycles, opening with, 148,
  157-159
Higher structures, 316
Hippuric acid spectral test for
  azlactones, 246
Histidine, activation with
  isoxazolium salts, 278-280
Hofmann reaction, 323, 324
Homogenous proteinoids,
  definition of, 333
Hormonal activity of thermal
  proteinoids, 338
Hydrazides, use as carboxyl
  protection with
  isoxazolium salts, 252
Hydrogen cyanide
  oligomer, 298, 314
  polymer, 314
Hydrolysis of enol esters, 279-
  280
Hydroxyl containing amino acids,
  activation with isoxazo-

lium salts, 252, 276-277,
  279-280
1-Hydroxypiperidine esters,
  synthesis with isoxazolium
  salts, 287
Hydroxyproline, activation with
  isoxazolium salts, 252

I

Identification of products of
  epoxide openings
  NMR, 171
  spectroscopic methods, 170-
    171
  synthesis, 168-169
Illustrative procedures involving opening of epoxides, 186-191
Imides, from rearrangement of
  enol esters, 233 - 237
Intermolecular transposition,
  165-168
Ionizing radiation, 298, 307
Isoxazoles
  preparation, 211, 212
  quaternization, 213-215
Isoxazolium salt method of
  peptide synthesis, 208-
    287
  mechanism, 217-233
  racemization, 242-250
Isoxazolium salts, preparation,
  211-216

J

Japanese acid clay, 318, 320

K

K; see N-Ethyl-5-phenylisoxa-
  zolium-3'-sulfonate
Kinetic energy, of meteorites,
  309
Kjedahl method, 335

L

Lactams, 325
α-Lactams, reactions of, 178
β-Lactams, 160, 198
Lactones, opening of, 181-182
β-Lactones, reaction of, 191

L. arabinosus, 330
[D-Leu⁴,⁴']-gramicidin S, 16
[D-Leu⁴]-semigramicidin S, 16
Liquid ammonia,
    [Lys²,²']-gramicidin S, 30

                    M

Mechanism of opening, 148-153
    solvent influence, 149-152
α-Melanocyte-stimulating
        hormone, 41, 64, 120
    activity of thermal polymers,
        338
p-Methoxybenzl-8-quinolyl
        carbonate, 106
Mono-opening, 152
Morpholone, 174
    N-substituted, 174
Multimolecular systems, 317
Mustards, 162

                    N

NEPIS, see N-Ethyl-5-phenyl-
        isoxazolium-3'-sulfonate
Neutrons
    fast, 310
    thermal, 310
Nonaqueous solvents, 335
Normal opening, 148
Nucleic acid, 316
Nucleopeptides, synthesis with
        isoxazolium salts, 285
Nucleophilic opening; see
        Opening of
Nutritive properties of acidic
        proteinoids, 329

                    O

Opening of 3 and 4 member rings
    catalysis of, 160-163
    with carboxylic group, 147,
        150-153
    electrophilic, 149-150
    influence of inductive
        effect, 149
    influence of mesomeric
        effect, 149
    influence of steric effect,
        149, 151-159

mechanisms of,
    electrophilic, 151, 152
mechanisms of
    nucleophilic, 149
with NHR group, 147-149
nucleophilic, 147-155
with SH group, 174, 186
Outgassing
Oxazolones; see Azlactones
Oxides; see Epoxides

                    P

Pentachlorophenyl trichloro-
        acetate, 93
Peptide acids, in isoxazolium
        salt couplings, 252, 264-
        276, 278-281
Peptide bond, formation with
        isoxazolium salts, 208-210,
        238-239, 250-281
Peptide modification with
        isoxazolium salts, 285-287
Peptides
    formation of, 316
    opening with, 193, 195, 199
Phenylacetaldehyde, 316
Phenylacetylene, 316
Phosphopantethein, 5
N-phthalylamino acids, in isox-
        azolium salt couplings,
        251, 277
Physiological activity of
        compounds, 162, 181
Piperazone, 177, 179
β-Pleated sheet structure, 8, 9,
        18, 24, 25, 26, 28
Polyalanine, 320
Polyaspartic acid, 322, 323, 340
    α- and β-linkages, 323, 340
    formation by transpeptidation,
        340
Poly-α-DL-aspartic acid, 323
Polycondensation
    of amino acids, 329
    mechanism of, 334
Polyfunctional amino acids, acti-
        vation with isoxazolium
        salts, 252, 253, 276-280
Polyglycine, 312, 322, 338
Polylysine, 326
    α- and ε-linkages, 326
Polymerization of amino acids,
        166

Polymerization of enol esters, 285
Polymerization of lactones, 185
Polypeptides
  formation of, 316
  synthesis with isoxazolium salts, 285
Polyphosphoric acid, 328
Polysaccharide, 316
Prebiotic syntheses of amino acids, 335
Primitive multimolecular systems, 336
Primitive proteins, 338
Primordial Earth, 336
Propiolactones, 183, 185
Propiotheolactones, 184
Protein-like material, 329
Protein modification with isoazolium salts, 285-287
Proteinoid microspheres, 336, 337
  dynamic budding phenomena, 336
  electron micrographs of, 336
  Gram staining of, 336
Proteinoids, 329, 330
  acidic; see Acidic proteinoids
  thermal; see Thermal proteinoids
Proteins, 316
Proteolyzability of acidic proteinoids, 330

Q

Quaternization of isoxazoles, 213-215
N-(2-Quinoxaloyl) amino acids, in isoxazolium salt couplings, 252, 263

R

Racemization, with isoxazolium salts, 242-251, 252, 276, 278-280
Reaction intermediates, 311
Rearrangement, 153-154
  of enol esters to imides, 233-237
Retroenantio-[Gly$^{5,5'}$]-gramicidin S, 20, 31

Retrogramicidin S, 20, 29, 30
Retrosemigramicidin S, 30
α_{11} Ribbon structure, 23

S

Salt couplings with isoxazolium salts, 279
[Sar$^{5,5'}$]-gramicidin S, 18
Seleniranes
  opening of, 190
  reactions of, 179-180
Selentenes, reaction of, 185
Semigramicidin S, 30
Sephadex LH-20 column chromatography, 9, 16
Serine, activation with isoxazolium salts, 252, 261, 274, 276, 280
Sesquigramicidin S, 21
Shock waves, 298, 309, 311
Solid phase peptide synthesis with isoxazolium salts, 280
Spark, 299
Stereomer formations, 151-152
Steroidal aziridines, 179
Steroidal episulfides, opening, 176
Steroidal epoxides, opening, 150, 171-173
Steroidal peptides, synthesis with isoxazolium salts, 284
Strecker-type discharge, 301
Styrene oxide, 169, 172, 174
Synergistic activity, 12

T

Tabular survey of opening reactions, 192
4,5-Tetramethylene isoxazolium cations
  activation of the carboxyl group, 227, 230-231, 233
  azlactone formation, 230, 231, 233, 245, 278
  peptide synthesis with, 278
  racemization; see Azlactone formation
  use with peptide acids, 278
Thermal copolycondensation, 325

Thermal energy, 298, 308
β-Thermal polycondensation, 321,
    322
Thermal proteinoids, 335
  catalytic activity of, 335
  hormonal activity of, 335
  microsphere formation, 335
Thietanes, 184
N-Thiocarboxyanhydrides, syn-
    thesis with isoxazolium
    salts, 286
Thiolactones, opening of, 181,
    184
Thiranes; see Episulfides
Three member rings
  reactions of, 163, 192-197
    of aziridines, 175, 177-179
    of cyclopropanes, 181, 191
    of episulfides, 174, 176-
      177, 183
    of epoxides, 163-165
    of seleniranes, 181, 190
  reactivity of, 146-147
  strength and reactivity, 146-
    147
Topochemical approach, 20
Topochemical investigation, 31
Tosylarginine, 56
Transpeptidation, 340
N-Trifluoroacetylamino acids,
    in isoxazolium salt
    couplings, 252, 262
Tripeptide, 325
Tryptophan, activation with
    isoxazolium salts, 252,
    261, 275
Tyrocidine A,
  structure, 4
  synthesis, 22
Tyrocidine B
  ion carrying activity of, 31
  structure, 4
  synthesis, 22
Tyrocidine C
  structure, 4
  synthesis, 22

Tyrocidine D
  structure, 4
Tyrocidine E
  structure, 4
  syntheses, 23
Tyrocidines, 2, 3
Tyrosine, activation with
    isoxazolium salts, 252,
    261
Tyrothricin, 2

                    U

Ultracentrifugation, 333
Ultraviolet rays, 298, 305, 308,
    316
Unprotected amino acids and
    peptides, reaction with
    isoxazolium salts, 285

                    V

Valine-gramicidin A
  structure, 4
  synthesis, 15
Valinomycin, 31
Volcanic activity, 308

                    W

Woodward's reagent K; see N-
    Ethyl-5-phenylisoxazolium-
    3'-sulfonate
WRK; see N-Ethyl-5-phenylisox-
    azolium-3'-sulfonate

                    X

X-Rays, 307

                    Z

Zürcher corrections, 172
Zwitterion structure
  in ethanol-water, 154-155
  in water, 154-155